P9-CCK-409

POLITICAL GEOGRAPHY
OF THE TWENTIETH CENTURY

A GLOBAL ANALYSIS

This book is dedicated to the memory of
Peter Taylor (1920–1980)
sailor, miner, teacher and socialist,
a man of his century

POLITICAL GEOGRAPHY OF THE TWENTIETH CENTURY

A GLOBAL ANALYSIS

Edited by Peter J. Taylor

with contributions from
Gerry Kearns
John O'Loughlin
Herman van der Wusten
R. J. Johnston
Stuart Corbridge
John Agnew

and

Vladimir Kolossov
Akihiko Takagi
Ghazi Falah
Bertha K. Becker
Chandra Pal Singh
C. O. Ikporukpo

Belhaven Press
London

Co-published in the Americas by Halsted Press,
an imprint of John Wiley & Sons, Inc., New York

72716

Belhaven Press
(a division of Pinter Publishers)
25 Floral Street, Covent Garden, London WC2E 9DS, United Kingdom

First published in 1993

© The editor and contributors 1993

Apart from any fair dealing for the purposes of research or private study, or
criticism or review, as permitted under the Copyright, Designs and Patents Act,
1988, this publication may not be reproduced, stored or transmitted in any
form or by any means or process without the prior permission in writing of the
copyright holders or their agents. Except for reproduction in accordance with
the terms of licences issued by the Copyright Licensing Agency, photocopying
of whole or part of this publication without the prior written permission of the
copyright holders or their agents in single or multiple copies whether for gain
or not is illegal and expressly forbidden. Please direct all enquiries concerning
copyright to the Publishers at the address above.

Co-published in the Americas by Halsted Press, an
imprint of John Wiley & Sons, Inc., 605 Third Avenue,
New York, NY 10158–0012

British Library Cataloguing in Publication Data

A CIP catalogue record for this book is available from the British Library.

ISBN 1 85293 196 5 (hb)
    1 85293 197 3 (pb)

Library of Congress Cataloging-in-Publication Data

Political geography of the twentieth century : a global analysis :
  edited by Peter J. Taylor; with contributions from Gerry Kearns . . .
  [et al.].
    p.   cm.
  Includes bibliographical references and index.
  ISBN 1-85293-195-5.—ISBN 1-85293-197-3 (pbk.)
  1. Political geography.   2. Geopolitics.   3. World politics—20th
century.   I. Taylor, Peter J. (Peter James), 1944–.   II. Kearns,
Gerard.
JC319.P586   1993   92–27642
320.1′2—dc20   CIP

ISBN 0-470-21965-3 (cloth – in the Americas only)
      0-470-21966-1 (paper – in the Americas only)

Typeset by Florencetype Ltd, Kewstoke, Avon
Printed and bound in Great Britain by SRP Ltd, Exeter

Contents

6/3/93

JC319
.P586
.1993

Contents

List of contributors

John Agnew is Professor of Geography at Syracuse University. He has written extensively on political geography including *The United States in the World-Economy: A Regional Geography* (Cambridge University Press) and *Place and Politics* (Allen and Unwin).

Bertha K Becker is Professor of Geography at the Federal University of Rio de Janeiro. She has published extensively on the political geography and geopolitics of Brazil including *Brazil: A New Regional Power in the World-Economy* (Cambridge University Press).

Stuart Corbridge is a Fellow at Sidney Sussex College and lecturer in geography at the University of Cambridge. He has written books and articles on India and development in general including *Capitalist World Development* (Macmillan) and *Debt and Development* (Blackwell).

Ghazi Falah is the founder of the Galilee Center for Social Research, Nazareth and a Visiting Associate Professor (1992–93) at the Department of Geography, University of Northern Iowa. He has written extensively on the political geography of land-use conflict on Israel/Palestine.

C O Ikporukpo is a lecturer in geography at the University of Ibadan. He has published on the dynamic political geography of Nigerian federalism.

R J Johnston, formerly Professor of Geography at the University of Sheffield, is Vice-Chancellor at the University of Essex. He is author of very many books and articles on human geography including *Geography and the State* (Macmillan), co-author of *A Nation Dividing? The Electoral Map of Great Britain, 1979–87* (Longman) and co-editor of *World in Crisis? Geographical Perspectives* (Blackwell).

Gerry Kearns is a lecturer in historical geography at the University of Liverpool. He has written on medical geography topics in nineteenth century Britain as well as a seminal paper comparing Halford Mackinder with Frederick Jackson Turner.

Vladimir Kolossov is Professor at the Laboratory of Global Geographical Problems, Academy of Sciences, Moscow. He has written extensively on the new electoral geography of USSR/Russia and has conducted comparative studies of local government areas across east and west Europe.

John O'Loughlin is Professor of Geography at the University of Colorado at Boulder. He is the author of many papers on political geography, is co-author of *Geography and International Relations* and co-editor of the series of that name (Belhaven), and is the American editor for *Political Geography* (formerly *Quarterly*).

Chandra Pal Singh is Reader in Geography at the Dehli School of Economics, University of Delhi. He has written extensively on the electoral geography of India and was the secretary of the International Geographical Union Study Group on the World Political Map.

Akihiko Takagi is an Associate Professor in the College of Liberal Arts and Science, Ibaraki University. He has translated *Political Geography: World-Economy, Nation-state and Locality* into Japanese.

Peter J Taylor is Professor of Political Geography at the University of Newcastle upon Tyne. He is author of many books and articles on political geography including *Political Geography: World-Economy, Nation-State and Locality* (3rd edition, Longman), is co-editor of *World in Crisis? Geographical Perspectives* (2nd edition, Blackwell), co-edits the series *Geography and International Relations* (Belhaven) and is editor of *Political Geography* (formerly *Quarterly*).

Herman van der Wusten is Professor of Political Geography at the University of Amsterdam. He has written many articles on political geography and is co-author of *Geography and International Relations* (Belhaven).

Preface

This volume of essays is an unusual one. Its genesis derives from my experience editing another volume of essays, *A World in Crisis* (Oxford: Blackwell, 1989), with Ron Johnston. We decided that the opening chapter of that book should be on the contemporary travails of the world economy and we invited Nigel Thrift to do the job. As it turned out this was an inspired choice and Nigel presented us with a splendid essay. But there was one problem, he overran his allotted space by about 100 per cent. We managed to trim a little here and there but the final chapter remained a very long one. Ron and I had no problem accepting this outcome since, with hindsight, we realized we had given Nigel such a huge task that the length of his essay was entirely reasonable. The book was a success and invariably commentators mentioned Nigel's chapter as one of the reasons for that success. In other words this unplanned outcome, the long chapter, worked.

This book consists of five long chapters with each author allocated 20,000 words plus to say their piece. The book is planned as a supplement to political geography texts that present the theoretical basis and overall content of the subdiscipline. Here we select five key topics and invite the authors to explore them in some depth and take the argument further than can be expected from conventional texts. Since we are approaching our *fin de siècle* we treat as our subject the twentieth century. This period is interpreted as the era when politics finally became global in scope, hence our subtitle 'a global analysis'. Selection of topics is inevitably a personal matter but I think I have chosen sensibly five areas of political geography that will satisfy a good amount of the curiosity of those who read political geography books. Changes are traced in geopolitical world orders, patterns of war and peace, the nature of the state, the status of the 'Third World' and the power of the USA. Within each of these topics the authors were given a free rein to describe their subject matter within a critical theoretical context but without dwelling on the theory: this is an empirical book not a theoretical one. In addition there is a prologue and epilogue devoted to the past and present *fin de siècle* which aid in defining the outer parameters of our time period.

I have dedicated this book to my father Peter Taylor whose life covered the core middle sections of the twentieth century. He was a man of this century in more ways than just timing, however. Growing up in the bad times of the 1920s and 30s (what we will later call Kondratieff IIIB), he came of age to fight in a world war. His contribution was made in the engine room of Royal Naval vessels protecting vital arms supplies in the Atlantic and Arctic Oceans as they made their way from the USA to Britain and the USSR. As part of the British 'electoral cohort of 45' he was ready to build the new post-war world and with the nationalization of the British coal industry he moved north to become a miner. Taking advantage of the educational training offered by the National Coal Board he changed course in mid-life and became a lecturer at a College of Further Education, a good example of those who can do, teach. He lived long enough to experience the return of the bad times (Kondratieff IVB) and a year or so of the Thatcher government beginning its dismantling of what he held so dear. But he lived the main constructive part of his life in the good times of the fabled 'post-war boom' (Kondratieff IVA) when the pessimism of what went before and was to re-appear was banished. As a child of Kondratieff IVA it was an exhilarating experience to grow up with the assumptions that with education everything was possible for both me and society as a whole. Struggle was not banished but there was no doubt that the world was getting better and better. As a man of his century my father imbued these happy thoughts into my being, I remain my father's son.

Peter J Taylor
October 1992
Blacksburg, VA.

Acknowledgements

The Editor, contributors and publishers would like to thank the following for permission to reproduce copyright material:

Blackwell Publishers, Oxford for Figures 7.2 and 7.3 from *Models of Democracy* by David Held; Prof. B.J.L. Berry for those figures from his book *Long-Wave Rhythms in Economic Development and Political Behaviour* (Johns Hopkins University Press, Baltimore, 1991); John Urry for Table 16 from his book *The End of Organized Capitalism* (Polity Press, Cambridge, 1987) and Prof. S. H. Franklin for Figure 3 from *Cul-de-Sac: The Question of New Zealand's Future* (Unwin, Wellington, 1985).

The lines quoted, in Chapter 4, from the poem *Suicide in the Trenches* by Siegfried Sassoon are reproduced by kind permission of George Sassoon.

In a few cases it has proved impossible to trace the copyright-holder of material reproduced here and the publishers would be grateful to receive information that would enable them to do so.

Introduction

A century of political geography

Peter J. Taylor

The idea that we are living in the 'twentieth century' derives from a purely arbitrary construct of a calendar time imposed on the world by the West as part of its relatively recent global pre-eminence. Whether we celebrate specific decades, centuries or millennia, the exact time periods we use are devoid of any direct social meaning. As Raymond Williams (1983, 3) has pointed out we are, in effect, 'time prisoners' of the sixth-century Scythian monk, Dionysius Exiguus, who first devised our 'western time' for Christendom.

If we take the view that the *raison d'être* of social scientists is to interrogate the taken-for-granted world they experience, then it would seem to be particularly inappropriate for us to focus our concern on something as artificial as the twentieth century. Surely our efforts should be directed at meaningfully defined 'social times', rather than at what the calendar tells us? And yet if in the timing of society we are all 'dupes of Dionysius', it follows that calendar time itself becomes part of society and will be reflected in social processes. This seems to be particularly the case in modern society where the promise of progress gives way to forebodings of concern for the future at the *fin de siècle*. This idea was prevalent at the end of the nineteenth century and seems to be repeating itself in our time. Of course, social change cannot be simply willed through contemplating the calendar, but the co-incidence of *fin de siècle* with ongoing material

cycles and trends can define a meaningful 'century' for us to study.

The twentieth century, defined by the two *fin de siècles*, is the period when world society finally became truly global in scope. The term 'international' had been coined a century earlier to denote the nature of the competitive state-system based upon Europe; by 1900, those states had all but incorporated the rest of the world into their world-system. This was expressed at the end of the nineteenth century by the notion of *global closure*. Forebodings arose because the end of frontier and imperial expansion – the 'safety valves' of nineteenth-century European and European-settler expansion – meant that in the new century aggressive expanding empires would face one another in a geographical zero-sum game. War seemed inevitable. In our times the notion of *globalization* is one of the key ideas for understanding contemporary social change. After political global closure, the world has become more and more economically integrated, so that by the end of our century forebodings centre upon the the dehumanized scale at which our lives are organized. Even the states, the focus of most political activity, seem to be marginalized as trans-state economic forces gather pace out of control. The twentieth century is *the period of the geographical global scale*, and the chapters in this book concentrate upon global analyses of its political geography.

The century we deal with in this book, there-

1

fore, is not a time of exact dates, but extends from one *fin de siècle* to the next. Beyond the 'global geography', the forebodings felt in these two periods were complex and multifaceted, but they did include key *political geography* components. At the end of the nineteenth century the 'long peace' was coming to seem fragile for the troubled empires of Europe. In 1898 the USA defeated Spain, in 1904 Japan defeated Russia: the end of Europe dominating the world was nigh. At the end of our century it is America, above all, that worries for its future. While the USA was 'winning the Cold War', Japan was winning the all-important economic race: the end of the West (as Europe and European-settler regions) dominating the world is nigh. The main chapters in this book take different key themes for telling the story of how the world got from the first *fin* to the second – that is our political geography of the twentieth century.

Fin de siècle geopolitics

The text begins and ends with a Prologue and an Epilogue on the *fin de siècle* geopolitics that set the limits to our century. The two sections could hardly be more different from one another. In the Prologue, Gerry Kearns concentrates upon the political geography concerns of the 'largest empire the world had ever seen', that of Britain. Here was the leading European global power concerned for its future. At this time, what Britain thought and did mattered hugely to the rest of the world. In the Epilogue, six different authors from beyond the North Atlantic traditional core-area of political geography studies give their views of world prospects. They represent a pluralistic world where the ideas and actions of the old areas of domination are no longer automatically what matter. If we really are one world community, as so many commentators of this century have insisted upon, then we need genuine 'geographical dialogue' between the peoples within what remains a highly hierarchical world. Our Epilogue aspires towards beginning such dialogue.

Kearns' chapter deals with contrasts between two leading British intellectuals grappling with the problem of how to maintain Britain's pre-

eminent position in the world. This is particularly appropriate in providing the setting for the remaining chapters for two reasons. First, we should not doubt the dominance of Britain in world affairs at the beginning of this century. As pioneer industrialist, free-trade champion, naval-imperial power and world's banker, Britain had done far more than any other country to create the world as it existed in 1900. Even as late as the 1930s, Britain continued to occupy 'the central position . . . in American maps of the world' (Watt, 1984, 44) and for all other states. British views mattered.

Second, the particular views discussed here – those of Halford Mackinder and John Hobson – do much more than define different positions within the British imperial debate; they introduce themes that were to reverberate throughout the international relations of the twentieth century. The geographer Mackinder envisioned a relatively self-sufficient protected empire from a realist perspective that has dominated subsequent international relations writings. A lineage can be drawn from Mackinder through the *geopolitik* of Hitler's Germany to US strategy to counter the USSR as the 'world's heartland'. The economist Hobson was a critic of imperialism and a supporter of a liberal world order. He is famous for being used by Lenin in producing the standard anti-imperialism text that remained a key orthodox theory in the USSR until very recently. Hence both Mackinder and Hobson appear politically as intellectual 'fathers' on opposite sides in the Cold War (and despite the fact that their economic arguments would reverse their positions in Cold War idolatry). Within the intellectual contest between Mackinder and Hobson, Kearns' Prologue shows us the seeds of some of the crucial debates of our century.

The Epilogue provides a voice for those outside the mainstream North Atlantic core. The contributions are loosely ordered in terms of how much an 'outsider' the author or his/her region appears to be. Each short essay is a separate statement highlighting very different concerns for where the world is going at our *fin de siècle*. Vladimir Kolossov's Russian view focuses on the negative impact of the new nationalisms that are fragmenting the world political map, but he is ultimately optimistic for the necessary establishment of world government. From a Japanese perspective, Akihiko Takagi concentrates on the problem of hegemony in the world-system, but concludes in a similar optimistic vein that a future world will be a very co-operative one. The pessimism begins with Ghazi Falah's Middle Eastern view where he sees population growth exacerbating current problems of inequality with a clear potential for increasing warfare. From Brazil, Bertha Becker considers processes of globalization from the point of view of the semi-periphery and supports Falah's increasing polarization position. In complete contrast, Chandra Pal Singh's Indian view is much more optimistic that through the right political structures and policies, although inequalities will remain, quality of life can be raised throughout all countries of the world. Finally, the view from Nigeria/Africa by C. O. Ikporukpo is by far the most pessimistic. In his realist account, the USA as lone superpower can coerce and cause instability across the world. This very disparate range of views coincides with some of the concerns for the future expressed in the main chapters of the text but they also add new emphases and themes. We cannot know which, if any, will provide reasonable accounts of what is in store for us, but we do know that in the future such views from beyond the North Atlantic region will contribute crucially to making the next century.

Moving between these two contrasting *fin de siècle* geopolitics in this book takes five substantial chapters. Clearly we cannot attempt a comprehensive coverage of the political geography of the century, but we do attempt to cover some key themes. They can be identified to some extent from the way popular commentators have labelled the twentieth century. Perhaps the most common designation is as the century of great wars. On the other hand there has never been more peace-making and international organization than in our times. The two elements of this contradiction between 'war and order' are the subject matter of the first two chapters. The twentieth century is also a time of unprecedented growth of the state. This has occurred in two ways. First, the world political map is now a global pattern of sovereign units in the wake of the 'revolt of the periphery' – decolonization – this century. Second, the state has never before

penetrated civil society to the degree it has in this century. The second pair of chapters covers these themes, concentrating on 'developed' and 'developing' worlds respectively. Finally, we should not forget that this century has often been referred to as 'the American Century'. The power of the USA in our world has been a major political geography fact of our time and is the subject matter of the fifth chapter. Let me now introduce these five contributions in more detail.

War and order

The twentieth century has been, so James Joll (1985, 5) tells us, 'a century of wars'. The clearest expression of this view can be found in Richard Natkiel's (1982) *Atlas of Twentieth Century History*. This contribution to understanding our century is totally dominated by war maps; it is quite a depressing portrait of our times. And yet there is an alternative view. Gaddis (1988), for instance, describes the period since 1945 as 'the long peace', and Boulding (1978) finds great hope in the oases of peace in what had been the great 'continent of war', Europe. There is what we may term a political paradox of the twentieth century: a curious mixture of war and order.

Why has war been so commonly accepted as the *motif* of the twentieth century? One important reason is historical: to contrast it with the nineteenth century and the original so-called 'long peace' from 1815 to 1914. The optimism of Victorian liberals with their faith in progress and rationality was rudely disrupted by the new century. War, the negation of liberal rationality, seemed to be at the heart of what was soon to be called the 'crisis' in the twentieth century (Churchill, 1923). Two basic strands of argument for indicting our century can be traced. First, war was transmuted into 'total war' with its mass mobilization and concomitant aggressive ideological beliefs. Second, technological advances were hijacked by the war-lords to produce unprecedented destructive capability culminating in the nightmare of nuclear war. Hence, whether war has been directly experienced or not, our world has been utterly transformed: the 'high politics' of the statesmen now impinge in one way or another on the ordinary daily existence of everybody. No wonder what had been thought of as the coming century of promise and progress came to be designated 'a century of wars'.

The paradox is that there is a contrary argument which seems equally convincing in lauding the peace-making of the twentieth century. More statesmen have devised more mechanisms and institutions to create a peaceful world, albeit often naively and with limited success, than in any other era. War is no longer normal; it has to be justified: witness the wholesale conversion of 'War Ministries' to 'Ministries of Defence' in this century. More substantial has been the marshalling of technology to enhance productivity to a degree where millions of ordinary people in one part of the world in the second half of the twentieth century have been able to live lives of peace and prosperity undreamt of by their nineteenth-century ancestors. For them at least, the new century finally kept its promise. This is the oasis of peace and stability in which I am composing these words.

War and order: what does the contradiction mean? Chapters 1 and 2 try and make some sense of all this. Beginning with my chapter on geopolitical world orders, an attempt is made to understand the changing overall patterning of political power across the world-system. Chapter 1 is a very factual account of the high politics of the twentieth century which has the purpose of codifying the events of traditional narrative history in terms of broader political geography concepts. A descriptive framework based upon the order and stability created by the political elites of the Great Powers is provided as a backcloth to the remaining chapters. Although long cycles of material production and world leadership are well known (Goldstein, 1988), related sequences of political order in the world have been less systematically studied. While not cyclical in pattern, definite periods of distinctive distributions of power can be identified as geopolitical world orders. Two are described covering most of the twentieth century: the Geopolitical World Order of the British Succession and the Cold War Geopolitical World Order. As the name of the former world order suggests, these relate directly to hegemonic cycles; the USA finally 'succeeds' Britain in 1945. At each *fin de siècle* a geopolitical transition is identified when international rela-

tions were, or are being, transformed as they were also in 1945, at the mid-century transition to the Cold War.

The world is not as simple as international political elites would wish it to be. Geopolitical world orders may provide an initial patterning, but the political geography is much more complex than I have portrayed it in Chapter 1. In some ways the remaining chapters deconstruct my neat political ordering in their very different ways. This is certainly the case with John O'Loughlin and Herman van der Wusten's chapter on the political geography of war and peace where our century is explicitly interpreted as a 'bloody century'. Here we are presented with a much messier reality in terms of where and when wars occur or peace prevails. Like the first chapter, this is a very factual one but preceding the empirical treatment of wars there is a review of explanatory approaches from among which the authors choose the 'historical-structural' method for their presentation. This means that the search for causes of geographies of war and peace takes the authors far away from conventional political geography and its narrow concerns for political practices alone. O'Loughlin and van der Wusten explicitly link war and peace cycles to economic and power cycles, typifying the approach of all the chapters in this volume: a broad political-economy viewpoint supersedes a purely political view of the world. We came across this in Chapter 1, it is taken further in Chapter 2, and thereafter it pervades the chapters that follow.

These two chapters by no means 'solve' the paradox of war and order, but they do reflect it. Their common historical political geography investigations do help us understand how and sometimes why the paradox has operated.

The rise of the state

States, even modern capitalist states, long predate the twentieth century, but our times can truly be designated the century of the state. The wars discussed in Chapter 2 are predicated on the assumption that the norm is for all individuals to be ready to 'die for their country'. And many millions have done just that. For the most part wars have been extremely popular, reflecting a strong loyalty to the state. An attitude of supporting 'my country right or wrong' has usually defeated peace campaigners who can be easily dismissed as 'disloyal'. The general popularity of wars by states despite great sufferings confirms the salience of the state in the twentieth century.

But the state in the twentieth century is much more than a war machine. The war-making functions of the state derive from the territorial state structure that developed in Europe in the sixteenth century. At the Treaty of Westphalia in 1648 this form of political sovereignty – a competitive inter-state system – was confirmed for Europe. At the end of the eighteenth and into the nineteenth century this spatial structure is given a crucial cultural dimension. The territorial state becomes the vehicle for the politics of cultural expression, resulting in the nation-state. Sovereignty of the state becomes vested in 'the people' who constitute 'the nation' that now legitimates the state.

It is with the transmutation of territorial state into nation-state that wars become popular, as 'state imperatives' become 'the national interest' which is 'the people's cause', even if it means killing other people. The classic case remains the outbreak of World War I in August 1914 when nationalist calls for war throughout Europe drowned out the rival peace policy of the Socialist International. But the state has had to pay for such colossal expressions of loyalty. State elites could no longer take the population of their territories for granted. As state subjects became state citizens they gained political rights. With these rights they could make demands on the state, a process culminating in the twentieth-century state. The result has been the massive growth in the state bureaucracy, in the proportion of gross national product going to the state, and generally in the number of state functions that impinge upon society. The nature of the state has changed fundamentally in the twentieth century.

Ron Johnston begins Chapter 3 with a series of quantitative indicators that illustrate the rise of the state in the twentieth century. Concentrating largely on the states of the developed world, he derives the basic functions of the modern state and shows how their operation has varied with the economic cycles state elites have had to cope

with. These functions have been expressed through the development of two parallel state structures, the welfare state and the corporate state. The former results from pressure for policies to provide for the economic security of citizens, the latter from policies to intervene in the economy to promote economic growth. Much of the twentieth century has involved the phenomenal growth of these two state structures. The latest downturn in the Kondratieff cycles – since about 1970 – has produced an important backlash against this growth of the state, and Johnston discusses the debate that was to result in a change of policies to curtail the state in the 1980s. In his comparative analyses Johnston is at pains to highlight similarities in responses across the countries he surveys, but also makes the point that every state's response is at the same time different. These differences magnify when we leave the developed countries of the First World and consider the poorer states of the world. Chapter 4 by Stuart Corbridge is about the states in this very different political world.

What Johnston's chapter illustrates is that, recent political attacks on the state notwithstanding, putting together the political and the cultural in the nation-state has produced a resilient but, above all, truly powerful social institution. The real mark of this power is that it has become hard to imagine a world that is not parcelled up into separate national sovereignties. States as nation-states appear to be almost natural entities beyond the realm of political discourse:

sovereign statehood . . . is now so ingrained in the public life of humankind and imprinted in the minds of people that it seems like a natural phenomenon beyond the control of statesmen or anyone else. When schoolchildren are repeatedly shown a political map of the world which represents the particular locations of named states in different continents and oceans they can easily end up regarding such entities in the same light as the physical features such as rivers and mountain ranges which sometimes delimit their international boundaries. It is nevertheless the case that not only the map itself but also the sovereign jurisdictions it represents are a totally artificial political arrangement which could be altered or even abolished. (Jackson, 1990, 7.)

Quite simply, the world does not consist of one or two hundred contiguous 'nations' waiting for states to put boundaries around them. Every nation-state is a particular political construction, very few of which are even close to being culturally homogeneous.

These observations are crucial for understanding the 'new states' created in the twentieth century out of the ashes of the European empires. Geographically one of the most interesting features of these new states has been that, by and large, they have maintained the boundaries imposed by the Europeans in the nineteenth century. According to Jackson (1990) they exhibit a 'negative sovereignty' which is imposed as a means for dealing with the outside world, but has little meaning for the populations living in the European-demarcated territories that are the new states. Hence African states, for instance, appear on the world political map and have joined the inter-state system (certified by membership of the United Nations), but they do not have any domestic citizenry articulating demands on the state. The 'positive sovereignty' of the people of nation-states in the 'developed' world in not found here. Hence there is no welfare state or corporate state, and nation-building and even state-building are extremely problematic in these circumstances.

Stuart Corbridge tackles these issues in his chapter on the new states of the Third World. As he points out, the contemporary pessimism of prospects for the Third World is in marked contrast to the naive optimism of the immediate postcolonial period. The 'failure' of the new states has brought to the fore the question of the legacy of colonialism and he deals with this in the first part of his chapter which he terms 'writing history'. The second part – his 'writing politics' – explores the reasons why the Third World has not lived up to expectations. In fact he is less pessimistic than many other observers and reminds us of the achievements of some Third World states. India, in particular, nearly made it to becoming a fully-fledged 'development state' according to Corbridge. But he concedes that the most common type of state to emerge has been various forms of patrimonial state which has not normally been an institution for progressive development. Nevertheless, there are new movements in civil societies in the Third World that both work with and bypass the state in em-

powering people to control their futures: the subalterns are speaking back and more. Corbridge concludes that this new postcolonial politics will have important lessons, not just for the political geography of the Third World, but for the First World also.

The American Century

The phrase 'American Century' is usually attributed to the publisher of *Life* magazine, Henry Luce, writing in 1941. The idea that the time was ripe for moulding the world in America's image was much older than this statement, however. The first sustained attempt came in the presidency of Woodrow Wilson who hoped to use US involvement in World War I to reorder international relations. With the failure of the USA to join Wilson's own League of Nations, American governments pursued a policy of minimal involvement – isolationism, always stronger in rhetoric than fact – until joining World War II. Lucy's revival of the idea that the century was indeed America's is part of the US political domestic debate on the road from isolationism to hegemony.

Hegemony occurs in cycles, and since US hegemony from rise through to the beginnings of decline occur in the twentieth century, the notion of the 'American Century' makes sense in political geography. Although US hegemony was finally confirmed by military victory in 1945, the concept of hegemony implies much more than military prowess. To define a state as hegemonic means that the country involved is qualitatively ahead of all rivals in economic performance, production, commerce and finance (Wallerstein, 1984). With this material leadership come opportunities to reorganize the world. Hegemonic powers are champions of liberalism both economically and politically; the USA promoted the United Nations, the General Agreement on Tariffs and Trade and the World Bank, for instance. In the event, US hegemony dominated about two-thirds of the world as the Cold War closed off a Soviet sphere of influence. The hegemony was not in doubt; the USSR was certainly never an economic rival and there were no such rivals for about a quarter of a century after 1945, the period sometimes termed 'high hegemony' at the centre of the hegemonic cycle. Hence, for a book on the twentieth century the USA warrants a chapter to itself, and this is provided by John Agnew as Chapter 5.

For Agnew, the USA has always been the 'archetypal liberal state', giving it key advantages over rivals in its rise to hegemony. The great irony of achieved hegemony was that the accompanying Cold War forced the US to become the great military power it had avoided being in the past. Agnew's chapter concentrates on these 'superpower years' culminating in the Reagan administration's attempt to force America 'back on top'. But fighting the Cold War has taken its toll, although winning the Cold War has left the USA undisputed geopolitical world leader it is now poorly equipped for the new geoeconomic competition. As the world's largest debtor and with an economic growth rate consistently below rivals, the USA has become the undisputed world political leader just as its hegemonic foundations have gone. This has led to what Agnew identifies as America's impasse: domestic prosperity and a world leadership role are no longer compatible.

In many ways America's impasse is similar to that of Britain in the decline phase of its hegemonic cycle. Certainly many of the problems of Britain's impasse as confronted by Mackinder and Hobson in Kearns' Prologue are reappearing in contemporary US debates. But we must be careful not to take the hegemonic historical analogy too far, which is why our Epilogue is so different from the Prologue. Agnew does not think that there is another state that will follow the USA as hegemon. The world is changing via contradictory trends of globalization and fragmentation. The former means that territorially-defined state economies have less salience in the operation of the world-economy, the latter that states are under threat from ethnic divisions below. In these circumstances the sequence of hegemonic cycles will conclude with the USA.

Agnew predicts a new era of 'transnational liberalism'. This American view of the future world order can be compared with those from beyond the North Atlantic in the epilogue. It seems closest to the view from Japan and furthest from the view from Africa. No surprise there; it confirms the need for geographical dialogue if America's impasse is not to become a global impasse in the next century.

References

Boulding, K. E. (1978) *Stable Peace*. Austin, TX: University of Texas Press

Churchill, W. (1923) *The World Crisis, 1911–18*. London: Butterworth

Gaddis, J. L. (1987) *The Long Peace*. New York: Oxford University Press

Jackson, R. H. (1990) *Quasi-states*. Cambridge: Cambridge University Press

Joll, J. (1985) Some reflections on the twentieth century. In N. Hagihara, A. Iriye, G. Nivat and P. Windsor (eds.), *Experiencing the Twentieth Century*. Tokyo: University of Tokyo Press, pp. 3–11.

Natkiel, R. (1982) *Atlas of Twentieth Century History*. London: Hamlyn-Bison

Wallerstein, I. (1984) *Politics of the World-Economy*. Cambridge: Cambridge University Press

Watt, D. C. (1984) *Succeeding John Bull*. Cambridge: Cambridge University Press

Williams, R. (1983) *Towards 2000*. London: Chatto and Windus

Prologue

Fin de siècle geopolitics: Mackinder, Hobson and theories of global closure

Gerry Kearns

Fin de siècle geoeconomics

The temptation to personify periods of time such as centuries is understandably strong when the calendar delivers the end of one and the start of another. It is a literary conceit to speak of the passage from one century to the next as a movement from one sort of world to another. Damnation or salvation may equally be expressed in these terms. The dominant tone among the educated bourgeoisie of Europe at the beginning of this century was pessimistic: mourning the twilight rather than celebrating the dawn. This is the structure of feeling usually termed *fin de siècle*:

For decades *fin de siècle* implied a 'go to the dogs' feeling that was thought to pervade European 'civil-ised' society in the years around 1900. This mood of malaise certainly affected individuals and sections of aristocratic as well as bourgeois social background towards the end of the nineteenth century. Underlying it was a cocktail of lamentations for the past and fears of the future, countenancing the notion that human progress was being brought to a halt, if not to an end. This evaluation now, as we ourselves approach another turn of the century and look back, appears distinctly simplistic. (Teich and Porter, 1990, p. 1.)

Simplistic or not, for one part of 'civilized'

Europe historians have strongly corroborated this sense of crisis. Bernard Porter's (1987) history of the British Empire terms the years 1895–1914 those of 'Crisis'. Paul Kennedy (1991) writes of the end of Pax Britannica' during the period 1897–1914. In many ways the crisis of the British Empire was seen in recognizably geographical terms: in terms of the relative prosperity of different economic arenas, in terms of the strategic reach of British power, in terms of how the British should cultivate their territories. The British crisis was presented as one of global economics, global strategy and global development. As we approach our own *fin de siècle*, perhaps we might learn something from a brief survey of the sense of crisis which coloured the early years of this century. I will do this by comparing two influential views of the crisis of British imperialism, but first I want to sketch in the general economic context of Britain's place in the world-economy at the end of the nineteenth century.

The crisis was essentially one of British imperialism. Imperialism is a term which is defined rather differently within various theoretical traditions. For our purposes we may take the following as a lowest common denominator: imperialism is the use of the resources of one place by people from another where the terms of the extraction are set by the inequality in power between the two sets of peoples. This, then, would include both the displacement of local peoples by European colonialism (taking the land by force) as well as the exploitation of weak countries today through what is often termed unequal exchange (the resources are taken on the cheap). The forms of imperialism have changed over time.

In the fifteenth century, several European nations, having reached the limits of their internal agricultural frontiers, given contemporary farming practices and prevailing price levels, began to look for areas overseas to 'plant' productive enterprises on for the purposes of expanding their tax revenue (Wolf, 1982). The Spanish, having expelled the Moors and recolonized southern Spain, led the European rush out of the Mediterranean into the Atlantic world, establishing plantations on the Canary Islands. This new imperialist machine was soon exported to Latin America where it served to suck out the silver which oiled European trade with South East Asia and brought Asian spices and tea back to stimulate jaded European palates. In the seventeenth century, European nations such as Holland and England had established trading posts in South East Asia, but exercised rather little control over these societies and we may think of trade in this region as being on a more or less equal footing. Similarly, in the earliest trade with the peoples of North America, the European settlers, be they French or English, were as desperate to trade with the local peoples to secure food as were those groups to get access to European goods. Very quickly, this era of mercantile capitalism saw the Europeans work to change the terms of exchange. In North America, they secured an independent supply of food and took land from the local peoples by force. In South East Asia, the British began to insert themselves in the political and tax-gathering structures of India and, while protecting their domestic market from Indian cottons, they redirected the production of raw cotton to the needs of British mills.

As Britain managed to fund more of its trade through the export of its own manufactures, world imperialism moved from the period of merchant capitalism to that of industrial capit-

alism. Increasingly, the terms of trade were set by the exchange of high-value manufactures for low-value primary products. Britain, as the workshop of the world, commanded the wealth of the world, and poor countries hoped to fund their own development through exporting primary-goods into the most buoyant and open world market – the British. Britain's domination was staggering. It produced six times as much coal as the rest of the world in 1800, and coal was the crucial precondition for industrial production in the age of steam (Pollard, 1981). Such a headstart was unsustainable, and when European markets began to recover from the torrent of British competition released by the end of the French revolutionary wars, first Belgium and then northern France and then the German coalfields began to eat into the British lead and provide the basis for their own domestic industrialization. In the early 1850s, Britain still produced twice as much coal as the whole of the rest of Europe. By the late nineteenth century, the British were relying rather heavily upon those industries on which their early lead had been based. That is, rather than moving into the new industries such as chemicals in which technological advance was now most rapid, the British continued to rely upon the cheapness of their coal to give them a competitive advantage in the steam-driven cotton textile industry. This threatened the British ability to continue winkling primary products out of other countries. Those countries would be as well placed trading with Germany or the United States. Furthermore, the British could no longer actually sell their own manufactures in these German and American markets where they were often undersold by more efficient producers whose development had been nurtured behind tariff walls.

In response to the failure of British industry to diversify and with the temptation of higher rates of return elsewhere, British investors swilled their spare capital overseas where it funded infrastructural development (notably railways) in the poorer countries and industrial competition in the richer countries. By 1913, one-third of all British wealth was invested abroad: a truly staggering amount and without historical precedent (Edelstein, 1981, 70). Furthermore, the reliance of the British on importing primary products was significant, with four-fifths of their wheat coming from abroad (Foreman-Peck, 1983, 106) – and this in a country which had been almost self-sufficient in food a century earlier. The importance of the exporting of investment has led many commentators to see a new era of world imperialism in the late nineteenth century, a period moving to the sway of finance capital.

The situation was heavy with portent. To understand the economic basis of the crisis of British imperialism, it is important to break the national economy down into its agricultural, industrial and financial sectors. They each saw the future in rather different terms. The agricultural depression of the mid-1870s was a serious shock. The completion of the railway links into the wheat belt of the United States and a series of excellent summers brought an unprecedented quantity of American wheat on to the market in 1873. This was a short-term phenomenon, as the urbanization of North America in the last quarter of the century would soak up much of this new bounty, but surplus American wheat elbowed British wheat out of a relatively open market for some years, sending the British landed interest back to examine the small print on the social contract which had marked the opening of the era of free trade with the repeal of the Corn Laws in 1846. But the true significance of the so-called Great Depression was that it signalled a quickening of competition in all primary-goods markets as a result of a dramatic fall in long-distance transport costs. The writing was on the wall for British coal as well as for British wheat.

British industrial capital and British finance capital shared common interests as long as Britain remained the workshop of the world and free trade would secure markets for the former and reliable money for the latter. The response of Britain's industrial competitors was crucial. With tariffs introduced in Germany in 1879 in response to the depression, and France, the United States and others following, the British were drawn into an entirely new game. Free trade and peace oiled the flow of capital to and from overseas markets. With sterling as the currency of international exchange, the City of London could basically levy a tax on world trade for administering the system (Ingham, 1984). As

Britain's industrial rivals went off the gold standard, they ensured that their imports would largely be balanced by exports of their own products, but the decreasing attractiveness of British products, combined with Britain's retention of the gold standard, threatened to lead to a drain on Britain's gold reserves as imports grew. The City was keen to defend a sterling trading area against the protectionist tendencies of Britain's industrial rivals so that a large volume of world trade would continue to be conducted in sterling, thereby enriching the City. Furthermore, the City was anxious to retain South African gold within Britain's trading orbit since it was now so crucial to the convertibility of sterling. Trade with India, which produced a net flow of sterling to London as well as a large quantity of goods for re-export, was likewise central to Britain's balance of payments. It seems likely that the increase in the size of Britain's formal empire in the late nineteenth century was largely a defensive measure to keep as much as possible of world trade under the terms of free trade conducted in sterling. This was the nature of the City's strategy.

The needs of industry went further than this. Industry wanted some measure of protection at home to prevent Britain's competitors underselling British producers, by fair means (greater efficiency) or foul (state-subsidzed dumping of surplus production). Industry naturally supported the extension of the formal empire as a defensive measure against the annexation of colonies by Britain's rivals, beginning in 1884 with Germany's dash for African territory. Far more disturbing, however, was the Boer War of 1899–1902. The British belief in peace and free trade had always been interpreted flexibly enough to allow the use of military muscle in the periphery to soften up markets or lever resources in the direction of British investors. Imperialist wars had largely been fought and won against black people. In the Boer War, Britain became drawn into a war against a rival group of settlers, the Dutch farmers. In a staggering display of 'fair play', both sides refrained from the 'dangerous' practice of employing black mercenaries to prosecute the war. The Dutch farmers gave the British the run-around. Liberal Britain set up murderously unhealthy concentration camps for captured Boer settlers. The patriotism of the industrial north delivered up a crop of soldiers, up to half of whom in some towns had to be rejected as unfit to die for their country. The domestic political spectrum focused on the need for social reform to breed a race fit to defend its empire, an empire which was becoming more formal and, therefore, more of an affront to jealous rivals.

For industrialists this also seemed to offer an opportunity to bury the class struggle with their employees in an orgy of patriotic social welfare. Social reform plus industrial protection appeared to have something to offer everyone: everyone, that is, excpt the City. The last thing the City wanted was any suggestion that the British economy become more autarkic. This would reduce the attractiveness of sterling as the means of international exchange and it would have to be accompanied by the abandonment of the gold standard and the convertibility of sterling. It would be to renounce economic interaction with the most dynamic economies and would reduce the net volume of trade conducted in sterling to defend an altogether smaller economic arena. The City's levy on world trade would plummet.

These economic relations may be illustrated through a brief comparison of Britain with its two major rivals, Germany and the United States (all data are calculated from tables in Mitchell, 1975 and 1983). By the end of the nineteenth century, Britain was still the most industrialized of these economies; only 7 per cent of its national product came from agriculture compared to 17 per cent in the United States and 30 per cent in Germany. Yet the two latecomers were growing more quickly. The British Gross Domestic Product was only one-third higher in the second half of the 1890s than in the first half of the 1880s. In Germany the increase was two-thirds and in the United States, four-fifths. Indices of industrial production tell the same story. Over the same period the British increase was only two-fifths compared to about nine-tenths in both Germany and the United States. In relation to the total volume of trade (imports plus exports of goods), the balance of trade for goods was consistently and increasingly negative for Britain but substantial earnings from services, banking, insurance, investments and shipping (the

invisible balance of payments) kept the overall current account balance in surplus (see Figure Prol.1). In many ways the German economy was growing to look more like the British with a deteriorating visible balance of payments matched by an improving balance on invisible services. The strength of the American economy in agriculture and manufacturing is shown by the extraordinary growth of the visible surpluses in the late 1890s which began to overtake the drain implied by an unfavourable balance on services.

The relations between the three economies may also be illustrated by calculating a set of figures for bilateral trade. In this case, we may look at the ratio of imports to exports. Transport costs are included in the value of imports but not exports, so that the value of exports from Britain to Germany, for example, is not the same as the value of imports from Britain received by Germany. Nevertheless, the picture shown in Table Prol.1 is reasonably clear Whereas the trade in goods was roughly equal between Britain and Germany, the United States had an increasingly positive balance of trade with the other two, and especially with Britain. However, as Table Prol.2 shows, the British market was far more important to American exporters than vice versa. In other words, the American penetration of the British market was far greater than the British penetration of the American market. No such gross inequality is revealed by the bilateral trade in goods between Britain and Germany.

These few statistics highlight the crises for British industrial and financial capital. The reliance of the British on invisible earnings is a testimony to the importance of finance capital in the national economy. The fact that Britain was falling behind Germany and the United States in industrial development is also clear, and the failure to compete in advanced markets is revealed most directly in relation to trade with the United States. In the absence of industrial supremacy, what might be the possible attraction for international traders in transacting their business in sterling? Would it only give them access to a market they were increasingly avoiding? Here, the idea of the Empire became increasingly important as a fillip to the British balance of payments, as a source of gold and a set of assets to sustain the convertibility of sterling. For the

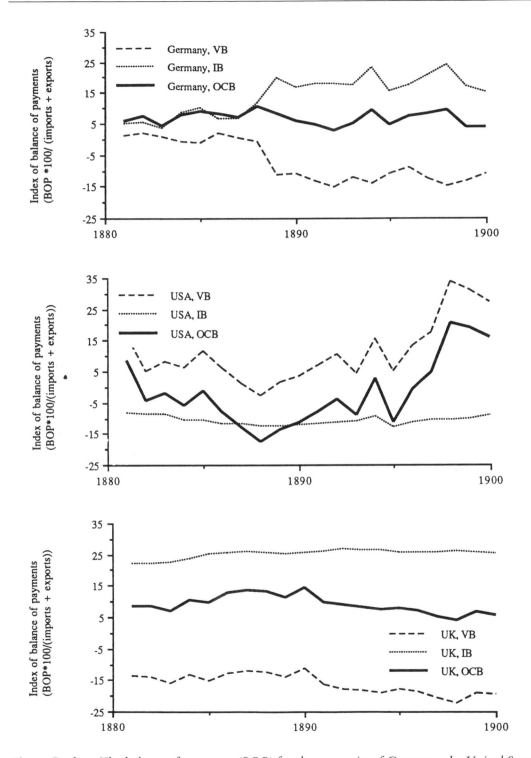

Figure Prol.1 *The balance of payments (BOP) for the economies of Germany, the United States and the United Kingdom, 1881–1900; distinguishing the visible balance (VB) on goods, the invisible balance (IB) on services and the resulting overall current balance (OCB)*

Table Prol.1: *Bilateral trade, the ratio of imports to exports, 1881–5 and 1896–1900*

	Bilateral trade with		
	Germany	United States	United Kingdom
1881–5			
Germany		0.76	0.89
United States	0.95		0.41
United Kingdom	1.39	3.42	
1896–1900			
Germany		2.06	0.80
United States	0.67		0.30
United Kingdom	1.19	6.54	

industrialists, the Empire appeared to promise protected markets, a strategy they wanted adopted at home too. Yet the City was uninterested in such restrictions on trade at home or abroad. Their aim was to keep the colonies trading openly and thus pre-empt the closure of these markets by Britain's imperial rivals.

The world capitalist economy does not develop evenly. National economic strategies are often negotiated between conflicting class interests against a background of depression, crisis or war. The particular sequence of crises punctuating the relative rise of Germany and the United States and the relative fall of Britain not only registered this seismic shift in geoeconomic power but were also instrumental in engineering the shift. Crises (be they wars or depressions) serve to redistribute economic and political value within the world-system of competing nation-states. Bankruptcy and defaults on loans devalue certain existing capital stocks revealing how chimerical the expectation of sales had been from certain sectors of an economy. War is even more significant. In the first half of the 1880s 31.5 per cent of the British government's spending went on the army and navy; by the second half of the 1890s this had risen to 39.5 per cent (calculated from figures in Mitchell, 1988). Maintaining the British position in the world was imposing a real strain on an economy which was not growing as quickly as those of its main rivals.

The First World War merely confirmed this weakness, since the loans which allowed the war to be fought had to repaid. The main beneficiary was the United States, and the massive share of British wealth held abroad was partly liquidated for this purpose. While Britain had only managed to supply one-quarter of the Allied troops in the war, it had met two-fifths of the Allies' costs. Indeed, the Central Powers taken together only spent 7 per cent than did Britain alone (Kennedy, 1991, 354). Europe had forgone an enormous quantity of industrial production to fight the war and had also lost a significant part of its labour force. The United States had not. Furthermore, the United States was now the world's greatest creditor nation and backed by the world's most dynamic economy, the dollar was increasingly the currency of first preference in international exchange. The weakness of the British economy, the evaporation of so many of its overseas assets, the ballooning national debt,

Table Prol.2: *Bilateral trade, the proportions of imports from and exports to various countries*

	Proportion to/from		
	Germany	United States	United Kingdom
Imports, 1881–5			
Germany		0.03	0.14
United States	0.08		0.25
United Kingdom	0.06	0.23	
Imports, 1896–1900			
Germany		0.15	0.12
United States	0.12		0.19
United Kingdom	0.06	0.26	
Exports, 1881–5			
Germany		0.04	0.16
United States	0.08		0.52
United Kingdom	0.06	0.09	
Exports, 1896–1900			
Germany		0.10	0.19
United States	0.11		0.41
United Kingdom	0.08	0.06	

and the pressure for increased social spending at home did not inspire confidence in the ability of the British to maintain a gold standard for sterling. The return to the gold standard in 1925 was the City's last desperate attempt to restore sterling's hegemony. But the essential contradictions of the British economy would not go away and the policy resulted in an overvalued pound, attractive to investors but crippling for industrial exports.

It is clear, then, that Britain's changing position within the world-system entailed a reorganization of the national economy. The resorting of the world-economy largely takes the form of the restructuring of national economies. Not surprisingly, then, contemporary views of Britain's geoeconomic and geopolitical future were largely coloured by this recognition of the need to address domestic political issues at the same time as the international question. In the remainder of this chapter I want to introduce two contemporary views of the British crisis: that of the geographer, Halford Mackinder (1861–1947) and that of the economist, John Hobson (1858–1940). They stressed very different aspects of this crisis and the strategies they proposed were in direct contradiction. Yet they not only spoke in terms of gross national product, of balance of trade, or of indices of industrial production; they also saw the British state as an organism with clear biological and territorial needs. They were the unconscious prisoners of the prevailing metaphors of their own day, just as we may, unwittingly, be of those of our day. I shall conclude with some brief remarks about the blindspots inscribed in these metaphors.

Mackinder and Hobson

Mackinder made an outstanding contribution to the development of British geography (Blouet, 1987; Kearns, 1985): as a lecturer – being the first Reader in Geography at Oxford in modern times and holding at one point academic positions simultaneously at three universities (Oxford, the London School of Economics, and Reading); as a writer of textbooks – most notably *Britain and the British seas* (1902), the first fruits of his New Geography; and as campaigner for the cause of geography both in intellectual and political circles. He was also a moderately important political figure, as a Member of Parliament between 1910 and 1922 and as part of an extremist group of ultra-imperialists in British politics united around Joseph Chamberlain's tariff reform, or protectionist, campaign (Semmel, 1960; Parker, 1982).

Hobson had little formal academic success, being offered neither lectureship nor chair at any British university, berating the increasing specialism of professional academia and presenting himself as an economic heretic (Hobson, 1976). He stood for Parliament once, in 1918 as an Independent, and lost. Yet his journalism and his books were important to political debate among the Left within the Liberal Party (which he quit in 1916) and later within the Independent Labour Party (which he ultimately joined in 1924). Hobson attacked the protectionism of Chamberlain and his band of Unionists (later Conservatives) and fought the creeping protectionism within the Liberal Party itself. His economic ideas were taken up enthusiastically by Keynes and his work on imperialism was central to Lenin's own writings on the topic (Townshend, 1990).

Mackinder presented his global vision in a number of places and it informed, as we shall see, the causes for which he struggled in the House of Commons. In 1900 he gave classes to trainee bankers on the historical geography of world trade (Mackinder, 1900) in which he set out the physical context of the distribution of nomadic and settled peoples in Eurasia and described the development of trading relations between the settled peoples of Europe and Asia via the zone of the nomads, the region inaccessible by river and thus beyond the reach of the settled powers of the moist coastlands. In 1904 he spoke of the 'geographical pivot of history' (Mackinder, 1904) as the pressure exerted on economic development in Europe and Asia by this nomadic zone, before drawing his audience's attention to the fact that the basket of resources contained therein was now available for mobilization by a settled power which would integrate the region by railway. This vast landpower might, organized either by Germany or (eventually) Russia, support a sea-power which could challenge Pax

Britannica and cut off the international links which alone could sustain Great Britain in the manner to which her former industrial supremacy had accustomed her. After the First World War, Mackinder (1919) argued that transport technology had advanced sufficiently that the region in Eastern Europe and western Russia which a coherent landpower could defend against sea-power was now bigger and included the whole basin of the Black Sea. The aim of the Allies at the Peace Conference in Versailles should be to so fragment the heartland among a number of independent nations that it could never fall under the sole sway of either Germany or Russia.

In this dangerous new world, Great Britain could not survive as a world leader if it had to rely on its industrial muscle alone. It had to forge alliances, either with the Anglo-Saxons in exile in North America or by organizing the Empire into a single political and economic force and throwing those vast resources into the balance against any continental power which might threaten the global reach of this sceptred isle. The unity of the Empire, argued Mackinder, could only be ensured if it was generally more worthwhile for its members to trade with each other than with the rest of the world. An imperial tariff which discouraged our competitors from stealing markets or resources from the Empire was, according to Mackinder, a crucial first step. Beyond that, the British should give serious thought to the creation of an Imperial Parliament which could take at least the Dominions (self-governing, white-settler colonies) fully into the confidence of His Majesty's Government on matters of war and trade.

Hobson's world-view was expressed in a truly daunting volume of work: 'He wrote over forty books, and published more than 600 reviews and articles on a large variety of economic, political and literary topics' (Townshend, 1990, 8). Hobson's point of departure was an attack on the idea that a capitalist economy was automatically self-balancing. According to Say's Law, all production created its own demand and the market will automatically clear all goods produced through balancing supply and demand, so long as economic agents act rationally. Important to this optimistic view of the free market was the claim that what was rational (and selfish) for the individual was ultimately rational (and altruistic) for the social unit as a whole. Hobson argued that this easy assumption was simply untrue (Rummery and Hobson, 1889) in the important case of savings. While thrift was frequently rational from the individual's point of view, for too many people to save too much deprived society of investment and reduced the effective demand for the goods which had already been produced. The latter was the more serious consequence and was, believed Hobson, the cause of the periodic gluts of apparent overproduction, idle capacity and unemployment which ought to lay to rest the spectral image of a harmonious and self-balancing capitalist free market. The tendency to crisis was endemic and not an occasional hiccup or blip. Hobson never departed from this under-consumptionist theory of capitalist crises and, to his mind, here lay the explanation for what he saw as the unnatural imperialist drives of the British and other Great Powers. Unable to employ their spare capital at home since no effective demand existed there, they strove to engage it under beneficial terms in the world's periphery. There, imperialist economic development sustained the infrastructural investment which guaranteed them a good return, and the panoply of costs attending an imperial policy (war, administration) soaked up more of their idle funds, making the British state the guarantor of their dividends.

In 1894, Hobson published *The evolution of modern capitalism*, where he argued that machine production exaggerated the problem of under-consumption by bringing unwanted goods to market more quickly and in larger quantities: 'The general relation of modern machinery to commercial depression is found to be as follows:– Improved machinery of manufacture and transport enables larger and larger quantities of raw material to pass more quickly and more cheaply through the several processes of production. Consumers do not, in fact, increase their consumption as quickly and to an equal extent' (Hobson, 1916, 286). But it was still the psychology of saving which was seen as preventing consumers exercising effective demand. By 1902, with the publication of *Imperialism*, Hobson, as Townshend (1988) points out, had

moved away from a purely psychological explanation of over-saving to an understanding of the
uneven distribution of demand in society in
terms of the unequal bargains struck between
poor worker and rich employer in the labour
market and between poor consumer and monopolistic capitalist in the goods market. This
inequality introduced such distortions into economic bargaining that the rich and powerful were
able to retain far more of the social product than
they could gainfully deploy, thus creating the
phenomenon of over-saving.

We might say that Mackinder wanted Britain
to be the world's greatest imperialists, whereas
Hobson wanted the British to abjure this strategy altogether. This coloured their world views
totally. While Mackinder celebrated people such
as Cecil Rhodes who urged imperialist expansion
in Africa, Hobson called Rhodes 'unscrupulous'
and 'short-sighted' (Hobson, 1988, 231). For
Mackinder it was precisely free-trade liberals like
Hobson who failed to take the long-term view.
Addressing the Liberals in his maiden speech in
the House of Commons, Mackinder drew a
contrast between short-sighted Britain and far-
sighted Germany:

> At the present time democracy itself is at stake, and the
> question is whether democracy, with the kind of argu-
> ments supplied to it by hon. Gentlemen opposite, will
> be able to take a forward view, to look into the future,
> and do as other countries are doing, notably those
> where there are bureaucratic and undemocratic
> Governments. Whether we shall as a democracy be
> able to hold our own with the results of those scientific
> policies which are possible in countries in which
> parties matter little . . . (*Hansard's Parliamentary
> Debates*, 5th series, vol. 14, House of Commons, 23
> February 1910, cols. 322–3.)

The contrast between Mackinder and Hobson
seems complete.

Global economics

Both Mackinder and Hobson were convinced
that very great changes were under way in the
world-economy. We might rather loosely refer
to their different conceptions of global closure.
By this I mean that they both thought that territorial conflicts between the Great Powers in the
periphery of the world-system were likely to

become more intense in the near future. This
development was thought by both of them to be
pregnant with dangers for Britain. Beyond this,
however, they saw things in very different ways.
I want to draw attention to two features of their
ideas which underwrite their contrasting prescriptions for the British economy. Mackinder
disagreed with Hobson about the nature of the
disjuncture in world-economic development
which ushered in the new perilous years. That
disagreement rested on a fundamental conflict
between their respective views of the nature of
economic wealth.

Mackinder and the protectionist empire

Mackinder spoke explicitly of a closed world
political system in which regional conflicts rever-
berated off the walls of a finite and largely
spoken-for set of territories (Kearns, 1984). This
territorial closure of the world was a fundamentally new condition of things, according to
Mackinder. It followed the 400 years of the
Columbian era when friction between the Great
Powers could be debouched onto the open
spaces of the 'uncivilized' periphery. It was unlikely that Britain would ever command more
territory than it already did. Furthermore, it now
not only had to defend this estate against its
neighbours, Britain had also had to watch its
rivals develop their own estates. As they did so
they would seek to deny Britain access to any
markets they could command. Britain must keep
markets it could not monopolize open, and it
must try to keep its rivals out of those it could
control. The crucial dimensions of this zero-
sum game were the implicit (and recognizably
ecological/geographical) claims that the basis of
wealth was territory and that the roots of power
lay in demography. Sentiment could unite and
organize people and their earth.

Let me illustrate this. One of the subjects on
which Mackinder spoke most frequently in the
House of Commons was that of the tariffs policy
of Canada:

> I think the House will agree with me that the geo-
> graphical position of Canada may be represented by
> saying that, geographically, Canada is an artificial con-
> struction. The frontier thrown across the Continent

runs at right angles to all the natural features, and Canada has been created in the face of the greatest natural temptations to fall into a great Continental State together with the United States – in other words, Canada is the greatest and most signal instance at the present moment of a state which has resulted from a settled policy acting under modern conditions, and the abandonment of the *laissez faire* policy, which was no doubt suitable to the time of Adam Smith. (Hansard, vol. 19, 21 July 1910, cols. 1469–70.)

In placing a high tariff against goods from the United States, the Canadian government was trying to accentuate the dependence of western Canada upon eastern Canada for its imports – imports ultimately derived from Europe. Eventually, western Canada would develop its own manufactures and the treacherous early years would have passed, leaving only those sentimental and communications filiations running east–west along Canadian, rather than north–south along American, lines. Tariff measures forge territorial relations.

The sentiment had to come first, and be nurtured. On this basis, Mackinder urged that a viable way to prevent the unification of the Eurasian heartland, which he so feared, would be to create a series of coherent spatial and ethnic units in south-east Europe out of the rubble of the Habsburg, Ottoman and, eventually, Russian empires. Speaking of his conception of Britain's intentions during the First World War, he said:

One of the foremost of those aims must be to secure a resorting of the peoples of South-Eastern Europe in such a way that South-Eastern Europe shall be built of bricks and not of pieces of bricks; that we may have a structure which is strong and based on nationality, and not a mere heap of peoples kept together by military control. (*Hansard*, vol. 93, 16 May 1917, col. 1652.)

Similarly, he believed that one way of constraining the Soviet Union's access to the Mediterranean would be by building a state of South Russia out of the Slav peoples who lived between the Russian and Polish peoples (*Hansard*, vol. 129, 20 May 1920, col. 1713).

Tariff reform was vital to the development of this sentiment. This was a truth which he believed Germany, with its 'scientific' scheme of tariffs, had grasped more firmly than any other Great Power. In the depths of the First World War Mackinder was perfectly sanguine about the

debts the British were incurring in the United States, seeing it as a 'mere accumulation of wealth' which would come scurrying back to fund the recovery in Europe after the war. Instead, 'What we have to face at the present moment is . . . still the very solid and serious fact of the economic strength of Germany, and that it will require the strength of our Empire to meet it.' He warned:

What is the position you have got in Germany, a great organising centre? She has, for practical purposes, more than double . . . her normal productive area. Germany has annexed, for the time being, all the region producing food in Hungary . . . She has Poland, Belgium, a portion of France, and now . . . to some extent she has annexed the resources also of an undefined area in the Near East.

Germany is treating all that as a great estate. (*Hansard*, vol. 77, 10 January 1916, col. 1340.)

After the war he never lost sight of this German threat. In 1921 he was warning that Germany would dump goods at low price in Britain simply to wreck British productive capacity, compromising Britain's economic independence and softening the country up for another bruising war:

Her whole policy has always been to equip herself as a complete economic unit, and the result will be that she will put out of action a certain amount of your skill and plant in those things which temporarily she was monopolising. Having done that, and having injured you to the extent that you will not be able to restore that skill and that plant, except after that lapse of a certain number of years, she will then return to the position she had before the War, and will attract to herself every trade which will enable her to work upon you a second time. (*Hansard*, vol. 143, 30 June 1921, col. 2407.)

In the new closed political system, nationalism and protectionism helped countries mobilize the resources of their earth and their people. The old ways would not work any more. The free exchange of products based on environmentally-determined natural advantage was inappropriate in a world where the transport revolution had rendered industry footloose and fancy free: 'in Adam Smith's day the theory was that the natural facilities, climate, food production of the nation and so forth, gave it an advantage which decided what should be its main product for the purpose of exchange. In the world today that is all changed. Today you can place your industry where you choose' (*Hansard*, vol. 141, 9 May 1921, col. 1565). A country without access to the full complement of modern industries was vulnerable, would be a push-over in a war, and thus would attract the bellicose attention of more well-balanced nations. Protection was strength and free trade was weakness, a 'wishy-washy philanthropy' (*Hansard*, vol. 93, 16 May 1917, col. 1565). For Mackinder this was the choice between wealth and efficiency. With the changed circumstances of the post-Columbian age, 'the time had come when defence was more important than opulence' (*Hansard*, vol. 114, 25 March 1919, col. 330). The free-trade policy would make the British merely a 'rich, small *bourgeois* nation' like the Dutch. '[O]n the other hand, you may determine to face the conditions of the future and you may determine that this nation will make its fortune afresh, that the strength of this nation shall lie, not in these fat *bourgeoisie*, living on the fortunes of the past, but in a great, vigorous, magnificent nation of workmen' (*ibid.*, cols. 333–4).

In this regard, Mackinder distrusted effete, cosmopolitan banking-capital, placed out in foreign lands beyond the empire or held in anticipation of those distant needs, 'kept relatively idle for the purpose of meeting contingencies in remote portions of the world' (*Hansard*, vol. 77, 10 January 1916, col. 1337). The City had little to do with vigorous, magnificent industrial Britain:

What money usually goes into industry? Not, as a rule, any large quantity of money that is acquired by staid people in the South of England. A few shares, usually wild-cat shares, may be bought by them, but the money that keeps the industries in the North going is what I may call money that is already in the industrial fund. If you want some thousands of pounds to put into a new factory, where do you go for it? Not to the City of London, you go to a manufacturer in your own neighbourhood . . . (*Hansard*, vol. 82, 18 May 1916, col. 1727).

Yet British industry could not hope to survive, even behind the most 'scientific' scheme of tariffs, unless it had access to all the crucial raw materials it needed. Herein lay the peculiar potential of Britain's imperial 'estate'.

Hobson and the atavistic Empire

Hobson, like Mackinder, recognized a quickening in the pace and a change in the character of the contest between the Great Powers over the fate of the Empires. For Hobson, however, protectionism was the cause of, rather than an intelligent response to, this world crisis. The crucial shift was seen as being from colonialism to imperialism. By colonialism, Hobson meant the transplanting of some section of the host society overseas: the basis of what we might call the white settler colonies – clearly Canada, Australia and New Zealand, more questionably South Africa. Imperialism, in contrast, was the control of new territory overseas by the home country (Hobson, 1988, 7). This was the extra factor introduced by the Scramble for Africa and the floating of the project for a new protectionist Empire: 'when Joseph Chamberlain set out to convert the Empire into a close preserve by his policy of tariffs and preferences, and the magnificent projects of Cecil Rhodes began to influence the mind and language of English politicians, the larger significance of our Imperialism became manifest' (Hobson, 1976, 59–60). This larger significance was not lost on Britain's rivals, and 'aggressive imperialism' must, therefore, 'defeat the movement towards internationalism by fostering animosities among competing empires' (Hobson, 1988, 11). In 1902, Hobson pointed to the imperialist wars in the periphery – 'most of these wars have been directly motivated by aggression of white races upon "lower races", and have issued in the forcible seizure of territory' (*ibid.*, 126). But he also noted that the Great Powers were girding themselves for the day when they would fight each other over the exclusive rights they claimed to their new territories. The New Imperialism was a repudiation of free trade and a desecration of the true principles of the British Liberal party:

By the close of the century this Imperialism, composed of force, finance and false philanthropy, now masquerading as defence, now as mission, now as commercial policy, had secretly eaten its way into the vitals of the Liberal Party. The connivance, consonance, nay the active participation, of Liberals in the wreckage of South Africa [the Boer War] was a revelation of the measure of this betrayal of Liberal principles. (Hobson, 1974, vii.)

This intensified, 'insane' imperialism marked the beginning of a new era:

The statement, often made, that the work of imperial expansion is virtually complete is not correct. It is true that most of the 'backward' races have been placed in some sort of dependence upon one or other of the 'civilised' powers as colony, protectorate, hinterland, or sphere of influence. But this in most instances marks rather the beginning of a process of imperialisation than a definite attainment of empire. The intensive growth of empire by which interference is increased and governmental control tightened over spheres of influence and protectorates is as important and as perilous an aspect of Imperialism, as the extensive growth which takes shape in assertion of rule over new areas of territory and new populations. (Hobson, 1988, 223.)

This new 'insane' stage in world capitalism resulted from a series of conscious choices by the Great Powers, most notably Great Britain. Mackinder's account of the sources of national wealth was, Hobson would surely have argued, an example of the self-seeking tendency of imperialists to present their economic strategy as natural rather than chosen. In Hobson's terms, the imperialists had a purely quantitative understanding of the relations between people, earth and wealth:

Imperialism is a depraved choice of national life, imposed by self-seeking interests which appeal to the lusts of quantitative acquisitiveness and of forceful combination surviving in a nation from early centuries of animal struggle for existence. Its adoption as a policy implies a deliberate renunciation of that cultivation of the higher inner qualities which for a nation as for an individual constitutes the ascendancy of reason over brute impulse. (*Ibid.*, 368.)

Yet, interpreted correctly, the state of the world-economy revealed the potential for alternative, qualitative economic strategies.

In broad terms, we might say that Hobson saw the emergence of a higher, qualitative form of economic growth and that he saw growing evidence of the potential for international co-operation rather than competition. In evolutionary terms, Hobson contrasted primitive organisms like the sponge which appeared to have no clear centre of conscious life (the sensorium) with more advanced organisms like human beings which did. Societies which refused to direct their development were the blind servants of lower nature, but in many societies there was

an 'educated' class which acted as the sensorium and which could direct the group will to purposive, national ends: 'Society is certainly a psychical organism, a moral, rational organism with common psychic life, character and purpose' (Hobson, 1974, 74). Hobson argued that as societies evolved they moved towards the cultivation of higher and higher needs. In this regard the share of the social product going to satisfy basic biological need would fall. Yet it was precisely these low-order basic goods that dominated colonial trade. Only about one-fifth of the British workforce was engaged in production for export markets of any description. The truly dynamic markets were those of our industrial rivals and not those of our own colonies. Imperialism was not worth the candle, it 'has been bad business for the nation' (Hobson, 1988, 46). However, one group clearly benefited from imperialism in terms of money, not goods. Here Hobson makes common cause with Mackinder in denouncing the *rentier* capitalists of metropolitan England:

Aggressive Imperialism, which costs the taxpayer so dear, which is of little use to the manufacturer and trader, which is fraught with such grave incalculable peril for the citizen, is a source of great gain to the investor who cannot find at home the profitable use he seeks for his capital, and insists that his Government should help him to profitable and secure investments abroad. (*Ibid.*, 55.)

Whereas foreign trade brought £18 million in profits to Britain each year, the profit of foreign investments was five times as high. This was a uniform development throughout the rich countries: 'The growing cosmopolitanism of capital has been the greatest economic change of recent generations. Every advanced industrial nation has been tending to place a larger share of its capital outside the limits of its own political area, in foreign countries, or in colonies and to draw a growing income from this source' (*ibid.*, 51). Over-saving operated as a uniform cause of imperialism throughout the rich world, but the efficiency of this connection depended upon the power of finance capital:

The stronger and more direct control over politics exercised in America by business men enabled them to drive more quickly and more straightly along the line of their economic interests than in Great Britain.

American Imperialism was the natural product of the economic pressure of a sudden advance of capitalism which could not find occupation at home and needed foreign markets for goods and for investments. (*Ibid.*, 79.)

Imperialism was, therefore, a class interest which diverted attention from the pursuit of higher things into a base struggle for brute reproduction. This perversion is the feverish outgrowth of the unnatural wealth falling out of what we would now term unequal exchange:

The over-saving which is the economic root of Imperialism is found by analysis to consist of rents, monopoly profits, and other unearned or excessive elements of income, which, not being earned by labour of head or hand, have no legitimate *raison d'être*. Having no natural relation to effort of production, they impel their recipients to no corresponding satisfaction of consumption: they form a surplus wealth, which, having no proper place in the normal economy of production and consumption, tends to accumulate as excessive savings. (*Ibid.*, 85–6.)

Somehow this 'sinister' class interest managed in country after country to pass itself off as the general interest. Here Hobson offered a rather simplistic conspiracy theory and a more sophisticated account of the deformation of public discourse. First, the conspiracy of finance capital:

United by the strongest bonds of organisation, always in the closest and quickest touch with one another, situated in the very heart of the business capital of every state, so far as Europe is concerned, chiefly by men of a single and peculiar race, who have behind them many centuries of financial experience, they are in a unique position to manipulate the policy of nations. No great quick direction of capital is possible save though their agency. Does any one seriously suppose that a great war could be undertaken by any European State, or a great State loan subscribed, if the house of Rothschild and its connection set their face against it? (*Ibid.*, 57.)

A sensationalist press, an overheated but uncritical public drugged by passive spectator sports and music hall and 'a sea of vague, shifty, well-sounding phrases which are seldom tested by close contact with fact' (*ibid.*, 206) – these mangled the ideological universe of imperialism. Jingoism was 'merely the lust of the spectator' (*ibid.*, 215), and the whole language of diplomacy connived in falsifying the shameful reality of war and imperialism:

Paramount power, effective autonomy, emissary of civilisation, rectification of frontier, and a whole sliding scale of terms from 'hinterland' and 'sphere of influence' to 'effective occupation' and 'annexation' will serve as ready illustrations of a phraseology derived for purposes of concealment and encroachment. The Imperialist who sees modern history through these masks never grasps the 'brute' facts, but always sees them at several removes, refracted, interpreted, and glozed by convenient renderings. (*Ibid.*, 207–8.)

The brute facts that these territorial metaphors obscured were: the class-interest of imperialism, the parasitic exploitation of the 'lower' races, the incalculable danger posed to Britain by the jealousy of its rivals, and the possibility of the educated middle classes directing national development away from the competitive development of lower life to the cultivation of higher instincts. The alternative may partially be glimpsed in Hobson's description of Chinese economic development. The preferred alternative was close to autarky: 'Possessing in their enormous area of territory, with its various climatic and other natural conditions, its teeming industrial population, and its ancient, well-developed civilisation, a full material basis of self-sufficiency, the Chinese, following a sound instinct of self-defence, have striven to confine their external relations to a casual intercourse' (*ibid.*, 305–6).

Global strategy

For Mackinder, the Empire proved its value during the First World War when the British prevented their colonies trading with Britain's enemies and tried as far as possible to meet a large share of British raw material needs from within the Empire. Mackinder saw no difference between the strategic priorities of peacetime and wartime. In 1912, while advocating the need for a strong navy to protect trade, Mackinder said: 'I firmly believe without any sense of panic that the German nation is forced to contemplate the invasion of this country because in no other way is it possible for her to remove the threat which would throttle her on the way to the oceans of the world' (*Hansard*, vol. 41, 22 July 1912, col. 921). 'Without any sense of panic' – yet this is precisely what Hobson was so distressed about.

Mackinder and trade as war

Mackinder professed to believe that if Great Britain was strong enough, Germany and all other powers would simply let the British be and accept that Britain should have privileged access to markets such as China and to resources such as those of tropical Africa. He criticized the House of Commons for failing to introduce Imperial Preference: 'Our markets in China, India, Africa, and even our North American markets, where we help to maintain the Monroe Doctrine, are ultimately held by force, and you refuse to widen the basis of your power' (*Hansard*, vol. 19, 21 July 1910, col. 1474). Britain's strength would come from a closer union within the Empire: 'something more organic', an 'organic union of this Empire' (*Hansard*, vol. 73, 21 July 1915, cols. 1526, 1528). The crucial precondition was a protectionist tariff around the Empire, and a set of tariffs within, which would ensure harmonious development and self-sufficiency. Even within the Empire, free trade was inappropriate:

I cannot help feeling that what would have resulted from the condition of things which John Bright and Cobden anticipated would have been a concentration of manufacturing here and hewing and drawing in the rest of the world. I cannot believe that our Empire is likely to be strong, to be many-headed, if you are to have concentrated here all the manufacturing of the Empire and elsewhere merely the production of our materials and foods. (*Hansard*, vol. 19, 21 July 1910, col. 1473.)

Britain's role was clear: 'we have got to do the thinking of the Empire, to a large extent' (*Hansard*, vol. 18, 29 June 1910, col. 980). Britain should at least share with the white settler colonies, the Dominions, the 'entire theory and position of our foreign affairs' (*Hansard*, vol. 24, 19 April 1911, col. 995).

The Empire was central to Mackinder's vision of a Great Britain which could take on all-comers in the global power struggle because the Dominions and colonies could support Britain's military strength and guarantee the country's lines of supply. However, it was also of strategic significance in sheltering our naval power in the different world geopolitical theatres. In particular Mackinder was concerned that 'the centre of world politics is shifting', and that: 'you have today in the Far East, as we all know, new problems which, at the rate at which things develop, may very quickly come to be the chief problems of the world' (*Hansard*, vol. 24, 19 April 1911, col. 996). The opening of the Panama Canal would shift the attention of the United States from the Atlantic to the Pacific and if Britain wished to exert an imperial presence in the new arena, it must work hard to associate Australia and New Zealand with its policies; otherwise (and here Mackinder had to speak against derisory laughter) those two countries would ally themselves with the United States in the Pacific.

Apart from using the Empire to promote Britain economically and militarily (in so far as one may distinguish these goals in Mackinder's strategy), Mackinder's global trading policy aimed at restraining the development of Germany. After the First World War, Mackinder proposed that Britain should require Germany to pay an indemnity for war damage solely in terms of 'low-grade materials' lest the fine stimulate German industry too successfully (*Hansard*, vol. 85, 2 August 1916, col. 450). This would be a reversal of the policy under which Germany had developed. Germany's growth before the war had been achieved by importing low-technology goods from Britain and sending sophisticated products in return. In the post-war world, warned Mackinder, competition in low-grade goods would be severe so Britain should retain control of, and a presence in, as many high-technology industries as possible (*Hansard*, vol. 141, 9 May 1921, col. 1568).

Hobson and liberal free trade

Hobson's global vision could not have been more different from Mackinder's when it came to questions of strategy. Hobson was concerned to follow L. T. Hobhouse and recognize international rights, to assert that the play of human sympathy passed outside national boundaries, to develop a community of interest across race and flag. He wrote:

There is of course a school which utterly repudiates such rights, a *real politik* which virtually regards weaker nations as legitimate prey of stronger ones, and considers that the sole moral duty of a statesman is to promote the strength and well-being of his own state, disregarding utterly the interests and so-called 'rights' of others. Under such a creed imperial aggression requires no justification, and admits of none. (Hobson, 1974, 255.)

The worst international strategy was aggression, which would bring its own nemesis. Even Chinese isolationism was preferable to this. However, best by far would be progressive federation. Here, Hobson saw Britain working initially with like-minded nations before moving on to embrace others. However, this coalescence of equals is very different to the imperial notions floated by Mackinder. The irreversible trend within the British Empire was for the more mature colonies to resent ceding authority to Westminster: 'As each colony has grown in population, wealth and enterprise it has persistently asserted larger rights of independent government which the Mother Country has, sometimes willingly, sometimes reluctantly, conceded' (*ibid*, 237). Mackinder and his colleagues were spitting into the wind: 'Those British Imperialists who, with the events of the last few years before their eyes, still imagine a closer imperial federation in any shape or form practicable, are merely the dupes of Kiplingesque sentimentalism' (*ibid.*, 238).

With or without protectionism, a united Empire was an improbable dream. The self-governing colonies would see the danger in being associated too strongly with Britain's aggressive stance unless Britain gave them a blank cheque to engage in imperial adventures of their own. This was, indeed, Hobson's analysis of the way Britain came to sponsor the Boer War: 'Such a spirit and such a purpose was plainly operative in South Africa for many years' (Hobson, 1988, 345). This only increased the chances of Britain becoming involved in costly wars. Furthermore, Hobson believed that free trade remained the cheapest, as well as the most moral and safest means for Britain to secure its share of the fruits of world economic development:

Free trade trusts for the increase of our foreign trade to the operation of the self-interest of other trading nations. Her doctrine is that, though it were better for us and for them that they should give us free admission to their colonial and home markets, their protective tariff, even if . . . it prohibits us from trading directly with their colonies, does not shut us out from all the benefits of their colonial development. (*Ibid.*, 68.)

These indirect benefits resulted from reduced competition for tropical goods in other markets, the reduced price of secondary goods from the protected market, and so on.

Finally, only through a federation of states could the rich and powerful countries faithfully discharge the moral responsibility they claimed on behalf of the poorer and weaker countries. Only where such development was under the control of some 'organised representation of civilised humanity' (*ibid*, 232) could there be any guarantee that interference would genuinely benefit world civilization or 'raise' the 'lower races'. Yet: 'What actually confronts us everywhere in modern history is selfish, materialistic, short-sighted, national competition, varied by occasional collusion. When any common international policy is adopted for dealing with lower races it has partaken of the nature, not of a moral trust, but of a business "deal"' (*ibid.*, 241). Such a moral trust would necessarily entail cooperation rather than competition between nations. Advanced civilizations might substitute a struggle with the environment for the more primitive battle between nations (*ibid.*, 183), and this higher struggle might move beyond mere animal existence to 'other ends of an enlarged and more complex life, for comfort and wealth, for place and personal honour, for skill, knowledge, character and even higher forms of self-expression, and for services to their fellow-men, with whom they have identified themselves in that expanded individuality we term altruism or public spirit' (*ibid.*, 172). There is certainly a grandeur in this vision altogether missing from Mackinder's territoriality.

Global development

It should by now be obvious that Mackinder and Hobson viewed world geopolitics in very different terms. Their contrasting analyses of global closure suggested alternative geopolitical strategies for Britain. Furthermore, their approaches to the whole question of economic development revolved around the very same contested biological and territorial metaphors that we have identified above.

Mackinder and biological imperalism

Mackinder's suspicion of finance capital had a basis in his conception of economic development as a question of the organization of biomass. For Mackinder, money was chimerical; the world turned on force, and force ultimately rested on organized 'man-power', not mere 'money-power' (Mackinder, 1905). After the First World War he responded to suggestions that Britain might simply reschedule war debts and defer payment: 'We have spent in this war goods, things. It is not a mere question of book keeping and of writing debts down' (*Hansard*, vol. 116, 21 May 1919, col. 456). At the height of the war, he told his colleagues: 'you have got to think as Germany thinks, and as we ought to think of it [the war effort], without money. It is the use of pounds, shillings, and pence which is so confusing in these matters, and which leads to all the fallacies which mislead the classes. Think of it in actual human work, in actual service to the nation' (*Hansard*, vol. 75, 10 November 1915, col. 1235). In this connection, foreign investment was foolish for it merely equipped Britain's rivals, attracting Britain's 'human capital' overseas and bleeding the country of the muscle it required to defend itself (*Hansard*, vol. 14, 23 February 1910, col. 320).

If the goal of development was the accumulation of force, the means was the organization of biomass. In a very real sense, Mackinder believed, as in fact did Hobson, that the economy-in-motion was more than the sum of its parts. It was an organism: 'The wealth of this country rests mainly on the country as a going concern' (*Hansard*, vol. 82, 18 May 1916, col. 1726). Industries can only be built up over a period of years and must be in continual production to remain efficient: 'machinery, if not kept in repair, will wear out in the course of two or three years' (*Hansard*, vol. 116, 21 May 1919, col. 460). In this sense an industrial nation cannot leave its assets idle for any length of time: 'Our assets, in the main, are wasting assets' (*Hansard*,

vol. 129, 11 May 1920, col. 320). Free trade, in leaving the rise and fall of industries to the vagaries of the international market, seemed willing to contemplate irretrievably writing off skills and organization on the mere whim of a price movement. The country should cultivate in its manpower that broad range of skills which would enable it to stand alone in time of war. The organic interdependence of the different sections of industry merely underlined this point.

The Empire, as we have seen, had a central place in this strategy. Yet, considered from the British point of view, the development of the Empire for the benefit of the home country was not only a duty to world civilization, but also an obligation which the government owed to the British people themselves:

> It is our bounden duty to see that steadily we have progress in the conditions of the savage and barbarous inhabitants of those regions . . . [in Nigeria]. But our duty is not limited to them. Our duty is towards our own race and mankind. Our duty is to develop those vast regions, and a portion of these results we are entitled to take for ourselves and for civilised mankind, for that civilised humanity which in the course of some generations probably the people who inhabit those regions will be able to join. We are entitled to take it because we have fought and spent that which cannot be estimated in cash and that ought to be taken into your economics. It has cost us the lives of our men . . . (*Hansard*, vol. 85, 3 August 1916, col. 573).

Hobson, scientific selection and mutual aid

Hobson stressed the organic unity of the economy as a 'going concern' in precisely the same terms as Mackinder. Indeed, Clarke has remarked that: 'The persistence and fervour of Hobson's almost mystical adherence to the organic view can hardly be exaggerated' (Clarke, 1974, xvii). Townshend notes that 'Hobson's organicism reversed the traditional liberal order of priority: the "whole", i.e. society, ultimately took precedence over the "parts", i.e. the individual' (Townshend, 1990, 138). This perspective, in which 'scientific selection' replaced 'natural selection', took Hobson a long way towards a set of very illiberal conclusions and left him relying on a purely pragmatic defence of individual rights. For example, Hobson's racism, which we have already seen in his conspiracy theory of Jewish finance capital, was given free rein in his eugenic theory of world development. He appeared to believe that weak human stock might simply evaporate on contact with higher civilization, but he was willing to contemplate helping this process:

As lower individuals within a society perish by contact with a civilization to which they cannot properly assimilate themselves, so 'lower races' in some instances disappear by similar contact with higher races whose diseases and physical vices prove too strong for them. A rational stirpiculture in the wide social interest might, however, require a repression of the spread of degenerate or unprogressive races, corresponding to the check which a nation might place upon the propagation from bad individual stock. (Hobson, 1988, 190–91.)

Although he believed that the highest intellectual qualities might be inherited, in the main he argued, as a Social Darwinist might, that the tropical environment had selected a lower-order person for survival than had Britain's temperate environment. He was, therefore, an enthusiastic member of the National Birth Rate Committee (1914), an official inquiry into the dangers posed by falling British fertility. He appears to have believed that it was only on moving to more temperate climates that African people could hope to raise themselves up:

Even the minority of thinkers who are persuaded that no inherent differences of racial values exist and that all the higher qualities of civilised life are due to differences of environment and education, would have to admit that these differences require a considerable time for their beneficial operation, and that a rapid decline of the more civilised peoples could not be compensated immediately by the fuller opportunities afforded to migrants from the backward countries. (Hobson, 1946, 149–50.)

There is certainly racism in Mackinder's writings, but nothing comparable to this international eugenicism. The point to note is that in both cases their organic perspective drove them to stress collective rights over individual rights: in Mackinder's case the right of the race/nation to insist on adequate living space and separate development (apartheid) in the periphery; and, in Hobson's case, the right of 'civilized humanity'

to manage the reproduction of different parts of the world organism.

There were other aspects of Hobson's view of global development which were more liberal than this. For example, he distinguished two main causes of poverty: 'waste of human power', and the 'inequitable distribution of opportunities' (Hobson, 1974, 162). Global development ought properly to address these issues. Here, on the pragmatic grounds of promoting economic growth, Hobson attacked the biological determinism of the New Imperialists. Human power is wasted because society does not bring science and co-operation to bear on matters of basic production. Acknowledging his debt to Kropotkin's theory of mutual aid, Hobson identified private ownership of resources as the main obstacle to the scientific management of subsistence industries. Hobson argued that all the industries which met uniform, biological, lower-order needs should be under collective, rational control. The higher-order needs which promoted individuality could more properly be left to the market.

The 'inequitable distribution of opportunities' needed to be addressed, not only because of its consequences for over-saving (as we have discussed above), but also because it locked up human initiative. Individuals needed access to productive resources if society were to benefit from their potential. This entailed collective control of crucial industrial and infrastructural resources and their provision to individuals on the cheapest terms. Hobson argued that equality of opportunity implied: mobility (cheap, nationalized railways), the option of self-sufficiency (cheap, nationalized agricultural land), access to power (cheap, nationalized electricity) and security (cheap, nationalized personal insurance) (*ibid.*, pp. 98–107).

Weaknesses of the ecological view of geopolitics

Because they saw the national units of the world economy as biological units, Mackinder and Hobson were led to accept what Marx once termed chaotic conceptions as the basic units of their analysis of global closure. In a much-quoted passage, Marx discussed how one might choose the point of departure for an analysis of a national economy; what were the basic concepts?

It seems to be correct to begin with the real and the concrete, with the real precondition, thus to begin, in economics, with e.g. the population, which is the foundation and the subject of the entire social act of production. However, on closer examination this proves false. The population is an abstraction if I leave out, for example, the classes of which it is composed. These classes in turn are an empty phrase if I am not familiar with the elements on which they rest, e.g. wage labour, capital, etc. These latter in turn presuppose exchange, division of labour, prices, etc. For example, capital is nothing without wage labour, without value, money, price etc. Thus, if I were to begin with the population, this would be a chaotic conception of the whole . . . (Marx, 1973, 100.)

This passage comes after a long discussion of the basically social character of production in which Marx argued on empirical grounds that production is always a collective act and that in modern societies the social character of production is structured by property relations which determine people's access to the means of production and the social product. These property relations, and their defence at law, form the basis of the economic classes in society, and it is on this basis that Marx argued that class should come before population in describing the laws of a capitalist economy. It is not, therefore, a question of class being more 'real' than population but, rather that, for the purposes of understanding the dynamics of the economy, the divisions of class rather than the biology of populations are of greater explanatory power. It is because the category 'population' conflates the dimensions which Marx believed to be empirically most important in explaining how economies developed that he referred to it as a 'chaotic' conception. For other purposes, for example in understanding genetically-transmitted disease susceptibility, class itself would be a 'chaotic' conception since the ethnic and biological dimensions must be considered to have empirical priority in such a case.

The point I want to make is that the choice of explanation in the social and economic sciences is a strategic one; a certain order in the hierarchy of concepts which makes up an explanation throws light on particular areas and obscures others. All theoretical accounts have some blind spots. So far, I have been drawing attention to the essentially contestable nature of the biological metaphors deployed by Mackinder and Hobson. I have suggested that they manoeuvred very differently within this biological world-view, and managed to sustain very different conceptions of the nature of the British crisis and possible responses to it on the basis of a fundamentally similar set of concepts. I want to conclude by turning the matter around and proposing that the biological units they saw in the world-system obscured the class divisions within nations and the individual basis of political rights. Although I do not have the space to explore this here, I think that some of these issues remain pertinent to current debates about what the fundamental building blocks of geopolitical analysis are, and it is exciting to find this recognized in recent work (see Balibar and Wallerstein, 1992).

The tariff reform strategy postulated a common interest between industrialists and their employees in which the former would receive protected domestic markets and the latter would benefit from social reform. Such a social contract would, they believed, defuse the class struggle and make the national interest a real, and efficient, agent in world geopolitics. In Mackinder's case, this national interest was itself an expression of a race-interest, and he spoke of 'the English race, the English blood' as 'valuable' in 'carrying a certain character' which 'is, it seems to me, something physical, and therefore not wholly transferrable except with the blood' (Mackinder, 1925, 726). In fact Mackinder disliked class politics (by which he meant working-class political movements). Regionalism was, for Mackinder, an alternative to class politics: 'I believe that whether we look at the matter from the point of view of the freedom of men or of nations we shall come to the same conclusion; that the one thing essential is to displace class organisation, with its battle-cries and merely palliative remedies, by substituting an organic ideal, that of the balanced life of provinces and of the lesser communities' (Mackinder, 1919, 241–2). Beatrice Webb noted in her diary for 1918 that although Mackinder 'still talks in continents and waterways, in mass movements and momentums . . . he has become uncomfortably aware of another kind of mass movement, of another type of momentum – the uprising of the manual workers within each state . . . it is an uncomfortable shadow falling across his admirable maps of the rise and fall of empires' (quoted in Parker, 1982, 46). Stressing the priority of a biological understanding of community rendered class politics illegitimate, parasitical.

Now Hobson, of course, based his whole theory of imperialism on the class conflict between the rich financiers and the poor workers. But he also postulated an 'organic' national identity whose interest was not well served by imperialism, but which nevertheless had an independent existence of its own. This existence was perceived in biological and not class-conflict terms. Imperialism was understood as a form of biological parasitism and seen as subject to universal biological laws which ensured the eventual collapse of all imperial powers. Hobson's *Imperialism* ends with a comparison between ancient Rome and modern Britain. He argues that Rome collapsed because parasitic classes living off the despoiling of the Roman Empire created such a degree of exploitation that they undermined the conditions for continuing production in the periphery:

The direct cause of Rome's decay and fall is expressed politically by the term 'over-centralisation', which conveys in brief the real essence of Imperialism as distinguished from national growth on the one hand and colonialism upon the other. Parasitism, practised through taxation and usury, involved a constantly increasing centralisation of the instruments of government, and a growing strain upon this government, as the prey became more impoverished by the drain and showed signs of restiveness. (Hobson, 1988, 366–7.)

British imperialism differed in a few particulars from that of Ancient Rome: 'But nature is not mocked: the laws which, operative throughout nature, doom the parasite to atrophy, decay, and final extinction, are not evaded by nations any more than by individual organisms' (*ibid.*, 367).

Both Mackinder and Hobson set organic needs over individual rights. This crucially affected their conceptions of democracy. Mackinder was quite blatant. He believed that the captains of industry exercised the most responsibility within society, and that: 'those who have the greatest stake in the nation and its business, in competition with the trade of other nations, should have a considerable and more than average voice in the management of the country' (*Hansard*, vol. 52, 30 April 1913, cols. 1277–8). He saw no general argument in favour of individual democratic rights: 'For my part I do not worship King Demos; I am of a more rationalistic turn of mind' (*Hansard*, vol. 34, 19 February 1912, col. 368). For Mackinder, the vote represented either physical or moral force. As physical force, the right of individuals to vote averted the recourse of men to arms in a successful revolt of the majority against tyrannical decisions. As moral force, the vote might at some future time represent the moral duty of responsibility towards the state placed upon differentially responsible individuals and interests. The physical force argument, argued Mackinder, told against admitting women to the franchise. Under present conditions, he said: 'A vote is a cheque or draft on power, and, ultimately, on physical power' (*Hansard*, vol. 25, 5 May 1911, col. 763). At some remote time in the future, men may not threaten to resort to violence in the face of collective decisions they disagree with, and the mob may cease to be the threat behind national politics: 'If men at any time come to approximate more to the condition of angels with the progress of society then I see no reason why we should not give women votes' (*ibid.*, col. 755). Universal male suffrage, then, rested on no fundamental matter of principle, but was simply a tactical concession in the face of the mob's threat of civil war. The moral argument for the vote would set the value of the working class even lower than this and would, in Mackinder's eyes, justify some sort of plural voting system based on a property franchise. This concept of responsibility would recognize the relative importance of different people to the organic efficiency of society: a utilitarian rather than a rights-based idea of representation. The individual was never the point of departure for Mackinder.

I have already remarked that Hobson presented a purely pragmatic defence of forms of political representation, although in his case those forms must be based on universal and equal suffrage. The purpose of government should be to serve the organic needs of society as a collectivity (Hobson, 1974, 74–84). Just as in the hive, the individual should serve a social end. All production is necessarily social, and it is this fact which establishes the organic rights of the group as prior to those of the individual. Similarly, all property is held in trust because its enjoyment rests upon the productive efforts of the group as a whole. Democracy is an efficient way of identifying the needs of the social organism since it gives all taxpayers a way of examining the way the social budget is used and gives them an opportunity to bring forward information about how it affects those interests with which they are most familiar as individuals.

The utilitarian defence of forms of political representation in Hobson and Mackinder presupposes some consensual yardstick against which to measure the performance of political systems. In their case, that yardstick was the organic efficiency of the state. Hobson was surely right to draw attention to the difference between quantitative and qualitative development. The grounds on which Mackinder dismissed the high standard of living of the Dutch – his celebration of the manly vigour of industrial Britain and his defence of restricting the franchise to men in recognition of their monopoly of physical force – underline what I have termed his treatment of power in terms of biomass. The class struggle of the socialists undermined this biological definition of the national interest. Hobson's treatment of the working class was quite different, but a biological conception of the polity led him to postulate a national organic interest which was peculiarly apparent to the educated middle classes. It is precisely this casual assumption of a common national interest bestriding the class divisions which appears rather simplistic in retrospect.

It is also clear that organic definitions of the national interest make it difficult to respect cultural diversity. Mackinder was quite clear about English racial superiority and Hobson was too. In Mackinder's case, this might not seem too surprising, but Hobson's commitment to international eugenics was of a piece with his own biological perspective. In Hobson's case, it went with a rather spurious distinction between colonialism and imperialism, which I would treat as

different imperialisms. The idea that the collapse of populations in the face of European diseases is a function of racial inferiority is indefensible; one only has to think of the early impact of bubonic plague or syphilis on Europe to see that there are no stable hierarchies of disease resistance. Hobson also underestimated the extent to which European nations undermined the autonomy of the societies of the periphery during the earlier period of mercantile capitalism.

Finally, the biological conception of geopolitics created certain strategic blindspots for these two important analysts. Mackinder conflated economic and military power in his ideas of biomass. They were so closely interrelated that he took assertion in the latter as reliable evidence of competitive advantage in the former. He persistently undervalued the virtues of restraining military expenditure, and imagined that Britain could in some fashion sustain a level of military preparedness which would terrify all other powers. This was an impossible dream. Britain's earlier economic lead was unsustainable. Military spending could not prolong it indefinitely. Mackinder also continually saw Britain's main economic competition as coming from Germany because of its military expenditure, and he was far too sanguine about the economic rise of the United States.

Hobson's treatment of imperialism as biological parasitism encouraged him to pay attention to the internal contradictions which would bring down imperialisms from within. In this he was followed by Lenin who also saw imperialism as a new atavistic level of capitalism, capitalism in a tail-spin dive. Rather than being a sort of terminal disease which all the Great Powers were contracting, the competition between the imperial powers incorporated the restructuring of separate national economies. In the process, the British lost many of their overseas assets. This was a crisis of British capitalism rather than a stage in the evolution of world capitalism. Cartels and monopolies were hardly as general as Hobson, and subsequently Lenin, believed. Unequal exchange is, indeed, an endemic feature of capitalist economies, both in domestic and international relations, but so is what Cohen (1991) terms 'geopolitical repartitioning', and this makes it impossible to establish the long-run dynamics of the world economy simply on the basis of the assumed functioning of a closed system. The study of capital-in-general is a dangerous basis from which to generalize about the likely functioning of a competing set of capitalist states.

This brings me to the last point I want to make about these conceptions of global closure. The idea of a closed system was crucial to the appropriation of the language of biology for understanding geoeconomics and geopolitics. The geopolitical gaze across the space of international competition only captures the central terms of economic restructuring where territory is the crucial precondition of economic growth. I have tried to show here that the territorial imperative of international competition at the start of this century rested upon a contingent set of circumstances, notably the British defence of a sterling zone of worldwide free trade and the protectionist policies of Britain's rivals. The British economic lead had evaporated, and for the British the world did indeed seem to be shrinking, to be closing in. Facing relative decline, ecological perspectives provided the basis for what we might term the paranoid style in *fin de siècle* British geopolitics.

References

Balibar, E. and Wallerstein, I., 1992, *Race, nation, class: ambiguous identities*, Verso, London

Blouet, B. W., 1987, *Halford Mackinder: a biography*, Texas A&M University Press, College Station, TX

Clarke, P. F., 1974, Introduction, in Hobson, J. A., *The crisis of liberalism*

Cohen, S. B., 1991, Global geopolitical change in the post-Cold War era, *Annals of the Association of American Geographers*, 81: 551–80

Edelstein, M., 1981, Foreign investment and empire 1860–1914, in Floud, R. and McCloskey, D. (eds.), *The economic history of Britain since 1700. Volume 2: 1860 to the 1970s*, Cambridge University Press, Cambridge

Foreman-Peck, J., 1983, *A history of the world economy: international relations since 1850*, Wheatsheaf, Brighton

Hobson, J. A., 1916, *The evolution of modern capitalism: a study of machine production*, new and revised edition, Walter Scott, London (original edition 1894)

Hobson, J. A., 1974, *The crisis of liberalism: new issues of democracy*, edited and with an introduction by P. F. Clarke, Harvester, Brighton (original edition 1909, P. S. King, London)

Hobson, J. A., 1976, *Confessions of an economic heretic: the autobiography of J. A. Hobson*, edited and with an introduction by Michael Freeden, Harvester, Brighton (original edition 1938, Allen and Unwin, London)

Hobson, J. A., 1988, *Imperialism: a study*, third edition, edited and with an introduction by Jules Townshend, Unwin Hyman, London (third edition first published 1938; original edition 1902)

Ingham, G., 1984, *Capitalism divided? The City and industry in British social development*, Macmillan, London

Kearns, G., 1984, Closed space and political practice: Frederick Jackson Turner and Halford Mackinder, *Environment and Planning. D. Society and Space*, 2: 23–34

Kearns, G., 1985, Halford Mackinder, in Freeman, T. W. (ed.) *Geographers: biobibliographical studies. Vol. 9*, Mansell, London

Kennedy, P., 1991, *The rise and fall of British naval mastery*, third edition, Fontana, London (original edition 1976, Allen Lane, London)

Mackinder, H. J., 1900, The great trade routes, *Journal of the Institute of Bankers*, 21: 1–6, 137–55, 266–73

Mackinder, H. J., 1902, *Britain and the British seas*, Heinemann, London

Mackinder, H. J., 1904, The geographical pivot of history, *Geographical Journal*, 23: 421–37

Mackinder, H. J., 1905, Man-power as a measure of national and imperial strength, *National and English Review*, 14: 136–45

Mackinder, H. J., 1919, *Democratic ideals and reality: a study in the politics of reconstruction*, Constable, London

Mackinder, H. J., 1925, The English tradition and the Empire: some thoughts on Lord Milner's Credo and the Imperial Committees, *United Empire*, 16: 724–35

Marx, K., 1973, *Grundrisse: foundations of the critique of political economy (rough draft)*, Penguin, Harmondsworth (written 1857–8 and first published in German in 1953)

Mitchell, B. R., 1975, *European historical statistics 1750–1970*, Macmillan, London

Mitchell, B. R., 1983, *International historical statistics: the Americas and Australasia*, Macmillan, London

Mitchell, B. R., 1988, *British historical statistics*, Cambridge University Press, Cambridge

Mummery, A. F. and Hobson, J. A., 1889, *The physiology of industry*, Murray, London

Parker, W. H., 1982, *Halford Mackinder: geography as an aid to statecraft*, Clarendon Press, Oxford

Pollard, S., 1981, *Peaceful conquest: the industrialisation of Europe 1760–1970*, Oxford University Press, Oxford

Porter, B., 1987, *Britain, Europe and the world 1850–1986: delusions of grandeur*, second edition, Unwin Hyman, London (original edition 1983)

Semmel, B., 1960, *Imperialism and social reform: English social-imperial thought*, Allen and Unwin, London

Teich, M. and Porter, R., 1990, Introduction, in Teich, M. and Porter, R. (eds.) *Fin de siècle and its legacy*, Cambridge University Press, Cambridge

Townshend, J., 1988, Introduction, in Hobson, J. A., *Imperialism*

Townshend, J., 1990, *J. A. Hobson*, Manchester University Press, Manchester

Wolf, E., 1982, *Europe and the people without history*, University of California Press, Berkeley, CA

1

Geopolitical world orders

Peter J. Taylor

The key issue in contemporary international politics, so President George Bush and his advisers keep reminding us, is the creation of a 'New World Order'. After the collapse of the communist regimes in Eastern Europe in 1989, it has become generally accepted that the Cold War – the 'Old World Order' – is finished, and international politics has to be reconstructed in another form. Hence the call for a New World Order from the remaining superpower, seemingly the only state with the power to embark on such a project.

It is the purpose of this chapter to explore this now popular notion of a 'world order' so as to set the current search for a new one into both a theoretical and an historical context. This requires the argument to be developed in two parts. The first develops a geopolitical analysis that interprets world orders as relatively stable distributions of political power across the world. To understand these 'geopolitical world orders' it is necessary to relate them to both other global patterns of social change, and the particular activities of governments that create the events that are the stuff of international politics. The result of this analysis is to produce a concept of world orders that is a framework for relating events to broader patterns of change. Two geopolitical world orders are identified in the twentieth century and they form the subject matter of the second and larger empirical part of the chapter. They are dealt with chronologically, first the Geopolitical World Order of the British Succession and then the Cold War Geopolitical World Order, as two subjects for describing the international politics of the twentieth century from a distinctively geopolitical perspective. With the concept of a world order so treated, we can return to George Bush's call for a New World Order and, in a short conclusion, lessons from our analysis and history are used to speculate on what we might expect from the contemporary reconstruction of international politics.

Geopolitical analysis

Geopolitics was born with the twentieth century. The German term *geopolitik* was first coined in 1899 by the Swedish political scientist Rudolf

Kjellen, and geopolitics' most famous model – pivot area, later heartland thesis – was presented to the Royal Geographical Society by Halford Mackinder in 1904. For many people geopolitical analysis came to mean the revealing of fundamental truths about global geographical constraints on the behaviour of states. Events were interpreted in terms of what were viewed as permanent spatial patterns of power. What came to be called 'traditional geopolitics' became a source of structural imperatives that determined the course of international politics. Although such thinking could still be found towards the end of the century, such geographical determinism is now generally considered to be inadequate as a social theory of international politics.

It is somewhat ironic that Mackinder's theory should be a stimulus for the identification of such ahistorical 'changeless geographical factors' when his analysis was imbued with historical insight. Mackinder's history is best interpreted in terms of Fernand Braudel's concept of the *longue durée*. The heartland model was originally devised to mark an important change in 'epoch' as Mackinder saw it. After 400 years of the 'Columbian era' dominated by sea-powers, Mackinder thought the twentieth century to be the beginning of the 'post-Columbian era' when global power would transfer to land empires: a key structural change of long-term importance was occurring. Mackinder feared for the new epoch since the expansion of sea-power had incorporated the whole world into a single system, eliminating colonial expansion as a 'safety-valve' in an increasingly competitive inter-state system.

Such *fin de siècle* thinking was relatively common, and in this respect we can link Mackinder's original geopolitics with other writers who were concerned with *longue durée* processes such as Lenin. In his revolutionary prescriptions for changing worlds, Lenin similarly focused on the new imperial competitiveness that worried Mackinder, identifying it as the 'highest', and final, stage of capitalism. Such long-term thinking is important in this chapter, not just as a recovery of Mackinder from ahistorical analysis, but because such ideas have been influential in the construction of geopolitical world orders in this century. The Cold War a half-century later, for instance, was built in part upon ideas from both Mackinder and Lenin to define the enemy for each side (threatening heartland = USSR, expansive capitalism = USA). And, of course, this competition between superpowers was launched by both sides in *longue durée* terms as the ultimate conflict between two incompatible civilizations vying to define a new epoch for the world.

In contrast to the time perspective of traditional geopolitics, the narrative history approach to international relations has concentrated upon the events themselves, the wars, the treaties and the political elites making the decisions. It was this concentration on the short term or *courte durée* that Braudel was rebelling against in his history. His social model of time went beyond 'eventism', not just in defining the *longue durée*, but by defining three levels of social time including a middle category, the *moyenne durée* of temporal sequences such as trends and cycles. Despite his attacks on the *courte durée* historical tradition, Braudel (1980) argued that for a comprehensive historical analysis, all three social times must be examined. In this chapter, as well as the *longue durée* claims of the Cold War described above, we will present a narrative of events, *courte durée*, but as part of an analysis of medium-term changes in patterns of international politics. Geopolitical world orders are creatures of Braudel's *moyenne durée*. In the analysis developed here we describe a geopolitics of the *moyenne durée*; we will be concerned with the changing political geography of 'periods' rather than 'epochs' and their (nearly) permanent geography or the particular transient worlds of individual events. But the medium-term time span is the most complex, dealing as it does with a very wide range of patterns of change. Hence our starting-point in this analysis must be to locate geopolitical world orders within the periods and cycles that form the twentieth century.

Periods and cycles

Distinctive historical periods can be conceptualized in several ways. They may be viewed as stages in a sequence of changes, as components of

cycles in which repetitions of circumstances are emphasized, or simply as different times that are not linked into any such patterning. Geopolitical world orders at first sight seem to fit into the latter category, but in fact can be related to other temporal sequences that are far more patterned. The international political elites and their governments making the decisions that collectively make up the the world orders do not operate in a material vacuum. The world-economy is notoriously cyclical in nature – what Wallerstein (1984b) calls its rhythms – and politicians have to accommodate to these systematically varying circumstances. Changing material contexts profoundly alter the circumstances in which politicians act by providing new agendas for action. Precisely how the world orders relate to these cycles is a complex and highly contested theoretical issue; in this chapter I am merely asserting the relationship, letting my empirical description of the world orders within a framework of material cycles justify this position. The starting-point of the analysis is a discussion of cyclical changes on a global scale, concentrating on the two longest cycles that are usually identified, Kondratieff cycles of approximately half a century and hegemonic cycles of about a century or so in length. (For the best and most comprehensive review of the literature on long social systemic cycles, see Goldstein (1988).) These two types of cycles and their component periods are compared across the last two centuries in Table 1.1 which we will refer to frequently in what follows.

Kondratieff cycles are usually described in strictly economic terms, but undoubtedly they have profound political impacts. The fifty-year cycle is divided into two approximately equal periods, an A-phase of growth and a B-phase of stagnation. There is a large debate on the causes of these long economic fluctuations, but their existence is now generally agreed upon. In conventional timing of these cycles the twentieth century covers the third and fourth Kondratieff cycles when dated from the industrial revolution in Britain (Table 1.1). In general terms B-phases are more competitive economically, although how this translates into politics is by no means simple. Probably the most successful interpretation of Kondratieff cycles in relation to political

Table 1.1 *Long cycles and geopolitical world orders*

Date	Kondratieff cycles	Hegemonic cycles	Geopolitical world orders
1790/8	*First Kondratieff Cycle* A phase (industrial revolution)	*British Hegemonic Cycle* Ascending hegemony (grand alliance)	(Napoleonic wars as French resistance to Britain's ascending hegemony)
1815/25	B phase (first long industrial depression)	Hegemonic victory (balance of power through Concert of Europe)	Disintegration *World Order of Hegemony and Concert* Transition (1813–15)
1844/51	*Second Kondratieff Cycle* A phase (mid-Victorian boom)	Hegemonic maturity ('high' hegemony: free trade era)	(Balance of power in Europe leaves Britain with a free hand to dominate rest of the world)
1870/75	B phase (late-Victorian depression)	Declining hegemony (age of imperialism, new mercantilism)	Disintegration *World Order of Rivalry and Concert* Transition (1866–71) (Germany dominates Europe, Britain still greatest world power)
1890/96	*Third Kondratieff Cycle* A phase (the Edwardian boom)	*American Hegemonic Cycle* Ascending hegemony (a world power beyond the Americas)	Disintegration *World Order of the British Succession* Transition (1904–7)
1913/20	B phase (the 'great' depression)	Hegemonic victory (not taken up: global power vacuum)	(Germany and USA overtake Britain as world powers, two world wars settle the succession)
1940/45	*Fourth Kondratieff Cycle* A phase (the 'post war' boom)	Hegemonic maturity (undisputed leader of the 'free world')	Disintegration *Cold War World Order* Transition (1944–6)
1967/73	B phase (the latest long 'slump')	Declining hegemony (Japanese and European rivalry)	(USA hegemony challenged by the ideological alternative offered by the USSR)
19??	*Fifth Kondratieff Cycle?*	*New Hegemonic Cycle?*	Disintegration *'New World Order'* Transition (1989–91)

processes has come through their linkage to hegemonic cycles.

Hegemonic cycles focus upon one state, the hegemon, that for a short period is pre-eminently powerful economically, politically and culturally. The cycle consists of the rise and fall from this position. Following Wallerstein (1984) we can describe the cycle as follows. The hegemon gradually gains a clear economic advantage in the realm of production and extends its leadership to the commercial and financial spheres. At the same time it becomes politically dominant after leading a coalition of states against its main political rival. Henceforth it is able to order the world to its advantage using such techniques as balance of power rather than outright coercion. This is possible in part because of its cultural leadership in 'universal' ideas – the hegemon is typically the champion of world liberalism. The period of 'high hegemony' is relatively short, and these leadership attributes from production to culture are progressively lost.

The relationship between Kondratieff A- and B-phases and the hegemonic rise and demise periods is illustrated in Table 1.1 as pairs of Kondratieff cycles coinciding with one hegemonic cycle defining British and US 'centuries' respectively. This means that the influence of Kondratieff phases will vary between rise and fall periods of the hegemonic cycles. The clearest relation between the two cycles in both centuries is the common coincidence of high hegemony with an A-phase to be followed by hegemonic decline and increased competition in the subsequent B-phase. For instance, the personal experience of many readers of this book has been of US high hegemony coinciding with the great 'post-war boom' (Kondratieff IV A) and our current worldwide recession (Kondratieff IV B) associated with recognition of US relative decline.

Although Table 1.1 covers two hegemonic cycles, we will be concerned with British hegemony only in its demise phase. At the beginning of the twentieth century Britain and its empire was clearly still the most powerful state in the world, but its undisputed hegemony was already a quarter of a century past. Its main rivals were Germany and the USA, with the latter becoming hegemonic after 1945. Hence we will be concerned largely with the US hegemonic cycle from its rise through the defeat of Germany, high hegemony at mid-century, and finally the debates over its relative decline in more recent years.

Geopolitical world orders are not as neatly related to the two material long cycles as they are to each other. As Table 1.1 shows, world orders have been of varying length so there is no cyclical pattern. Nevertheless world orders do generally begin and end at roughly the same time as Kondratieff phases; in fact, I would argue that both cycle phases and the world orders between them constitute the nature of our modern global times. For instance, the processes making up Kondratieff cycle IV, US hegemonic practices and the Cold War are impossible to disentangle in understanding the recent past and our contemporary situation. However, for pedagogic reasons we are going to concentrate largely upon geopolitics in the remainder of this chapter. We will be concerned with the practical geopolitical reasoning and actions of political elites in their creation of the two world orders of the twentieth century.

Geopolitical codes

Geopolitics is avowedly state-centric in its premises. By practical geopolitical reasoning I mean the way in which governments conceptualize the distribution of political power beyond their boundaries as a precondition for conducting foreign policy in their special national interest. This is the way state elites make sense of the world in order to respond to or create events to their state's advantage. Through studying such statecraft we can identify the geopolitical codes that are the building blocks of geopolitical world orders.

Following John Gaddis (1982), we shall use the term 'geopolitical code' to describe the output of practical geopolitical reasoning. These are the codes or geographical frameworks by which a government deals with the outside world. A national interest is defined and other states are evaluated in terms of whether they are real or potential aids or obstacles to that interest. A change of government in a state may change details in a code, but foreign policy in the twentieth century has come to be located within the consensus realm of most state politics. Bipartisanship is the norm, which means that we can usually identify general codes that transcend several governments but with specific codes incorporating perhaps different emphases for each of those governments. For instance, Gaddis (1982) identifies a general containment code for the USA after 1947 but with distinctive features for successive Democrat and Republican administrations.

One important reason for the relative continuity of codes across governments has been the formal democratization of politics in many states in the twentieth century. Bipartisan policies insulate foreign affairs from democratic pressures. For instance, in Britain there has been much discussion, when the Labour Party have been in opposition, of what a 'socialist foreign policy' should be, but no such phenomenon has been forthcoming in any of their periods of office. 'National interest' cannot be easily changed back and forth; 'friends' and 'enemies' once so identified have to have some degree of permanency to make sense. The definition of national interest and with it the general geopolitical code cannot be reduced to a mere political football like domestic policies. This is what is implied by the phrase 'high politics', being 'above' the partisan concerns of domestic politics. But in this matter the twentieth century differs from definitions of state interests in earlier periods, when 'high politics' was solely the concern of the monarch and his advisers. Winston Churchill at the beginning of this century saw the dangers of democratic input into foreign policy: geopolitical codes would lose their flexibility and politicians would be thus constrained in periods of inter-state tension (Bartlett, 1984, 86). Certainly wars have become 'patriotic', more total and exceedingly popular in the twentieth century. Of course, the influences on codes have by no means been all one way. Popular sentiment has been very efficiently mobilized when state elites have found it necessary to do so. This is obviously required when the government wishes to change the geopolitical code and designate new enemies. The conversion of the USSR from ally to enemy between 1945 and 1947 is a good example, and

was accomplished by a virulent anti-communist campaign (Adler and Paterson, 1970).

At a general level, the description of a government's geopolitical code is relatively straightforward and can be inferred from its foreign policies. Alliances, lesser agreements, overseas bases and levels of diplomatic status are all obvious indicators of the code. Some researchers have tried to quantify codes or groups of state codes through analysis of policy-makers' journeys (Henrikson, 1980), diplomatic linkages (Nierop, 1989), voting in the United Nations (Brunn and Ingalls, 1983) and the counting of place references in policy statements (Grant and O'Loughlin, 1991). All of these are valuable in defining the salience of different countries for each other. However, probably the most explicit source of information on any codes is to be found in the activities of the General Staffs of the armed forces. Parallel with the intellectual emergence of geopolitics, the same competitive pressures led to governments' creating new departments where the leading military elites could officially plan wars for the first time in peacetime (Kennedy, 1978). In such statements there is simply no room for any diplomatic ambiguity on who are targeted and who are not. We will use such materials frequently below.

Finally, we must recognize the geographical scale component in geopolitical codes. In general the salience of one state to another will decline with distance (O'Sullivan, 1986), but this will vary greatly between states. For all states their immediate neighbours are crucial components of their code either as friends or enemies. Most peaceful interactions such as trade generally occur between neighbouring states, but also most wars are border wars. Every state, therefore, has its own local code. For the majority of states, the small ones, this constitutes the effective whole of their practical operations. For medium and large states, though, there is a wider range of salience which is termed regional. Regional powers throughout the world define their national interests beyond the narrow confines of their borders. Brazil in South America, India in South Asia and Nigeria in West Africa are the three clearest examples of states which include domination of their region as integral to their national interests. Finally there are world powers whose codes are global in extent. Their governments consider events across the world as being of potential relevance to their national interest as a Great Power. The USA and USSR as superpowers in the second half of the twentieth century are classic examples of states claiming such extravagant interests. States can move between these categories. British withdrawal from 'east of Suez' in 1967 to concentrate on Europe is an example of reduction from world power to regional power. This also happened to the USSR in the final two years of its existence as it disengaged around the world, leaving its major successor state, Russia, a regional power.

Geopolitical world orders

The three-scale hierarchy of geopolitical codes defines relations between codes. At its simplest, the local codes of small states have to fit into the regional codes of medium states which in turn should fit into overarching global codes of world powers. This is what phrases such as 'sphere of influence' and 'backyard' imply as expressions of geographical power relativities. This leads to a series of bilateral and multilateral patterns of associated codes across the world. The concept of a geopolitical world order asserts that the geographical organization of power across the world is more than an aggregation of these interlocking hierarchies of codes. It is a whole that is more than the sum of its parts. Beyond any individual code, however powerful the state, there is a geopolitical order that defines the basic parameters of the international politics of the time. Such orders represent relatively stable patterns of geographical power-distributions over distinctive periods of time. During that period the geopolitical codes of most, though not necessarily all, states will accept the defining parameters of the order.

Geopolitical world orders will vary in terms of the degree of conformity required of states. We can tell from the name of the latest such order, the 'Cold War', that stability was based on very strict adherence to an antagonistic alliance system. Previous world orders have been much more flexible, although a general global structure of power was always easily discernible. All

orders may incorporate contrary codes, however. These may reflect vestiges of past world orders or may represent attempts to define new orders. For instance, Spain and Portugal vigorously pursued policies akin to pre-World War II colonialism for two decades after such activity had been de-legitimized by the Cold War. In contrast, other states, such as India, refused to be forced into taking sides in the Cold War. They formed a nonaligned bloc to proclaim their independence from the prevailing order. Notice that this policy is not one of neutrality as typically practised by some of the smaller European states. Neutrality is a policy that is effectively a local code, a strategy against more powerful neighbours. Nonalignment is a global strategy, a policy against a world geopolitical code: by its violation of the premises of the Cold War it constitutes a challenge to that world order. Refusing to choose between East and West, nonaligned states defined themselves as South in an attempt to change the key geographical fracture of world power from East/West to North/South. But the Cold War geographical pattern prevailed until its recent demise which was not precipitated by the South.

Geopolitical world orders are relatively stable patterns of power. They represent what many regard as 'surface features' of the world-system compared to the more basic material changes expressed in Kondratieff and hegemonic cycles. But this is a rather unhelpful way of looking at world orders, since they cannot be independent of these cycles. For instance, the Cold War is how US hegemony has been expressed as a world order. Cycles and world-order periods should be interpreted together.

Periodization

We are in a position now to define our periodization of the twentieth century in terms of geopolitical world orders as shown on Table 1.1. Geopolitically the century has consisted of two distinctive world orders plus very small parts of two others. These world orders are separated by relatively short geopolitical transitions when the assumptions on which the global power distribution exists are overturned. We identify three such transitions in 1904–7, 1945–6 and 1989–91; these transitions encompass two world orders which we will term the Geopolitical World Order of the British Succession and the Cold War Geopolitical World Order.

Our treatment of each of the two world orders will concentrate on three main themes. First we consider the precursors of each new order. Although geopolitical transitions constitute changes that surprise contemporaries, no new order can be constructed out of nothing. Hence, using our considerable powers of hindsight we can find pointers to the new order in the old. In the case of the first world order we describe, this will mean delving back into the late nineteenth century to consider its precursors in what we shall term the World Order of Imperial Rivalry and Concert. It also means that the contemporary search for pointers to a new world order should scour the recent history of the Cold War.

Secondly, we deal in some detail with the geopolitical transitions themselves. Turning worlds upside-down is a complex business; erstwhile friends become enemies and vice versa. Although we can identify a period of disintegration in the old world order prior to a transition that in hindsight we see as vital, when the change around comes it is rapid and decisive. A new international political logic is constructed and every state has to review its geopolitical code and change accordingly. In this way geopolitical transitions define our periodization. Notice that this periodization does not use World War I and its associated events – the Russian Revolution and the Treaty of Versailles – as a break point. Certainly contemporaries acting in these events very consciously thought they were creating fundamental transitions – witness the phrases 'the war to end all wars', 'world revolution' and 'peace with justice'. In terms of diplomatic organization and activities, an important change did occur as Clark (1980) has illustrated in his periodization of the period, but geopolitically this is not the case. In our interpretation we follow Wallerstein (1984) and Modelski (1987) in considering both World Wars I and II as essentially one prolonged 'thirty-year war', 'the German Wars'; hence we follow Lentin (1984) in treating the grand events at Versailles as 'a mere

truce', and we follow Dukes (1989) in viewing the Russian Revolution as a precursor of the Cold War. Geopolitically there is one world order from 1904 to 1945.

The third element in our discussion of each world order emphasizes the variety within each period. Although there is one dominating logic throughout the period of a world order, this does not mean that international politics stops still. Within the parameters of the world order different phases of political activity can be discerned. It is the identification of such phases that enables us to transcend World War I in our periodization. Certainly international politics changed after 1917/18/19, but the dominant logic of the World Order of the British Succession remained: how to control German expansion as British decline continued. More recently, identification of the 1970s detente as a phase within the logic of the Cold War enables us to pinpoint the essence of the transition at the end of the 1980s. This transition was not a new detente, a reform of the old order as originally thought (Falk and Kaldor, 1989), but a genuine change which turned international politics upside-down for the third time this century.

The Geopolitical World Order of the British Succession

As the new century dawned, political observers in different countries talked of a forthcoming 'German Century' or an 'American Century', but the idea of a further 'British Century' seemed most unlikely. Britain was still the most powerful empire in the world, but the signs of relative decline were uncontestable. The decline might have been obscured by the successes of empire as symbolized by the imperial pageant for Victoria's Diamond Jubilee in 1897 (Morris, 1968), but the increases in Britain's territory compared to other states had run counter to comparative economic indicators in the late nineteenth century. And there was little doubt which changes were of most long-term importance: Britain was being overtaken by both Germany and the USA as the 'workshop of the world', and no imperial aggrandizement could hide this fundamental fact of the world-economy. Which of these two challengers would take over Britain's erstwhile role as hegemonic power was settled in the Geopolitical World Order of the British Succession of 1904 to 1945.

Precursors of a new order

Imperial policy could reveal as well as obscure Britain's declining world position. Such was the case in 1900 as the small Boer republics continued to resist the greatest empire the world had ever seen. The possibility of defeat was not a realistic worry, but the exposure of Britain's international position was. Hostility to Britain was widespread, and the perennial British nightmare of a continental alliance of powers in opposition to Britain appeared briefly as a real possibility. Certainly the Franco-Russian alliance was explicitly anti-British at this time, and Germany was very publicly pro-Boer; if the Russians had succeeded in their attempt to bring together the arch-enemies France and Germany, then the twentieth century would have looked even less rosy for Britain. But in the event this Russian ploy turned out to be just another episode in the three decades of geopolitical manoeuvring that typified the World Order of Imperial Rivalry and Concert. We need to consider this world order to understand the stabilization in alliances that was to occur after 1900 to produce a diplomatic turnabout which left Germany more isolated than Britain.

The establishment of the German Empire in 1870–71 coincided with the earliest British worries about decline. In fact, the emergence of this dominant power in Europe had mixed blessings for Britain. For Disraeli, the British Prime Minister, German military successes had produced 'a new world' where the 'balance of power has been entirely destroyed' (Dukes, 1989, 94–5). Britain's ability to balance European powers to her own interests was most definitely curtailed after 1871, but initially at least the result did not have to be disadvantageous to Britain. What Britain required in Europe was political stability; since this was German policy after 1871, the changing power distribution had less effect on Britain than was first feared.

Diplomatically the late-nineteenth-century world order was one of immense fluidity. Alliances, agreements and understanding among the major states abounded, but with no consistent pattern. Russia, for instance, made agreements with Germany and France at different times indicating contrary geopolitical codes. But this fluidity should not be interpreted as chaos. There were two certainties that held this first post-British-hegemony world order together. In Europe, Germany was the status quo power wishing to preserve its recent rise to the top of the European hierarchy. In the rest of the world, Britain was the status quo power attempting to preserve its global pre-eminence. It is this combination of two status quo powers without overlapping interests that defines the order of this period. In their two spheres of operation, both states had similar strategic concerns. In Europe, Germany was between Russia and France and feared a European war on two fronts. Outside Europe Britain's two major adversaries were also Russia and France, and she feared an imperial war on two fronts.

Notwithstanding this common pattern of strategic foes, the British–German power duopoly was never a formal arrangement despite such suggestions from both sides. But an alliance proved to be unnecessary; the status quo powers, operating for the most part separately, were able to preserve the peace among the major states for over three decades. Ultimately the world order was an unstable one since the power trajectories of the two 'guarantors' pointed in opposite directions – one up, and one down. But in the meantime an order prevailed.

The peace of this world order was a very tense one and it is in this context that we can find precursors and preconditions of the warring world order that was to come with the twentieth century. In response to the perennial tension and associated suspicions, Europe became highly militarized in the last three decades of the nineteenth century. By the end of the century the German, Russian, French and Austro-Hungarian armies each numbered between 2.6 and 4 million men (Bartlett, 1984, 6–7). Even Britain had increased its military expenditure by 45 per cent before the Boer War. The remarkable feature of the old Pax Britannica of British hegemony had been its 'cheapness' (Kennedy, 1981, 32), but in this new world this was no longer the case. The great additional drain on the Treasury by the Boer War finally confirmed that the new militarization had finally caught up with Britain in its extra-European activities.

Associated with the great new armies came the integration of General Staffs into the state apparatuses. This followed the successes of the Prussian Army from 1864 to 1871 and the expectation that future European wars would be equally as swift. Hence in this period, and for the very first time, we get the development of detailed war plans in what is peacetime (Kennedy, 1979, 2–3). This one sector of the state apparatus translates each state's political and economic interests into friends and foes in battle. This state innovation was to be vital to both world orders of the twentieth century.

All this pent-up aggression was premised on one fundamental geopolitical assumption. This is what we may call an 'ideology of bigness'. Although Europe remained the political cockpit of the world throughout the nineteenth century, there were many commentators who saw the future in terms of large continental-scale states nominating the USA, Russia or even China as harbingers of new worlds. With efficient mobilization of resources, these giant extra-European states would be able to dwarf even the major powers of Europe. The response to this widely-held scenario came in the form of two geographical global models that challenged the 'one world' assumption of British hegemony. Between them they were to dominate geopolitical thinking in the twentieth century; we will term these important precursors the geostrategic and geoeconomic models.

The geostrategic model derives from Britain's response to the bigness thesis. From at least the 1840s onwards, Britain had been 'playing the Great Game' in Asia to counter the expansion of Russia (Edwardes, 1975). Friction occurred between the geopolitical codes of these two empires in an 'arc of conflict' from Afghanistan/North-West Frontier through Persia/the Gulf to the Turkish Straits/Mediterranean. There was no actual war in any of these zones in the last third of the nineteenth century, but pressure rose with the increased tension of imperial rivalry in the world order. A crucial new factor appeared in the

equation with technological advances in transport and communications. Russian railway-building into Asia seemed to presage the continental state beginning to realize its great-power potential. On the other side steamships, telegraph cables and the Suez Canal similarly made global mobilization of distant lands a new possibility. Kennedy (1979b) notes that telegraphic cable-laying was largely commercially led from 1837 to 1870, but thereafter strategic considerations dominated. In particular the British constructed their 'all-red' network, avoiding locations outside their direct control (Christopher, 1988). This was completed by 1900. It spawned dreams of a world-state based upon a new imperial unity. Many political commentators now believed that the culmination of the Great Game would be a new power constellation that pitted global sea-power against continental landpower. This geostrategic model was to be codified by Halford Mackinder as his Heartland Theory which has been so influential this century.

The geoeconomic model derives, in part, from Germany's response to the bigness thesis. It related initially to the 'new mercantilism' that followed the end of British hegemony, rather than the 'new imperialism'. As the free-trade arguments of the British came under attack after 1870, it was the German Empire that became the leading advocate and practitioner of protectionism. The logical consequence of this for a Great Power was autarky so as to become as economically self-sufficient as possible. But it was becoming more and more difficult to reduce economic vulnerability in an increasingly interconnected world. The solution to the impossibility of 'capitalism in one country' was to create economic zones of dependence beyond the state. Outside Europe these would be colonies or spheres of influence: inside Europe, dominated trading spheres. For Germany, central Europe (*mitteleuropa*) and the whole Danube basin constituted the zone of expansion linking the state irrevocably to Austro-Hungary (Schultz, 1989). Hence when Max Weber in his inaugural lecture of 1895 dismissed German unification as 'youthful folly' unless followed up by some grander scheme (Bartlett, 1984, 20), he was pointing towards the theory of pan-regions, the division of the world into a few large self-sufficient regions centred on the Great Powers (O'Loughlin and van der Wusten, 1990). Of course, such a strategy by Germany would bring her into conflict with Britain and doom the World Order of Rivalry and Concert that depended on the separation of their political interests.

Changing codes, changing world orders

In January 1896 Kaiser Wilhelm II proclaimed that Germany should pursue a policy of *weltpolitik*. At a stroke, Germany was converted from a status quo European state to a deprived global power. If the sun never set on the British Empire, then Germany too should have its 'place in the sun'. By giving notice of a global geopolitical code, Germany was directly challenging Britain's extra-European supremacy. Bartlett (1984, 80) has argued that the influence of *weltpolitik* has been overemphasized. Our interpretation is that the Kaiser's proclamation set the conditions for a new world order – early premonitions of transition.

Exactly the same point can be made for the Franco-Russian alliance of two years earlier. Kennan (1984) calls this the 'fateful alliance' that led to World War I. He argues that the mobilization provisions were crucial, since they meant that limited war would no longer typify European conflicts and a war *à outrance* (255) was set for the future. Once again Bartlett (1984, 13) argues against overemphasizing this event – no deep divide resulted from it initially. Nonetheless it is true that this treaty finally separated Germany and Russia and set the geopolitical codes in Europe into the pattern that was to be so important in the next century: this was a precursor of transition to a new world order.

There were other important political developments in the 1890s that were equally setting out the conditions for the transition to come. Britain was revising its geopolitical code, albeit much less radically. In the 'Great Game' it was reassessing its position with regard to the Turkish Straits. With German influence now extending through the Balkans to the Ottoman Empire, Britain had to reconsider its strategy of confronting Russia in the Black Sea. Egypt and the Suez Canal was the new strategic point on the route to the East, and became the focus of the imperial geopolitical code replacing the straits. But this had further implications for Africa as the British embarked on a conquest of the Sudan to protect their position in Egypt. At the same time the French had an east–west Sudanese African strategy and the countries clashed famously at Fashoda in 1898 where the French retreated. Hence, the considerations bringing the French and Russians together were as much anti-British as anti-German; no new order had yet arrived. Rather, these changing codes of the 1890s are best interpreted as the disintegration of the old order – not yet its replacement. And we can add to the list of code revisions: first the emergence of Japan as an Asian regional power after its defeat of China in 1894, and second, the extension of the USA's geopolitical code beyond the Monroe Doctrine in the Americas and eastern Pacific to include interests in the Far East, after the defeat of Spain in 1898 and the subsequent retention of the Philippines (Grenville, 1979).

The old order was definitely disintegrating in the late 1890s, but the actual transition was not immediately forthcoming. There was still one final important event of the World Order of Rivalry and Concert to occur which was quintessentially of the old order. After the Boxer rebellions in 1900, China was humiliated and brought into line by a joint force of imperial powers, with Japan and the USA alongside the Europeans. The major powers of the world were not to act in concert again until the Gulf War of 1991. To understand the creation of a new world of 'deep divide', we have to return to the activities of what was still the most important power – Britain.

While British forces were joining with their rivals in China, they were fighting a much larger and diplomatically lonely war in South Africa. This conflict precipitated a rethink of Britain's code. Joseph Chamberlain, the Colonial Secretary, called first for an end to Britain's isolationalism, and second, for tariff reform to end British free trade. In this challenge to traditional British thinking he was successful in the former, but he failed in the latter. In his campaign Chamberlain was actively supported by Halford Mackinder, whose development of the Heartland Theory should be interpreted in this light. The final years of diplomatic isolationalism are associated with Foreign Secretary Lord Salisbury and his appeal to retain Britain's 'free hand' in world politics. In his famous defence of this policy in 1901, he still talked in terms of France and Russia being the main enemies. But in this his preferred code is not dissimilar to Chamberlain, who favoured alliance with Germany – same code, different means. 'It is a matter of supreme moment for us,' stated Landsdowne, Salisbury's successor at the Foreign Office, 'that Germany should not be squeezed to death between the hammer of Russia and the anvil of France' (Bartlett, 1984, 42). It is the overturning of this traditional British geopolitical code focusing on the enmity of France and Russia that marks the core of the geopolitical transition to a new world order.

The end of the 'free hand' came in the Far East with the Britain–Japan naval agreement in 1901. The crucial importance of this was not that it ensured British neutrality in the subsequent Japanese–Russian war in 1904–5, but that it broke almost a century of British isolationism – the first major political retreat from British hegemony. In fact, we can see this as part of a general pattern, starting with the transition, of Britain securing its position with each of the other powers in turn in order to combat the threat of Germany. We may see this transition process as Britain's first round of appeasement in its political decline. For instance, later in 1901 Britain conceded USA predominance in Central America in the Hay-Pauncefote treaty and made concessions on the Alaska–Canada border. But the crucial changes came in relations with European states with what Langhorne (1981, 85 and 93) has termed the first and second 'impossible agreements'. In 1904 Britain settled its differences in Africa with France – they accepted each other's positions in Egypt and Morocco respectively – to sign the Entente Cordiale. This was a diplomatic revolution that surprised other states, especially Germany. Equally surprising, after Russia's defeat by Japan, Britain and Russia agreed on the partition of Persia between them and in 1907 joined with France in a triple alliance. The final effect of all this diplomatic activity was a 'rigidification of alliance blocs' (Kennedy, 1987, 249) into the geopolitical

pattern of the World Order of the British Succession: confronting Germany on two sides in Europe, with Britain controlling access to extra-European arenas, plus the USA being politically neutral while generally in sympathy with the Allies.

Phases of the succession

The first round in Britain's appeasement of its rivals should not be seen as a voluntary withdrawal on the part of the British from their global predominance. The British succession was to be contested; appeasement was merely a policy to hold the line at special times of perceived threat, such as after the Boer War. The actual succession by the USA at the expense of Germany took four decades of threat and real conflict involving two further rounds of appeasement by the British. There were five distinct phases in the unfolding of this world order.

(i) A fragile divide, 1907–14. The Anglo-French *entente* was tested by Germany almost immediately and before the addition of Russia to the alliance. In 1906 Germany called a international conference at Algeciras to discuss French activities in Morocco, but in the event Britain backed France with the added bonus of USA sympathy. Germany was humiliated and its isolation exposed. But Germany continued to probe the alliance ranged against it right up until 1914. And it was by no means a hopeless task. Although in hindsight we can see a geopolitical pattern that was to last for many decades, this was by no means so obvious to contemporaries. Rivalry and concert in the manner of the previous world order continued and sometimes cut across the new alliance system. France and Germany came to a bilateral agreement over Morocco in 1909, for instance. Even more ominously, the carefully-arranged division of Persia was failing to prevent renewed rivalry between Britain and Russia there in 1911. Certainly Germany had not given up prising Russia from the alliance and made several approaches along these lines during this period. Also in 1911 the Japanese–British naval alliance came up for renewal which was by no means automatic being

only agreed by the British with misgivings. There even seemed to be the beginnings of a *rapprochement* between Britain and Germany between 1912 and 1914, so much so that early in 1914 it has been claimed that a war was just as likely between Britain and Russia as between Germany and Russia (Bartlett, 1984, 73). We may conclude that the 'deep divide' which was constructed in the geopolitical transition remained fragile in this, its first phase of operation.

But there was a more fundamental basis to the new world order than these continuing diplomatic manoeuvrings suggest. This is clearly illustrated by documents of the two main protagonists at either end of this phase. In 1907 a British Foreign Office document, the 'Crow memorandum', identified Germany as the leading revisionist state in the inter-state system, and concluded that even major concessions by Britain would not satisfy her demands. There could be no appeasement here; rather, Britain and Germany were destined to conflict by 'the form of a law of nations' (Bartlett, 1984, 57–8). The converse of this can be found in 1914 with the 'Rathenau programme' for Germany's last chance to 'catch up'. This argued that Germany could never be a world power by accepting 'the charity of the world market'; rather, what was needed was 'territory on the globe' (Bartlett, 1984, 75). There could be no compromise with Britain; a redistribution of the world's lands was a necessity. Similarly, the Balkan Wars of 1911–13 confirmed the incompatibility of German and Russian interests in central and eastern Europe. Something would have to give.

Such global strategic thinking was complemented by war planning by the Chiefs of Staff, and by popular sentiment in the various countries. In Germany the famous Schlieffen Plan convinced the government that it could win a war on two fronts by knocking out France before Russia had had time to mobilize fully. On the other side, counter plans were in hand. Britain, for instance, had reorganized its imperial defences to support France on her left flank, thereby preventing German access to the channel ports. As one newspaper put it in 1912: 'We are in the position of Imperial Rome when the barbarians were thundering at the frontiers . . . We have called home the legions' (Kennedy, 1981,

131). Defending north-west Europe meant the army coming to the fore instead of traditional British military reliance on the navy; from their impossible task in the 'Great Game', they now welcomed a feasible continental engagement and planned for it accordingly. All the major powers on both sides had their plans for a great European war, and it is hardly surprising that many contemporaries thought it inevitable.

Armies were not all that were mobilized in 1914. Foreign policy had become an issue of popular concern. The rivalries of the period were producing mass phobias against the inhabitants of other countries. While the Kaiser could rely on Anglophobia in his political practices, there was equally an anti-German sentiment building in Britain through the popular press. Such processes put pressure on 'high politics' from below, and were occurring in all European states – even Tsarist Russia. Hence, when war came in 1914, it was very popular. Declarations of war were accompanied by mass expressions of patriotic fervour throughout Europe. This ensured that this war would be like none that had gone before.

(ii) The first round, 1914–19. The mobilization of armies across Europe in August 1914 ensured the era of limited wars, the minor adjustments of boundaries in favour of the victor, had come to an abrupt end. This was a war about the nature of the geopolitics of the world-system. It was a total war where the victor would design the pattern of power in the post-war world. Although there were a variety of motives from the different participants, the essence could be distilled into two contests. First there was the question of the mastery of Europe: would Germany be able to maintain her pre-eminence held since 1871 and indeed extend it? Second, there was the question of the global balance of power: would Germany be able to force Britain into a redistribution of colonies and influence across the globe?

Expectations on all sides had been of a quick war, but as it settled down to a stalemate in late 1914 and the belligerents decided to see it through to the end, the relative levels of power within each camp became clarified. A long war of attrition required a large and productive industrial base. Very soon a pattern of dependence was revealed. For the Central Powers, Austro-

Hungary needed German armaments and the finance to buy them with in order to stay in the war. For the Allies, Russia and France were similarly dependent on Britain. But this was only part of the story. From 1915 Britain was financing its war effort through loans from the USA. The war was depleting British overseas assets which had been built up over a century of hegemony and its aftermath. Britain was using up the proceeds of past dominance to preserve its current situation. By staying in a military war with Germany, it was effectively losing a financial war with the USA: World War I marked the transfer of the centre of world finance from London to New York. The final tangible element of Britain's financial hegemony was gone, whatever the outcome of the war.

The USA was the real winner of the war even before she joined with the Allies in 1917. But Britain remained the status quo power throughout. For instance, despite the American rhetoric of national self-determination, Britain was not necessarily against the maintenance of the Austro-Hungarian Empire as part of a new balance in Europe. But that was not to be. The irony was that the Austrian and Russian empires collapsed, but the German one did not. After the Armistice in November 1918, the German state remained intact while awaiting the Allies' terms. The problem for the victors meeting in Paris in 1919 was that Germany remained the largest and economically the most advanced country on continental Europe. Stripped of her colonies and with no remaining allies, she remained a formidable foe. The Versailles Treaty humiliated Germany politically as a major power and imposed some economic constraints on her immediate recovery, but left a potentially great but wounded state in the heart of Europe. The challenge was not over.

The discussion at Paris revealed the geopolitical codes of the remaining major powers. France was most concerned with her northern boundary with Germany and was to insist on the most severe conditions to maintain her position in Europe gained through the war. For instance, German sovereignty, not violated in the war itself, was now subject to external oversight in the Rhinelands. France also took her share of Germany's colonies to maintain her world role.

Britain and her empire got the lion's share of the dispossessed colonies, enhancing her world imperial role – this was to be the greatest extent the British Empire was to reach (Christopher, 1988). As the status quo power she presided over no major changes in the world outside Europe; certainly, national self-determination was not to be interpreted as a valid concept beyond Europe. The USA and Japan joined in sharing out the spoils of the German empire in the Pacific and China.

The position of the USA in the aftermath of the war is most interesting. Potentially the most powerful country, she took on the role of the honest broker with no major interests at stake. President Wilson maintained a position above the fray in a hegemonic-like stance to impose new institutions on the inter-state system. We will deal with this in more detail in the next section as a precursor of the future world order. The important point here is that the President's idealistic institutional arrangements were agreed by the other victors in return for satisfying their very realist demands. With the domestic defeat of Wilson and the failure of the USA to join the League of Nations, the new institutions set up to preserve the peace were doomed. The League of Nations, without the USA and the pariah states of defeated Germany and revolutionary Russia, was a mask for a power vacuum in the world-system.

(iii) Power vacuum interlude, 1920–31. This phase of geopolitical order was superficially like the Rivalry and Concert Order after 1871. There were two status quo powers, one in Europe and one for the rest of the world, but this is where the similarity ends. As a result of World War I and the Versailles settlement, France was the leading power on continental Europe. But the position was quite artificial and therefore short-term; it was only a matter of time before the larger and more industrial German state would be in a position to reassert her leadership of Europe. France understood this, accounting for her acute anxiety during this period. In the wider world Britain remained the status quo power, but her relative decline since 1871 meant that she was only a shadow of her former self as a world power, despite the size of her empire. Britain favoured

sharing global responsibilities with the USA through the League of Nations, but this was not to be. The USA was certainly not isolationist in this period, but she was highly selective in her interventions so as to preclude any overall construction of a Pax Anglo-America. In short, within and beyond Europe the inter-state system experienced a power vacuum.

The situation in Europe centred on providing France with security. This took three forms. First, France was the power most insistent on war reparations from Germany, both to pay off her own war debts and to retard German economic recovery. By 1923, when German economic recovery seemed to be overtaking that of France, the French army marched into the Rhinelands to force its reparation grievances to the fore. As the major debt-holder in Europe, the USA became involved in rescheduling debt and reparations, but this remained the most short-term of France's policies. Second, France formed a little *entente* with the new Eastern European states, continuing the old policy of confronting Germany on two fronts, but this was no substitute for the power of Russia. Third, France sought security guarantees from other Great Powers. The initial guarantee from Britain and the USA negotiated at Paris fell with the US's failure to join the League. This need became more obvious when the two pariah states, Germany and Russia, came to an agreement at Rapallo in 1922. France supported Britain's proposal for Germany to join the League in 1924 to counter this ultimate French nightmare of a threatening German–Russian alliance. At Locarno in 1924, general territorial guarantees were agreed among the European states (minus Russia), and this ushered in a short period of detente that lasted until the end of the decade. Briefly the hope of peace and stability promised by Versailles seemed attainable. But it was all based on the artificial position of French predominance in Europe.

Beyond Europe, Britain was equally as insecure as France was within Europe. Britain needed the USA as co-sponsor of any world order. She had tried to get the USA to accept a mandate in the Middle East, for example, but the United States limited her interests to the oil in the region, eschewing any territorial involve-

ment. Britain's initial concerns were in Asia and the continuation of the Great Game. Taking advantage of the revolutionary turmoil in Russia, Britain first sponsored a break-up of the old territorial state by supporting independence movements in south Russia – Mackinder was sent there as Britain's representative (Blouet, 1976). Britain also attempted to turn Persia into a puppet state now Russia was not on hand to balance her power. But all this changed rapidly as the communist regime consolidated itself. At Baku in 1920 the Bolsheviks went on the offensive, calling the First Congress of Peoples of the East where a holy war against British imperialism was proclaimed. But in the event, the new state had too many domestic problems to pursue traditional Russian expansionist policies at this time, and the Great Game faded from the picture except in the anti-communist rhetoric of such imperialist politicians as Winston Churchill.

Outside Europe, Britain's chief rivals during this phase were the USA and Japan. Britain coped with them in what we may see as its second round of appeasement in its long political decline. With the Anglo-Japanese naval agreement coming up for ratification, negotiations in Washington in 1921 and 1922 led to a new broader naval agreement incorporating the USA, France and Italy as well as Japan and Britain. The key point is that Britain bowed to reality and conceded naval parity to the USA. The principle of British global naval supremacy – 'Britannia rules the waves' – was ended. Japan was not conceded parity, but to compensate, a 3,000-mile non-fortification zone around the country was agreed. This meant that Britain could build a base no nearer than Singapore, and the USA had to abandon similar plans for the Philippines and have its nearest base at Pearl Harbor in Hawaii. Both decisions illustrate how far Britain had declined from its nineteenth-century position as the only global power.

Although the Washington Treaty was largely confirmed in London in 1930 despite some Japanese pressure for major revision, by this time the world situation was rapidly changing. The financial collapse in New York in 1929 was undermining the popularity of governments throughout the world. In 1931 two events marked the end of this artificial phase based upon two inadequate status quo powers: Britain took her currency off the gold standard, marking a new intensity in economic competition; and in China, the Manchuria crisis marked the beginning of a new intensity in political competition to be led by Japan. The power vacuum was ready to implode.

(iv) Reconstituting the divide, 1931–9. The implosion took the form of developing pan-regions. The nineteenth-century world that Britain had presided over was finally disintegrating into a deeply divided world.

The most developed pan-region, the one most commonly quoted by contemporary theorists, was the Americas under US leadership. Although originating from the Monroe Doctrine of 1823 warning off European powers from the western hemisphere, the major practical beginnings of this pan-region were at the turn of the twentieth century with the British concessions – the first round of British appeasement discussed above – leading to irregular US armed interventions into the 1930s (Pearce, 1981). In the reassessment of the Roosevelt administration after 1932, this was replaced by the 'good neighbor' policy consisting of seventeen bilateral trade agreements. This single-power dominance of a world region was becoming so clear-cut that pan-regions became commonly referred to as 'monroes' (Taylor, 1990).

In the Far East an analogous process was operating, but here the armed intervention by the Japanese was more permanent. During World War I when the Allies were diverted by events in Europe, Japan began the process of consolidating her position as the leading regional power. After seizing Manchuria in 1931, Japan continued to pressurize the Chinese for further concessions, leading eventually to war in 1937. By this time Japan was attempting to marshall Asian nationalist feelings against rival western powers under the umbrella of a 'Greater Asian Co-prosperity Sphere' proclaimed in 1938. By the end of the decade the Japanese were well on their way to carving out their own pan-region.

Other divisions were being forged. In a parody of the Roman Empire, Fascist Italy started to define its imperial goals in Africa with the invasion of Abyssinia in 1934. At the other end of the political spectrum Stalin was embarking on the USSR's autarky programme which he termed 'socialism in one country'. But the key actors remained Britain and Germany. In the former case, the economic collapse after 1929 precipitated the final revision of its nineteenth-century foreign policy principles – free trade was finally abandoned. In 1932 Britain agreed to adopt a trading policy of 'imperial preferences' aimed at both the USA and its trading rivals in Europe. Whether the British Empire dotted across the globe could have ever become a viable trading bloc is debatable, but certainly it was not a good time to try in the 1930s at the end of over half a century of British decline. Canada's growing trade dependence on the USA and Australia's vulnerability to Japan were the two glaring examples of other cross-cutting pan-regional projects. If the Americas represented the 'classic' pan-region, the British Empire was the epitome of an illogical, geographically-fragmented imperial inheritance that pan-region theory was designed to undermine. Britain and her Empire were out of tune with the times.

In contrast, it was in Germany that pan-region theory was developed as part of this revisionist power's legitimation of its expansionist foreign policy (O'Loughlin and van der Wusten, 1990). With the coming to power of Hitler, a four-stage plan came into operation (Stokes, 1986). First, Germany recovered its territorial sovereignty by militarizing the Rhinelands against the provisions of the Treaty of Versailles in 1934. Second the need for *lebensraum*, literally 'living space', led to the takeover of lands in central and Eastern Europe. The ultimate expression of this process was the war on the USSR – Germany's 'India' – in the carving-out of a Eurasian pan-region in the manner feared by Mackinder at the beginning of the century. Third, there would have to be war with the oceanic powers to force a redistribution of the world's territories in Germany's favour. This was to be more than the recovery of colonies lost at Versailles; a second German pan-region, Eurafrica, was envisaged. Finally, from this vantage point Germany could dominate the world. This *Stufenplan* was not necessarily attainable in a short time, but defined the geopolitical code and hence directed foreign policy.

These pan-regions began to come together as

two antagonistic world blocs as early as 1936, with Germany and Japan joining in the Anti-Comintern Pact against the USSR. Italy joined in 1937, and her help for the rebels in the Spanish Civil War alongside Germany solidified the growing world division. In 1938 at Munich, British Prime Minister Neville Chamberlain attempted to paper over the geopolitical divide by territorial concessions to Germany in central Europe. This region had never been strategically important to Britain and accounted for only about 1 per cent of British trade, but by now international politics was a zero-sum global game; a gain for Germany was a loss for Britain. In terms of Britain's long decline, this was the third and final round of appeasement. Appeasement in Europe, rather than in remote parts of the Empire, showed how weak Britain had become as British politicians surmised that another war would end any remaining ability of their state to play the role of a Great Power, whatever the outcome (Bartlett, 1984, 191). War had to be avoided even if it meant effectively conceding Europe to an enemy. The world Britain had created in the nineteenth century had finally come to an end. All that was left was the final showdown.

(v) Final showdown, 1939–45. As with World War I, this second episode of the 'German Wars' provides us with the explicit revelations of the geopolitical codes of the period. There was again a mixture of local, regional and world wars, and we will concentrate on the latter geographical scale. In fact, the alliance of Germany with Japan, coupled with the joining of the USA with the Allies, meant that this was the first truly global conflict with major confrontations outside Europe.

In 1939 the British policy of appeasement failed. Although Britain had no strategic interest in Eastern Europe at a regional level, it was clear that as Germany's expansion became perceived as a global challenge to Britain, the strategic equation changed dramatically. Any major territorial gain for Germany tipped the balance against Britain. There was a point, therefore, when appeasement would be counterproductive. The British government in 1939 chose Poland as the point too far for Germany.

Germany's invasion of Poland, unlike the earlier dismemberment of Czechoslovakia, triggered off the final showdown between Britain and Germany.

The German–Russian pact before the outbreak of war in Europe ensured that Germany would not be fighting on two fronts as in 1914. Germany and Russia divided Eastern Europe, allowing the former to concentrate its forces in western Europe, leading to the defeat of the arch-foe France in 1940. At this stage we can discern a German plan of four pan-regions. Germany largely ignored the USA and conceded the Americas to that quarter. From a different starting position but with the same outcome, east and south east Asia were conceded to Japan by the 1936 alliance. Russia was encouraged to develop its Eurasian pan-region with the ultimate prize of India. This left Germany with Eurafrica including the Mediterranean and the Middle East. The end-result of this world plan would be the dissolution of the British Empire, with Germany gaining African colonies, the USA inheriting Canada and the Caribbean, Russia conquering India and Japan likewise Australia and the Pacific islands.

There were three major problems with this seemingly neat division of the world. A pan-region scheme can only work when the geopolitical codes of the Great Powers do not overlap. In this case the only reliable participant who had no claims beyond its quarter was Japan. But east Asia had been a US area of economic interest from the 'Open Door' policy at the beginning of the century. Hence the USA refused to allow Japan its own pan-region and employed a trade embargo which precipitated the war in the Pacific. Similarly it could not be expected that the USSR would concede Europe to Germany and pursue a purely Asian strategy. Its ultimate security rested on its ability to control events on its European borders, and therefore it was never a feasible proposition that it would re-engage in the 'Great Game' with Britain while neglecting the more fundamental 'European Game' with Germany. But most important of all, Germany would not be limited to its Eurafrican pan-region. After the fall of France there was even concern in the USA lest Germany use French West Africa as a stepping-stone to the Americas to consolidate its already growing influence there, especially in Argentina. But the real threat of further German expansion was not a violation of the Monroe Doctrine, but a return to an anti-Soviet stance. The two pan-regions meeting in Eastern Europe were inherently unstable for both geostrategic and ideological reasons. The 1939 pact was generally interpreted as a short-term expedient by both sides. With the defeat of France, the western front was effectively neutralized, leaving Germany to turn again to her eastern front. In 1941 Germany invaded Russia and thereby converted her global strategy from a four to a three pan-region plan. In fact, this was a move towards the Heartland world model, since if successful it would leave Germany in possession of most of Mackinder's 'world-island', ready to strike at the remaining two smaller pan-regions.

By the end of 1941 the division of the world into a power struggle between the two alliances was in place; the Axis powers of Germany, Japan and Italy versus the 'Grand Alliance' of Britain, the USA and USSR. The former was the less co-ordinated alliance, with Germany and Japan effectively fighting separate wars, merely agreeing to a division of activities along the 70° longitude east. This allocated almost all of British India to Japan. Although both Japan and Germany made massive territorial gains initially, Japan never reached India and Germany never reached the Urals or Siberia. The Heartland survived, and the USSR finally defeated Germany on the eastern front. Both the three pan-region model and subsequent stages towards German world domination were brought to an end with the unconditional surrender of the two major Axis powers in 1945. Britain was on the winning side in this second great test of strength of the twentieth century, but this time its contribution was clearly secondary in both the European and the Pacific theatres of war. Germany may have been stopped a second time, but now the British succession could no longer be postponed. The time was ripe for building a very different geopolitical world order.

The Cold War Geopolitical World Order

The term 'cold war' was popularized by US political commentator Walter Lippmann in 1947

(Steel, 1980). It was born of disappointment in the new post-war era: the 'hot' war with Germany and Japan was over, only to be replaced by new international tensions as the Grand Alliance broke up. For the USA, the USSR soon replaced Germany as a great ideological enemy threatening the building of a liberal world order anchored in the United Nations. As the USSR slipped into Germany's role, the only change seemed to be the lack of military conflict itself, hence 'cold' war.

Lippmann's phrase has outlasted its original context because it conveys a second important feature of the post-1945 world order. The particular distribution of power that emerged seemed to be more permanent than any that had gone before. As E. P. Thompson (1987, 14) has remarked, continuing the climatological analogy, the world was 'glaciated into its Cold War form . . . like an immutable fact of geography'. As well as the two chief protagonists never coming to blows, the Cold War represented a freezing of international relations into a solid structure contrasting with the relative fluidity of previous world orders.

The freezing of the inter-state system into two antagonistic blocs had an ideological basis. The new world order was proclaimed in civilization terms. For the USSR this meant that the Cold War was just a step on the road to world revolution to create a new civilization. Western politicians were thinking in similar terms. In his famous speech to Congress of 1947, US President Harry Truman talked of the world having to choose between two 'ways of life', positing freedom against totalitarianism. Nazi Germany had been the totalitarian foe, so by branding the USSR with the same label the recent mobilization of resources and peoples for freedom could be continued: it justified the conversion of ally into enemy. In Britain a 1948 Foreign Office memorandum was entitled simply, 'The threat to world civilization'. For both sides the world was divided into more than mere blocs; two contrary 'systems' were facing one another. This is the language of the *longue durée*, of epochs rather than periods.

We will interpret the Cold War as a world order of the *moyenne durée*, but its civilizational pretentions remain a vital aspect of its character to be investigated. The Cold War covers the period of US hegemony in the world-system. Superficially this world order seems very straightforward, with an 'East–West' geographical pattern of power conflict dominating the inter-state system. But hovering in the background throughout there is an alternative 'North–South' geographical pattern. Originally benefiting from the anti-imperialism of both the USA and the USSR, proponents of this other interpretation of the global pattern of power came to question the relevance of the Cold War to most of the world's peoples. Hence, US political hegemony came to be challenged by a new 'Third World' as well as the USSR (Krasner, 1985).

Precursors of a new order

The idea of a great ideological contest between bourgeoisie/capitalism and proletarians/communism predates 1945. What happened after 1945 was the translation to the inter-state system of the century-long domestic socialist challenge. This process actually began at the end of World War I with the establishment of the USSR from the ruins of the old Tsarist state. The intervention of the 'west' – Britain, France, Japan and the USA – in the Russian civil war in 1920–21 was the first war with communism at the inter-state level and may be interpreted as a prologue to the Cold War world order.

Ideologically the Cold War contest has been traced right back to 1917 as a clash between Wilsonism and Leninism. President Wilson's administration represented the first tentative steps by the USA towards hegemonic policies, as opposed to the protectionist and relatively isolationist Republican administrations before and after. Wilson wanted to use the new-found financial power that World War I provided for the USA to manoeuvre international relations towards a more liberal order. He claimed a role for the USA as 'champions of the rights of mankind' (Lentin, 1984, 6) and 'trustee for the peace of the world' (Dukes, 1989, 85) – words very reminiscent of later Cold War statements. And this was at the same time that Lenin was supporting revolutionary movements in Europe as a

move towards a socialist world order. Dukes (1989) interprets this as the 'Great Conjuncture' of Leninism/Wilsonism when the Cold War was born.

We have to be careful reading back from the Cold War in this way. Dukes 'conjuncture' is a precursor to the next world order, but no more. To think otherwise is to project a Cold Warrior view of the world back on to the conflict over the British succession. In fact, USA–USSR relations were little developed after 1921 until they came together as allies twenty years later. The seeds of ideological conflict most definitely existed, but a great East–West global conflict remained very much a secondary concern while the German threat to both sides existed. This precursor of a new order lay dormant for most of the previous world order.

One effect of Wilson's rhetoric on US entry into World War I was that it 'transmuted a sordid imperialist war into a war of liberation' (Lentin, 1984, 6). At the Paris peace conference, the European imperial powers were able to limit this liberation by restricting national self-determination to Europe. Nonetheless, the share-out of defeated states' extra-European territory was in the form of mandates from the League of Nations which presumed development to self-government. The implication was that formal European imperialism would end; only the timing was in dispute. The first mandated territory to become independent was Iraq in 1932. In non-mandated possessions the anti-imperialism movement was led by India. The final successful threat to this 'jewel of the British Empire' came not from Russia in the Great Game, but internally through a radical national mobilization. New nationalisms were emerging outside Europe that would prove to be irresistible. The fall of Singapore to Japan in 1942 symbolically represented the end of European political superiority over Asia. As well as eliminating the organization of the world into pan-regions, World War II struck the death-knell of the European empires they were designed to replace. The precursors of the North–South challenge in the Cold War are to be found in the anti-imperialist movements of the previous world order.

How many worlds – one, two or three?

Because of the depth of change in 1945, this geopolitical transition is the classic case of its genre (Taylor, 1990). The world order was totally transformed; the USA replaced Britain as leader, and the USSR replaced Germany as challenger. The nature of international politics was turned upside-down. The British succession was settled, and a different politics would have to be built to replace it, but it was by no means obvious that the new world order would take the form of the Cold War.

The origins of the Cold War has spawned a very large and controversial literature (McCauley, 1983). The original 'orthodox' view in US writings emphasized the special nature of the USSR as an inherently expansionist state that could never be accommodated in a stable world order. The only feasible policy was to contain the enemy by encircling it with pro-USA states. This view of a benign USA holding back the destroyers of order was widely disputed during the US war in Vietnam, where the hegemonic state looked anything but benign. In the resulting 'revisionist' school, the Cold War is blamed on US demands for a free world market to suit US business. This forced the USSR to revert to its strong autarky policy to prevent its economic domination. This choice of interpretations between Russian political imperialism and American economic imperialism has been superseded by a 'post-revisionist' literature that attempts a more subtle analysis, emphasizing the interaction of policies by both countries in producing the Cold War.

The debate goes on, but there is one feature of it that we can challenge through our analysis. The British succession may have been completed, but Britain was not immediately finished as a major influence on world events. When the term 'superpower' was coined in 1944, it was applied as much to the British Empire as to the USA and the USSR (Watt, 1984, 11). At the end of the war, the peace was in the hands of the 'Big Three', with Britain accorded equal status in negotiations. It was only in 1947, after Indian independence and another British economic crisis, that the Big Three became definitively reduced to two and we enter the bi-polar world of the Cold War.

The key point is that the geopolitical transition predates this bi-polar world. Much of the literature on the origins of the Cold War suffers from the propensity to read the 'certainties' of the period backwards to before their existence. No account of the geopolitical transition that took place in 1945–6 should ignore the third major participant, Britain. In fact, this can be viewed as Britain's last fling of its world power dice as it strove desperately to retain its geopolitical status. The result was that Britain, despite its rapidly diminishing power, was surprisingly influential in creating the new geopolitical world order (Taylor, 1990).

Geopolitical transitions are pre-eminently fluid periods of international relations when different geopolitical options are vying for construction. Not all options are equally likely, of course, but the chief ones should be considered in an analysis that does not treat the outcome as inevitable. In this case if we limit our 'constructors' to just the Big Three, we can define five potential patterns of power in 1945: (i) *one world*, where the Grand Alliance survives to lead an undivided and peaceful world; (ii) *three monroes*, where the three superpowers split apart and each concentrates on their division of the world – a latter-day pan-region plan; (iii) *an anti-imperial front* producing two worlds where the USA and USSR combine to oppose Britain and other European empires; (iv) *an anti-hegemonic front* producing two worlds where Britain and the USSR combine, perhaps as socialist states after Labour's 1945 election victory in Britain, to confront the overwhelming economic power of the USA; and (v) *an anti-communist front* producing two worlds with Britain and the USA confronting the USSR. How did option (v) become the next world order?

As World War II ended, the expectation was that the Grand Alliance would be able to produce a peaceful world without deep divisions. In 1944 and 1945 a new world organization was agreed, and this United Nations was designed to overcome the weaknesses of the failed League of Nations. In particular the Big Three (along with the other two victors, France and China) were awarded permanent positions on the Security Council, the most powerful organ of the new institution. At Yalta and Potsdam the Big Three

51

met to agree the post-war order. Through this mixture of idealism and realism it was hoped the one world ideal would emerge. In the event the two peace conferences failed to come to any lasting agreement and delegated the negotiations to a series of Foreign Minister meetings whose breakdown signalled a divided world. Hence there was no peace treaty after World War II, leaving the victors largely in possession of what their armies had taken in war.

The demise of hopes of one world did not in itself mean the emergence of the Cold War. Bevin, the British Foreign Secretary, in particular feared a resurgence of autarky among his allied partners. This could be expected from the USSR, but what if isolationist forces came to dominate US foreign policy again? Bevin's 'nightmare' was what he termed a 'three monroes' outcome, where both the USA and the USSR possessed relatively compact and contiguous spheres of influence, leaving Britain with a ramshackle zone of leftover western colonies that would be impossible to defend in any future conflict (Taylor, 1990, 51). Since Bevin had little faith in the United Nations as an defence umbrella to shelter Britain, it follows that his policy was to promote a two-world solution to cope with Britain's vulnerability.

Of the three two-world options, the least likely was probably the anti-hegemonic front. Although there was a strong body of opinion in the British Labour Party in favour of closer relations with the USSR, this was not reflected in foreign policy making. Bevin was a long-term anti-communist from his trade union days and he easily transferred his antipathy to the international stage. The British general election of 1945 was held during the Potsdam conference, and when the Labour ministers returned to the conference they were generally considered to be more anti-Soviet than their Conservative predecessors (Shlaim *et al.*, 1977, 40). We can identify Bevin, therefore, as a major architect of the Cold War.

But the Cold War was not the only two-world option, as we have seen. Why should the USA and USSR be antagonistic rivals as Britain would wish? In terms of their immediate post-war geopolitical codes there was effectively no overlap to generate friction. The civil war in China had not

yet come to a head, and in Europe the USA was originally willing to concede the special interest of the USSR in Eastern Europe. Although the USA was against closed spheres of influence, the idea of 'open' spheres was proposed whereby Soviet political dominance would be conceded as long as economic transactions were not impeded (Harbutt, 1986, 131). For the new hegemonic state, it was such economic matters that were important initially. Hence in 1945 it was the British Empire with its imperial preferences enclosing the largest 'unfree' market in the world that was viewed as a potential enemy to US interests (Kolko and Kolko, 1973). Combined with traditional US anti-imperialism, there was certainly some potential for the isolation of Britain within the Big Three in 1945. The US solved the problem of closed British markets through its loan to Britain that effectively tied Britain to a new liberal US-led world-economy (Taylor, 1990). But there was still no basis for a USA–USSR split.

Although the British feared the USSR as a threat to its empire, it was hard to see how she could possibly get the USA to help defend it. Despite the acknowledged imminence of Indian independence, Britain persevered with its traditional imperial geopolitical code with the route to the east at its heart. USSR demands for access to the Mediterranean at Potsdam fuelled Britain's doubts on her ability to combat a resurgent Russian attack on the 'lifeline' of its empire. For Britain the problem was that if a new Great Game was to be initiated, how could the USA be involved on its side? The answer was to turn the Great Game into an ideological contest and make it 'universal' in scope (Taylor, 1990). If the USA could be marshalled to save the world from communism, then in the process the British Empire might yet be saved. In early 1946 we can see this strategy operating. At the foreign ministers meetings Bevin is hard at work driving a wedge between the USA and USSR (Deighton, 1987) while ex-Prime Minister Winston Churchill puts his immense influence behind the campaign with his famous 'iron curtain' speech positing the dark forces of communism against Anglo-American liberties. The mixed reception this speech received in the USA shows that the Cold War was not yet in place (Harbutt, 1986), but the rise of anti-communist feeling in the USA soon turned the tide. Ironically, the first real confrontation between the USA and the USSR occurred in Iran – one of the centres of the old Great Game – after the USSR was slow in withdrawing its troops (Harbutt, 1986). Within a year, Britain was to precipitate the Truman Doctrine, committing the USA to defend all countries against the spread of communism: Britain declared her inability to afford continuing the defence of Greece and Turkey, and the USA stepped in as new guarantor. The Cold War was now firmly in place.

In this interpretation the Cold War is just another stage in the Great Game, but with a new team leader, the USA. It is hardly surprising, therefore, that the geographer who codified the nineteenth-century Great Game as world strategic model should have a second 'life' as a major geopolitical theorist in the Cold War. Mackinder's Heartland thesis concerning the importance of the world-island and the inevitability of an era of seapower versus landpower conflict derived a new resonance with the coming of the new world order. The British imperialist Sir Halford Mackinder's two worlds finally came into being as the Cold War Geopolitical World Order.

Phases of the Cold War

Although the most stable of all geopolitical world orders, the Cold War nonetheless has exhibited a variety of international relations within the single pattern. The standard approach to differences over time has been to contrast different degrees of enmity in East–West relations (Halliday, 1983). We define four phases in this manner, but with the added ingredient that we show how these run parallel with developments in North–South politics.

(i) The freeze, 1947–53. By 'freeze' I do not mean that this phase experienced no 'hot' war as we shall see, but rather that the geopolitical world order stabilized into the form it was to take for four decades. Further, this initial phase represents the deepest the Cold War division was to reach: there were to be later crises in East–West relations, but the nadir of this politics is

usually dated around 1950. In addition, the domestic populations of the major participants were being mobilized to the new way of thinking through anti-communist and anti-imperialism programmes that eliminated government opponents (Truman's loyalty oaths and McCarthy 'witch hunts' in the USA, and Stalin's final purges in the USSR).

The Cold War begins with the division of Europe into two blocs. The Truman Doctrine promising military help to defeat communism was followed later in 1947 by the Marshall Plan, through which US capital was made available to reconstruct Europe. Since the USSR refused to allow states it controlled to accept such funds, the operation of the plan in 1948 effectively divided Europe into two economic regions. In addition the US, British and French occupation zones of Germany now came together to form 'West' Germany, leaving the Soviet zone to become 'East' Germany. Both Europe as a whole and Germany in particular were thus divided. This was confirmed by the formation of the North Atlantic Treaty Organization in 1949 that committed the USA to the defence of Western Europe, that is, to the liberal democratic capitalist region its capital was helping to construct (Taylor, 1991).

It is in this first phase that the geopolitical codes of the chief protagonists become clear. For the USSR, Eastern Europe is paramount (McCauley, 1983). Since it had been invaded twice through this region in the first twenty-five years of its existence, the Soviet state insisted on political control of a ring of buffer states from the Black Sea to the Baltic. The other two powers at Yalta had agreed to a Soviet special interest in this region, although not necessarily in the form it finally took. Other Soviet interests along its boundaries through Asia were relevant but remained secondary.

The USA code was originally just as selective in its scope. George Kennan, the first architect of containment, identified four advanced industrial regions that had the potential to sustain a war against the USA: Britain, Germany, Japan and the USSR (Gaddis, 1982). Containment of the USSR would consist, therefore, in ensuring Britain, Germany and Japan remained in the US camp, thus avoiding the situation of World War

II when the US faced an alliance of two of these key strategic regions. Hence the rapid conversion of the erstwhile enemies (West) Germany and Japan to friends, plus the Marshall aid for Britain and the rest of Europe. With the defeat of the communists in Greece, the removal of communists from government in Italy and France, plus the formation of a communist government in Czechoslovakia, the division of Europe was complete. In 1948 the divided city of Berlin, deep in East German territory, came under communist pressure by blockade, but the Berlin airlift from the West preserved this western outlier in the East. This crisis solidified the division, but importantly did not lead to military conflict. This was to typify the Cold War for its whole life. Europe was the main East–West front, with the greatest build-up of armaments ever facing one another, but without the massive arsenal ever being fired in anger: deterrence, not war, was the game.

In 1949 and 1950 the focus of international relations switched from Europe to Asia. The communists won the civil war in China in 1949, and a year later the Korean War began. At the same time the US code changed significantly from selective to blanket containment. In 1950 the policy document NSC 50 committed the USA to massive rearmament to face the communist threat across the world (Gaddis, 1982). This meant committing US troops to fight in Korea against Chinese, but not Russian, soldiers. The Cold War 'hotted up' in Asia as it never did in Europe.

As well as this first extra-European 'hot' outcome, the switch in concern to Asia is important because it mixed the East–West conflict with the emerging North–South confrontation. Soon after World War II the Philippines, India, Pakistan, Burma and Sri Lanka became independent, but colonial conflict continued against the French in Indo-China, the Dutch in Indonesia and the British in Malaya. Hence, the civil wars in China and Korea, while viewed primarily in East–West terms in the 'North', could be interpreted as part of a 'South' resistance pattern as well. In 1950 India convened the first 'South' caucus at the United Nations for countries that were soon to be labelled 'Third World' to distinguish them from the bi-polar world of the North. Although

the USA had secured UN support for its Korean policy, this institution had lost its original conciliatory role with the demise of the 'one world' scenario. India was beginning the process that was to make the UN the prime vehicle for Third World dissent. This phase comes to a close with the armistice and division of Korea, an East–West outcome, but carried out through the offices of Nehru, the Indian Prime Minister, to bring peace to the South.

(ii) Conflict and concert, contest or conspiracy? 1953–69. This rather awkward title of the second phase reflects the complexity that was emerging in the world order after the relative simplicities of the initial freeze. Fred Halliday's (1983) description of the period as 'oscillatory antagonism' sums up the mixture of despair interspersed with windows of hope that is our subject matter here. The first thaw in the Cold War occurred in 1955 and this period is differentiated from the previous one by the willingness of the protagonists to negotiate their differences. This is when superpower summits first began. But overshadowing all was the threat of nuclear war. The USSR detonated its first atomic bomb in 1949 and a nuclear bomb in 1953. With the US monopoly of this means of mass destruction terminated, the Cold War became dominated by the nuclear arms race with the USSR attempting to make up lost ground on the USA. The traditional concept of balance of power became translated into a balance of terror by the end of this phase, as the USSR could retaliate a US attack to guarantee 'mutually assured destruction' or MAD. This represents the nadir of Cold War thinking.

The events of this phase generally confirm the geopolitical codes of the superpowers. Both consolidated their positions in their own contiguous spheres of influence – the Soviet's buffer satellites in Eastern Europe, and the US 'backyard' in central America and the Caribbean. This was by armed intervention when necessary; the USA invaded Guatemala in 1954 and the Dominican Republic in 1965, whereas the USSR army put down revolts in Hungary in 1956 and in Czechoslovakia in 1968. There was a final showdown over Berlin in 1961 which resulted in the construction of the Berlin Wall to prevent migration from East to West, but generally the East–West conflict remained dormant in Europe after the USSR formed its satellites into the Warsaw Pact in 1955 to confront NATO. The USA continued its blanket containment policy and formed two further alliances to complement NATO which stretched from Norway to Turkey in the West: CENTO (Central Asia Treaty Organization), formed in 1959, stretched from Turkey to Pakistan, and SEATO (South East Asia Treaty Organization), formed in 1954, stretched from Pakistan to the Philippines. Together with the US defence treaty with Japan, the USSR was 'contained' from the north Atlantic to the north Pacific.

But the world is never that simple. Civil war in Cuba resulted in a revolutionary government coming to power there. The subsequent Cuba-USSR alliance meant that the USSR was not only breaking out of its containment, but entering its opponent's sphere of influence. The attempt by the Soviet Union to place missiles there resulted in the Cuban crisis of 1962 when it is often argued the world was closer to nuclear war than on any other occasion. In the event the USSR backed down, but Cuba remained a communist satellite off the coast of the USA.

The Cuban revolution illustrates the fact that political events could not be universally controlled by the superpowers. The collapse of CENTO after a coup in Iraq provides a similar message. Support for Israel in the Middle East made it difficult for the USA to keep allies in this region. All radical Arab regimes – Egypt, Syria, Iraq, later Libya – distanced themselves from the USA to become friendly with the USSR. The region became enmeshed in a complex pattern of overlapping superpower geopolitical codes which became recognized as the classic case of a 'shatter belt' (Cohen. 1982; Kelly, 1986). The ominous feature of this was that the USA was becoming associated with conservative regimes, leaving the USSR to reap a harvest of radical states into its camp. This process was to be repeated throughout the South, with catastrophic effects in the other shatter-belt of the period – South East Asia.

In terms of the South, this phase represents the great victory of decolonization: the rest of Asia and the Americas plus almost all of sub-Saharan

Africa became independent of the old European empires. This is what the British Prime Minister Harold Macmillan termed the 'winds of change'; in 1960 it turned into a hurricane, spawning no fewer than seventeen new states. Led by the Non-Aligned Movement, the United Nations in the 1960s became the world forum for Southern demands on the North. The movement had been formed in 1961 at a meeting in Belgrade and was originally led by Yugoslavia, India and Egypt (Willets, 1978). Yugoslavia had broken with the USSR in 1948; India formed a non-western gap in the containment arc between CENTO and SEATO, while Egypt had survived the last intervention of the European imperial powers – an Anglo-French invasion of Suez in 1956. India, despite its continued membership of the British Commonwealth, supported Egypt in the latter conflict. The movement's members emphasized their right not to choose sides in the East–West conflict, and promoted North–South issues instead. But the two geopolitical axes could not be kept apart, usually to the detriment of the USA. In Vietnam after 1964 the USA argued that its army was containing communism, but for most of the rest of the world she was opposing a national liberation movement. US-supported conservative regimes in South Vietnam were unstable and the USA had to fight a second war on the Asian mainland while the USSR stood back and, with China, provided military support for North Vietnam. The realization by the USA that she could not win the war and the beginning of peace negotiations signalled the end of this phase by enabling a long thaw – detente – to begin.

Before we leave the discussion of this phase, one further important development requires consideration. The Non-Aligned Movement was not the only major challenge to the assumption of a bi-polar world. Within each bloc there arose differences that exploded the myth of East–West as two frozen monoliths. The most important split occurred between the USSR and China in 1960, but in some ways equally symbolic was the decision of President de Gaulle of France to eschew US leadership in the West: France withdrew from NATO's military command in 1965, leading to the removal of NATO headquarters from Paris. These changes exposed the superpowers to more general criticisms: the Chinese

accused *both* the USA and the USSR of imperialism. The Cold War was coming to be seen by
critics on both sides as a convenient arrangement
for both superpowers to keep their allies under
continued control. Rather than the 'great contest'
of opposing ways of life as expounded when the
world order emerged, it had come to look to
some as a 'great conspiracy', a power duopoly
enjoying condominium over the globe (Cox,
1986; Taylor, 1989). The worldwide students'
and workers' resistance to authority in 1968 confirmed the importance of this interpretation. This
revolution may have incorporated a major anti-
American component, but it was most certainly
not pro-USSR in nature (Wallerstein, 1991).
After two decades of dominating their respective
realms, both superpowers were having to face for
the first time a relative decline in their powers.
This was the key prelude to detente.

*(iii) Detente and demands for a new order,
1969–79.* Moves towards a more extensive
thaw in superpower relations were stimulated by
the fear of nuclear war generated by the Cuban
missile crisis. This stimulated both the beginnings of negotiations on controlling nuclear weapon testing and dissemination, plus improved
communication between the USA and USSR: the
'hot line' linking White House and Kremlin was
instituted at this time, for instance. But if there
was to be detente in the 1960s it was delayed by
US involvement in Vietnam. With this obstacle
removed by the beginnings of peace negotiations,
both superpowers could begin their moves
towards mutual accommodation in 1969. This
represents their joint repudiation of ideas of
competitive incompatible civilizations and their
adoption of the roles of two status quo powers.

The change in geopolitical code was greatest
for the USA following on from the trauma of its
Vietnam experience. Under the tutelage of
Secretary of State Henry Kissinger, the blanket
containment code was replaced by a more
pluralistic balance of power model (Gaddis,
1982). A pentagonal distribution of power was
envisaged, with the two superpowers being
joined by China, Europe (meaning the European
Community) and Japan. This realistic reappraisal
recognized the economic achievements of the
latter two and the long-standing potential of

China. It is during this period that a new power
constellation is identified as the 'Pacific Rim',
drawing the USA's concern away from Europe
as the traditional focus of concern. President
Nixon visited China in 1971 and the communist
regime was finally admitted to the United
Nations. In an explicitly even-handed policy, the
USSR was not neglected and negotiations on
nuclear arms continued, culminating in the
SALT (Strategic Arms Limitation Talks) agreement of 1972. This represents the apogee of
detente.

There was also an important economic dimension to the new pluralism. In 1971 the USA
withdrew the dollar's convertibility to gold, thus
ending its special role as reserve currency
throughout the non-communist world. In effect
the USA was reverting to ordinary competitive
status within the world-economy (Corbridge,
1984). Thus began a new period of what Kaldor
(1978) termed 'West–West conflict'. There were
attempts to control the potentially destructive
effects by co-operation, resulting in new institutions. The Trilateral Commission was an
informal organization bringing together the
political and economic elites of the USA, Europe
and Japan to encourage the development of a
global management class (Gill, 1988). More
formally, this is the time when the leaders of the
seven largest capitalist economies – USA, West
Germany, Japan, France, Britain, Italy and
Canada – began their regular series of G7 meetings. In short, US economic hegemony had
ended in the West.

Pluralism was extended down the power hierarchy by the recognition of regional powers with
special responsibility to maintain stability in
their region. In the Middle East, for instance, the
US chose Iran under the Shah to play this role.
But the superpowers continued their repression
of opposition in their sphere of influence: in 1970
the USSR connived in the repression of a
workers' revolt in Poland, and in 1973 the USA
connived in the coup removing the socialist government in Chile. It was 'business as usual' –
repression – for the vast majority of peoples of
Eastern Europe and Latin America throughout
detente. Furthermore, the 1973 Arab–Israeli war
found the USA and USSR in their familiar positions on either side of the conflict. It is testi-

mony of the strength of detente that this latter event did not lead to a new 'freeze', but it also shows detente to be a reform of the existing world order rather than its replacement.

Nevertheless this phase of detente did provide a window of opportunity for forces that were demanding a new world order. By this time Third World countries formed a large majority in the United Nations and they were able to use this forum to generate a very different agenda for world politics. Stimulated by the economic success of OPEC (the Organization of Petroleum Exporting Countries) in the aftermath of the 1973 Middle East war in raising their commodity price, a New Economic International Order (NEIO) was demanded (Addo, 1984). This was a challenge to both the USA and the USSR with a new image of 'one world', where the needs of the South would take precedent over the Cold War and its insatiable demand for wasteful arms. The UN sponsored several major international conferences on such global issues as the environment, food, technology transfer and agrarian reform as part of this new world politics. The USA and its allies in particular found themselves as a permanent minority in the UN and at many of these conferences. This, more than superpower detente, seemed to offer a glimpse of the future. But where was the power to bring these ideas into effect? The OPEC success proved to be an isolated event.

As events unfolded during this phase, more and more opposition to detente grew in the USA (Dalby, 1990). In addition to the UN being out of control, it seemed to many critics that the USSR was gaining more from superpower accommodation than the USA. For instance, in the final wave of decolonization, the group of most resisted independence movements became, not surprisingly, the most radical. As colonies turned into new states, they were invariably Marxist in orientation, as in Angola and Mozambique, for example. In addition, new revolutions produced other Marxist regimes, as in Ethiopia. There was also suspicion concerning the Soviet policy on nuclear weapons. Hence, in the 1979 the US Senate refused to ratify the SALT II treaty. But there were three key events in the last two years of the decade that aided the 'cold warriors' in overturning detente. First, the

popular overthrow of the Shah in Iran led to a radical Islamic regime that was explicitly anti-American. Second, the radical Sandinista revolution in Nicaragua brought fears of a communist regime on the American mainland. Third, the USSR sent its army into Afghanistan at Christmas 1979. This latter event, ironically at the location where the original Great Game began, was to mark definitively the end of detente.

(iv) A process of freeze–thaw with deadly side-effects, 1979–89. Like detente, this phase began with the US in a state of trauma. The occupation of the US Embassy by revolutionary guards in Iran was a symbolic representation of American weakness – like Vietnam, albeit on a much smaller scale. However, the effect was to be very different. A demand for renewed strength brought President Reagan to power as a right-wing cold warrior 'to make America great again'. The Cold War political agenda was reaffirmed and Third World demands simply ignored in a more aggressive approach to the United Nations.

The geopolitical code of the USA returned to an earlier time of simple bi-polarity. The USSR, referred to by President Reagan famously as 'the evil empire', became the target of increased military expenditures and new missile deployment as the USA attempted to 'catch up', as the government saw it, after the disaster of detente. The most controversial decision came in 1983 with the go-ahead for the Strategic Defense Initiative (SDI) which proposed the militarization of space. All this amounted to a new freeze almost as total as the original phase of the Cold War: it is commonly referred to as the 'second cold war' to indicate this affinity (Halliday, 1983). However, in terms of the geography of the US policy the 'Reagan doctrine' went further even than blanket containment. Given the success of radical regimes in the Third World during detente, a new initiative was required to combat communism wherever it occurred. The invasion of Grenada in 1983 and the support for the rebel contras in Nicaragua was consistent with past sphere-of-influence policy, but the large-scale aid given to anti-communist rebels in Angola, Ethiopia and Afghanistan represented a new departure. It was found that communism could be confronted worldwide by sponsoring anti-communist groups; this was much cheaper than involving US troops. The Cold War had finally come to dominate the politics of the South.

The result for the Third World was nothing less than catastrophic. International wars, notably the Iran–Iraq conflict from 1980 to 1988, and civil wars on all three southern continents meant that the purchase of armaments received top priority in Third World states. A world of millions of refugees, large-scale famines, plus increasing poverty was created, as massive debt problems sucked capital from South to North. There is no doubt who the losers were in this final phase of the Cold War. With the demise of the UN as a relevant tool of redress, the major political reaction to this change has been intensified 'terrorism' – the war of the weak on any available targets of the strong (Herman, 1982). This has taken the form of aircraft hijacking and hostage-taking especially associated with the longest-serving group of refugees, the Palestinians living in exile since 1948.

Promoting the Cold War agenda had other important implications. By bringing political issues to the fore again, it confirmed the USA's standing as world leader. Although this was achieved at the expense of the US economy, in this phase military preparedness took precedence over concerns for economic decline. This process was confirmed by the re-emergence of that other relatively economically declining major power, Britain, as 'America's deputy policeman' under the cold warrior leadership of Margaret Thatcher, the 'Iron Lady'. But as in the period of British decline, political leadership can never wholly mask relative economic decline and its concomitant competition. The processes underlying the West–West conflict that emerged during detente did not abruptly end. Hence, with a new thaw in East–West relations in the second half of the decade, voices could be heard in the US identifying Japan as more of a threat to US global pre-eminence than the USSR.

For the USSR with its much more severe economic difficulties, the new freeze was potentially catastrophic. Without the means to enter a new arms race, the Soviet government reached a crucial dilemma in its policies just as a new leader came to power in 1985. President Gorbachev

seized the opportunity to change freeze into thaw with the unlikely support of both Reagan and Thatcher. Major new initiatives resulted in the first major destruction of nuclear arsenals after the INF (intermediate nuclear forces) agreement of 1987. The following year USSR troops withdrew from Afghanistan, removing the original cause of the 'second cold war'. Commentators began to speak of a 'new detente' (Kaldor *et al.*, 1989), but it soon became clear that this language was inadequate: the process in train was not merely one of reforming the Cold War; it was destroying it. This was confirmed by the revolutions in Eastern Europe in 1989. The USSR indicated that it would no longer employ force to keep its sphere of influence, and the communist regimes collapsed one after another, starting in Poland and ending in Romania – the only collapse accompanied by major violence. Economic crisis within the Soviet Union had occasioned a complete revision of their geopolitical code made possible by the thaw in East–West relations engineered by Gorbachev. On 9 November 1989 the Berlin Wall was breached by the new forces: this stands as the symbolic event of the end of the Cold War, leading to the unification of Germany the following year. And finally, as an aftershock of what the media had termed a 'geopolitical earthquake', the USSR was excised from the inter-state system in 1991.

And so the Cold War Geopolitical World Order began and ended in Europe, and benefits are expected from the healing of divisions there. But between this beginning and ending, the remainder of the world was incorporated into this world politics, initially to its benefit through decolonization, but latterly with disastrous consequences. The joy in Europe, and in the North generally, at the ending of the Cold War must be severely tempered while the vast majority of humanity in the South struggle just to survive. Will a new world order address their problems?

A new geopolitical world order

And so we return to the contemporary search for a new world order. After a transition, a new world incorporating new geopolitical assumptions is in the making, but it is not yet constructed and we cannot be certain what it will look like. With the demise of the USSR, there is a political power vacuum which accounts for the new order being generally associated with the USA and George Bush. Of course it is not that simple; the Cold War will not be replaced by a Pax Americana. The irony is that this power vacuum has occurred, not as US economic strength is at its height, but after two decades of relative decline. There is a crucial mismatch between political and economic trends in the world-system.

A lack of congruence between political and economic processes at the international level is not new. This is what we reported above at the end of the nineteenth century during British decline. And like Britain at that time, it would be wrong to write the USA off now. It remains not only the leading political power, but the largest economic force in the world, as critics of reading too much into US decline have emphasized (Strange, 1988). Hence it would seem premature to envisage a 'World Order of the US Succession' just yet, if at all. We must be careful not to use our historical analogies too uncritically. But on the other hand we are in a period of lost hegemony and we may expect the USA to employ similar strategies to Britain to maintain its power. Appeasing rivals will be a policy we can expect.

Making sense of where international politics is going must be speculative at this stage so close to transition. We can, however, progress a little beyond crude historical analogy by trying to identify those elements of the last world order that may continue to be important in the future. In a quarter of a century or so from now, perhaps some geopolitical analyst will be identifying the precursors of her or his stable contemporary world order. To predict what these might be I think we have to go back one phase to detente – the time of the Cold War most like our current situation. There were three very important tendencies within detente that survived the early 1980s freeze: the attempt at a new concert of Great Powers; the new economic rivalry, and the challenge from the South. Let us consider the possible relevance of each for the post-Cold War period.

The first major international crisis after tran-

sition, the Iraqi take-over of Kuwait, suggests that we may be entering a new period of 'rivalry and concert'. The United Nations emerged as a vehicle for great-power imposition of order as originally envisaged in 1945. Unlike the 1970s when the General Assembly dominated the scene, it is now the Security Council back in the driving seat and with the five permanent members acting in concert as status quo powers. In fact, this is just one of two key instruments of concert, with the so-called G7 group of leading advanced economies operating a second and complementary attempt to order the world. But it is in this economic arena that the rivalry continues. Hence, while 1991 witnessed agreement on military action in the Gulf, there was fundamental disagreement at the General Agreement on Tariffs and Trade (GATT) meetings. The question is whether the economic rivalry can be contained by the processes of concert.

If the economic rivalry comes to dominate the new world order, what form might it take? We can be reasonably confident that it will not take the form of several autarkic pan-regions as envisaged in the past. But this does not mean that the close interconnections among the economies of the North in the late twentieth century automatically precludes separate economic spheres if political elites deem this to be necessary for their particular ends. Following our previous discussion of increasing size of political-economy units, if a new divide is to arise it is likely to be based upon just two global zones. Wallerstein (1991) predicts a divide through the Atlantic, with Japan and the USA leading a Pacific Rim bloc against a 'greater Europe' incorporating the USSR and dominating the Middle East and Africa. Notice that this geopolitical arrangement, while still being bi-polar, completely turns the Cold War pattern upside down in terms of who sides with whom.

The two scenarios above are similar in that politics remains as before for the South; it is still the Great Powers who are calling whatever tune. The condition of the South in the 1980s illustrates how much their challenge of the 1970s failed. If there is an upturn in the Kondratieff cycle in the near future this will give opportunities to alleviate the worst problems of the South. However, given that domestic crises are going to dominate the states of the South into the foreseeable future, it remains unlikely that any of these states can provide the sort of leadership necessary for a new political challenge. Certainly there is no state like India in the 1950s and 1960s with a foreign policy to counter whatever the North decides for the South. If there is to be a challenge, it is more likely to be from the resurgence of Islam than from any one state. This is the most intriguing precursor within the Cold War order, and one that was briefly activated during the Gulf War by the surprising popularity of Saddam Hussein among Moslem peoples (Taylor, 1992). Geopolitically the main part of the North, on the world-island, has as its Southern fringe a long crescent of Islamic peoples from Morocco in the west to Indonesia in the east. It would be the ultimate irony if Mackinder's most famous geopolitical pattern were to be finally relevant as the South threatened the Heartland – the North.

References

Addo, H., 1984, *Transforming the world-economy*, Hodder and Stoughton, London

Adler, L. K. and J. G. Paterson, 1970, Red Facism: the merger between Nazi Germany and Soviet Russia in the American image of totalitarianism, 1930s–1950s, *American Historical Review*, 75, 1046–64

Bartlett, C. J., 1984, *The global conflict, 1880–1970*, Longman, London

Blouet, B. W., 1976, Sir Halford Mackinder as British High Commissioner to South Russia 1919–1920, *Geographical Journal*, 143, 225–40

Brunn, S. D. and Ingalls, G. L., 1983, Identifying regional alliances and blocs in United Nations voting, in Kliot, N. and S. Waterman (eds.), *Pluralism and political geography*, Croom Helm, London

Clark, I., 1989, *The hierarchy of states*, Cambridge University Press, Cambridge

Cox, M., 1986, The Cold War as a system, *Critique*, 17, 17–82

Dalby, S., 1990, *The coming of the second cold war*, Pinter, London

Deighton, A., 1987, The 'frozen front': the Labour government, the division of Germany and the origins of the cold war, 1945–7, *International Affairs*, 63, 449–65

Dukes, P., 1989 *The last great game: USA versus USSR*, Pinter, London

Edwardes, M., 1975, *Playing the great game: a Victorian Cold War*, Hamilton, London

Gaddis, J. L., 1982, *Strategies of containment*, Oxford University Press, New York

Goldstein, J. S., *Long cycles: prosperity and war in the modern age*, Yale University Press, New Haven, CT

Grenville, J. A. S., 1979, Diplomacy and war plans in the United States, 1890–1917, in Kennedy, P. (ed), *The war plans of the great powers, 1880–1914*, George Allen and Unwin, London

Halliday, F., 1983, *The making of the second cold war*, Verso, London

Henrikson, A. K., 1980, The geographical 'mental maps' of American foreign policy makers, *International political science review*, 1: 495–530

Herman, E. S., 1982, *The real terror network*, South End Press, Boston

Kaldor, M., 1979, *The disintegrating west*, Penguin, Harmondsworth

Kaldor, M., Holder, G. and R. Falk (eds.), 1989, *The new detente*, Verso, London

Kelley, P. L., 1986, Escalation of regional conflict: testing the shatterbelt concept, *Political Geography Quarterly*, 5, 161–80

Kennan, G., 1984, *The fateful alliance: France, Russia and the coming of the first world war*, Pantheon, New York

Kennedy, P., 1979a, Introduction, in Kennedy, P. (ed), *The war plans of the great powers, 1880–1914*, George Allen and Unwin, London

Kennedy, P., 1979b, Imperial cable communications and strategy, 1870–1914, in Kennedy, P. (ed), *The war plans of the great powers, 1880–1914*, George Allen and Unwin, London

Kennedy, P., 1981, *The realities behind diplomacy*, George Allen and Unwin, London

Kennedy, P., (1988) *The rise and fall of the great powers*, Random House, New York

Kolko, J. and Kolko, G., 1972, *The limits of power*, Harper and Row, New York

Krasner, S.D., 1985, *Structural conflict: the Third World against global liberalism*, University of California Press, Berkeley.

Langhorne, R., 1981, *The collapse of the concert of Europe*, Macmillan, London

Lentin, A., 1984, *Guilt at Versailles*, Methuen, London

McCauley, M., 1983, *The origins of the cold war*, Longman, London

Modelski, G., 1987, *Long cycles of world politics*, Macmillan, London

Nierop, T., 1989, Macro-regions and the global institutional network, *Political geography quarterly*, 8: 43–66

O'Loughlin, J. and Grant, R., 1990, The political geography of presidential speeches, 1946–87, *Annals of the Association of American Geographers*, 80: 504–30

O'Loughlin, J. and van der Wusten, H., 1990, The political geography of panregions, *Geographical Review*, 80: 1–20

O'Sullivan, P., 1986, *Geopolitics*, St Martin's Press, New York

Pearce, J., 1981, *Under the eagle*, Latin American Bureau, London

Shlaim, A., Jones, P. and K. Sainsbury, 1977, *British Foreign Secretaries since 1945*, David and Charles, Newton Abbot

Steel, R., 1980, *Walter Lippmann and the American century*, Little Brown, Boston, MA

Strange, S., 1987, The persistent myth of lost hegemony, *International Organization*, 41: 551–74

Taylor, P. J., 1989, *Political geography: world-economy, nation-state and locality*, Longman, London

Taylor, P. J., 1990, *Britain and the Cold War: 1945 as geopolitical transition*, Pinter, London

Taylor, P. J., 1991, A theory and practice of regions: the case of Europe, *Society and Space*, 9, 183–95

Thompson, E. P., 1987, The rituals of enmity, in Smith, D. and E. P. Thompson (eds.), *Prospects for a habitable planet*, Penguin, Harmondsworth

Wallerstein, I., 1979, *The capitalist world-economy*, Cambridge University Press, Cambridge

Wallerstein, I., 1984a, *The politics of the world-economy*, Cambridge University Press, Cambridge

Wallerstein, I., 1984b, Long waves as capitalist process, *Review*, 7, 559–75

Wallerstein, I., 1991, *Geopolitics and geoculture*, Cambridge University Press, Cambridge

Watt, D. C., 1984, *Succeeding John Bull: America in Britain's place, 1900–1975*, Cambridge University Press, Cambridge

Willets, P., 1978, *The non-aligned movement*, Pinter, London

Political geography of war and peace

John O'Loughlin and Herman van der Wusten

The blood-dimmed tide is loosed, and everywhere
  The ceremony of innocence is drowned:
The best all lack conviction, while the worst
  Are full of passionate intensity.

William Butler Yeats (1865–1939),
The Second Coming (1920)

Measured by the numbers of military and civilian deaths in war, the twentieth century has been the bloodiest century in history. Shortly after the century began, the 'war to end all wars' put paid to any notion that wars could be confined to the belligerent armies, or that wars were useful in settling disputes. As the Yeats' quotation indicates, World War I drowned the 'ceremony of innocence'. But the twentieth century has also seen more peaceful years than any recent century, and since the end of World War II much of the world has been free of inter-state and civil conflict. In this chapter, we not only present the spatial and temporal trends of global and local wars since 1890, but we also attempt to account for their geographic distribution and occurrence using a historical-structuralist perspective.

It is easy but foolhardy to equate the political geography of the century with the geography of the hundreds of wars, large and small, that have been fought over the past 100 years. Wars cast large shadows, but war by no means totally dictates the political geography of the century. While international relations researchers and political geographers have focused their attention on conflict, millions have prematurely died because of inadequate health care, poor diets and exploitation. To write about war and peace should not mean the denial of the fate of those whose quiet suffering brings no headlines; additionally, behavioural violence (war) and structural (silent) violence (premature deaths from disease, famine, etc.) continue to be intimately linked in many parts of the developing world as we enter the last decade of this bloody century.

Introduction

In this chapter, we attempt to understand the causes of wars in this century: to link war and peace cycles to economic and power cycles, to explain why peace seemingly reigns in some places and not in other parts of the world, and to anticipate what lies ahead for the world-system over the next half-century. In order to tackle this intimidating list of goals, we need to move away from political geography as usually practised under the guise of geopolitics. By this, we mean that we pay little attention to the usual concerns of geopolitics *sensu stricto*. There is little in this chapter about borders, battle plans, terrain analysis, geostrategies, military equipment or armament capabilities; this is the usual stuff of strategic manuals, military atlases and war games. Military analysts examine the characteristics of the states themselves, in effect adopting what is known as a state-centred approach to the study of international conflict. By contrast, we look to a global-centred view, in which we search for the causes of wars in the structure of integrated world political and economic relations.

Like other chapters in this book, we share the view that political geography is a specialist branch of a wider research stream, that of a structuralist approach to the analysis of the changing world-system. We share the behaviouralist science principles that: (a) war and peace are forms of human behaviour; (b) they can be studied scientifically, and (c) they must be viewed from a 'systems' perspective (Kelman, 1991, 246). Geography adds the specifics of place and the contingencies of contexts to that perspective. It is primarily in the relations between the members of this world-system and the resulting effects on global structures that we look for the conditions of wars, and only secondarily in the specific circumstances of the war outbreaks themselves.

War is sometimes portrayed as a social institution, a type of action based on internationally accepted rules. In feudal times, elaborate etiquette regulated collective fighting as a series of quasi-duels. Codes of conduct for modern warfare were agreed around the turn of the twentieth century at the Hague peace conferences, finally resulting in the Geneva Convention of 1925. At the same time, war is large-scale indiscriminate violence striking at the roots of civilization. In modern times, and particularly since 1914, weaponry has become so powerful, the battlefield so immense in size and in deadly risks, and the war effort so integral to society, that the spatially-discrete character of war has disappeared over huge territories, and the humans trapped in war have lost all sense of their physical environment and social surroundings. If these statements seem only apt for global conflicts like World War II, consider the ways in which the supposedly surgical Gulf War of 1991 wreaked havoc on Iraq and its various populations.

Explanations of war have been hindered by the disparate character of war. We can distinguish between four kinds of wars. *Global wars* are defined as wars involving most or all Great Power members of the world-system, that is, those powers with global reach, whose foreign policy activities are significant for the whole system. This concept is the same as Levy's (1983, 75) idea of 'General War', which he defines as 'wars involving at least two-thirds of the Great Powers and an intensity exceeding 1,000 battle-deaths per million population'. What is noteworthy here is that the world-system had few members up until the twentieth century. The Napoleonic War, 1792–1815, was a global war because it involved all the Great Powers of the day, both leaders and challengers, as well as many of their smaller allies. Since 1494, there have been five global wars, namely: the Italian wars, 1494–1517; the Spanish wars, 1581–1609; the wars of Louis XIV, 1688–1713; the revolutionary and Napoleonic wars, 1792–1815, and World Wars I and II, 1914–1945 (Modelski, 1983, 119). *World wars*, a second kind of conflict, are a twentieth-century phenomenon. This kind is defined by geography, suggesting earth-wide conflict, rather than by the character of the states involved. Wars with a worldwide battlefield reach define this category. Of course, global wars (World Wars I and II) of the twentieth century became world wars because of the ability of the major powers to fight on all oceans and in faraway places after the technological leaps in weaponry and mobility after 1890. It is interesting that the First World War was called the 'Great War' until the advent of the second 'Great War' of 1939–45. World War II was much more global than the 1914–18 war, as the two sides fought in battlefields stretching from North Africa to the Soviet Union, to Burma, to the Caribbean and to the North, Central and South Pacific ocean.

A third category of wars are *local wars*, confined to a few countries and geographically limited in range. Most wars since 1815 have been of this type and many have involved Great Power states against smaller opponents. Examples of this genre would include the Russo-Japanese war of 1904–5 (130,000 battle-deaths), the Chaco war of 1932–5 (Paraguay and Bolivia with 130,000 battle-deaths), the Six Day War of 1967 (Israel, Egypt, Syria and Jordan with 19,600 battle-deaths), and the Football War of 1969 (Honduras and El Salvador with 1,900 battle-deaths) (data from Small and Singer, 1982). Most local wars concern either independence movements from a colonial power or border wars between neighbours. A final type of war is *civil war*. Over the past century, it is this type that has grown as a ratio of all wars. By definition, civil wars are fought between participants in a country, usually on ideological, religious, ethnic or economic grounds. The reasons for civil wars are often mixed and not only concern which ethnic group will retain power in the state, but also offer an ideological choice (capitalist versus communist, pro-West versus traditional society). After 1945, civil wars have become concentrated in the Third World as state-making continued there in the wake of European decolonization. Recently, the collapse of the multi-national states of the Soviet Union and Yugoslavia have led to widespread civil conflicts. Increasingly, the participants in civil wars look to regional and global allies for help, in troops, economic aid, weapons and training. Other recent examples include the war in Afghanistan (1980–92), the turmoil in Somalia after the overthrow of Siad Barre in 1990, and the unsettled situation in Iraq on the long-term fate of the Kurds. Military forces in the Third World concentrate increasingly on repression of their own populations rather than protection against outside enemies.

The defence dilemma

State elites frequently have to expect efforts to unseat them and are constantly worrying that others are making inroads into the domain of their authority. In the final analysis there is no overarching international institution to protect them. Only in countries with a firm constitutional tradition or a long parliamentary tradition, as in the United Kingdom, can elites be unworried about infringements from inside the state territory. As a consequence, state elites have a strong incentive to protect themselves against attack. The weapons stocks that the elite builds up can eventually be threatening to themselves in times of civil unrest. Nuclear arms are an extreme example of the problem of weapons falling into the wrong hands, as is currently the fear in four ex-republics of the former Soviet Union (Kazakhstan, Ukraine, Russia and Belarus). The collapse of totalitarian regimes, mostly of the right in the 1970s and early 1980s and of the left after 1989, in Southern Europe, Latin America and Eastern Europe can be interpreted as the result of a crisis of legitimacy within the state elites. A minority can retain control of the state apparatus for a long time in the face of hostility from the majority of the population (Assad's Alawite minority support in Syria is a current example), but challenges and questionings of the legitimacy of the ideology on which the state is based cannot be chased away by military power alone – as the Shah found out in Iran in 1979, the Moscow coup leaders discovered in August 1991, and the white minority government recognized in South Africa after 1989 (Fukuyama, 1992).

National defence forces, organized initially for the protection of state borders against outside attack, can often turn their attention to domestic issues. Many countries in Africa have armed forces whose main and sometimes only role is to protect the state elite. But what if the armed forces turn to politics and become involved in an attempted destabilization of the regime? A useful strategy is to divert their attention to external matters. A case in point is the quite formidable Moroccan army. It tried to overthrow the monarchy in a *coup d'état* in 1973, and after that it was diverted to a long war in the former Spanish Sahara against the indigenous guerrillas, Polisario. The war had the useful function of keeping the troops busy outside the state borders. An element of this strategy was also involved in the Argentinian decision to invade the Malvinas/Falkland Islands in 1982, which provoked the losing war with the UK. These are obvious examples of what is called the 'defence dilemma'.

Defence dilemmas arise because the dynamics of the development of the institutions of the armed forces and of the armaments industry do not necessarily reflect the functions and needs for which they were ostensibly created. Defence measures may be irrevelant to the security that they are supposed to serve. If external or internal threats are economic in nature, a strong defence force does not help much. Defensive weapons may also be useless because offensive weapons on the other side have too much of an advantage. And, as noted in Morocco, defensive measures may be counter-productive because they threaten the society and state of the defenders directly.

The power–security dilemma

At the same time, the existence of armed forces is deeply ambiguous from the point of view of outsiders. They can perceive these forces as expressions of a surge for security or as an expression of the surge for an improved power position. This is the power–security dilemma. The pre-World War I naval arms race between Britain and Germany revolved around that ambiguity on both sides. In the inter-war years, the emphasis on security in Britain and France was the basis for widespread feelings of appeasement. Relative power positions based on armaments were dismissed as much as possible. Since 1945, the course of the long debate about nuclear weapons brought out new issues of the security dilemma. Security came increasingly to depend on deterrence, while the actual capability to defend was, in fact, undermined by offensive weapons development. At the same time, power positions were widely evaluated in terms of the possession of nuclear weapons. Efforts to find a legitimate doctrine for the possession of this awful weaponry were most openly debated in the United States. It obviously was also discussed in the other Great Powers who held them (the former USSR, the UK and France), and the debate was anxiously joined and followed by people in other countries who would have to share the burden of living with nuclear weapons on their soil (West Germany) or who have to share the consequences of a nuclear exchange (Scandinavia).

The power–security dilemma can only be mitigated by a clear distinction between offensive and defensive capabilities. This type of distinction remains, however, a contested notion. The dilemma is driven by the perceptions of political actors who see another state arm. Obviously, there is a strong element of reciprocity between actors. Weapons technology has accelerated very quickly over the past 100 years, and in the process this has deepened both the power–security dilemma, and the defence dilemma (Buzan, 1983).

Twentieth century, bloody century

The twentieth century has been both the best of times and the worst of times. By Tilly's (1990, 67) count, the world has seen 237 new wars since 1890 (extrapolated to 275 to the year 2000), resulting in 115 million deaths in battle, with civilian deaths probably equal to that enormous figure. By comparison, the nineteenth century had 205 wars with 8 million battle-deaths, while the eighteenth century had sixty-eight wars with 4 million battle-deaths. If we control for the size of the world's population, we see that the death rate was 5 per 1,000 population in the eighteenth century, 6 per 1,000 in the nineteenth century and 46 per 1,000 in the twentieth century. In another estimate, Singer and Small (1982) calculate new wars as occurring every 2–3 years from 1480 to 1800, every 1–2 years from 1800 to 1944, and every 14 months since 1946. These numbers suggest that the twentieth century has been the most bellicose of the modern era (generally dated from around 1500). However, using a relative measure of conflict, controlling for the number of states in the international system, the amount of war is not increasing. Holsti (1986, 368) shows that, while the severity of war has increased over time, the incidence of war has not. With a system of 160 states, the potential number of conflicts is 5.3 times greater than at the time of the Congress of Vienna in 1815, when there were just twenty-three states. Calculating that the average state has 2.5 neighbours (this is necessary since war between non-contiguous states is highly unlikely and impossible for weak powers), Holsti estimates that the possibilities for war are now far greater than they were 200 years ago. The growth in the number of wars has lagged the growth in the number of states. By Levy's (1983) results, we can also clearly see that the involvement of Great Powers in wars has declined over the past four centuries. The nineteenth century experienced wars involving Great Powers during only 40 years, compared to 78 years of war from 1700 to 1800, 94 years of war from 1600 to 1700 and 95 years of Great Power wars from 1500 to 1600. However, to this point in the twentieth century, Great Power war years have shown an upswing to 67. Levy (1983, 124) concludes that 'Great Power wars are increasing linearly in extent, magnitude, severity, intensity and concentration', though not in incidence.

There seems then to be general agreement that war incidence is down, if we control for the number of states, though war severity is up because of the increased destructiveness of weapons and the range of military power. What the numbers hide is two very significant developments in the conflict experience of the twentieth century. First, wars between major powers are becoming less frequent, especially since 1945. The world until 1945 was frequently wracked by Great Power war; World Wars I and II were just the latest manifestations of this kind of imperial conflict. Since 1945, though Great Powers have come close to conflict (e.g. Cuba 1962, Berlin 1948), they have instead fought their disputes through proxy states and forces. The most common kind of war since 1945 has been about state-making, both as a cause of and as a result of the decolonization of the erstwhile European empires after 1950 (van der Wusten, 1985). Civil wars between rival political and/or ethnic groups (e.g. Nigeria, 1967–70), conflict over borders between contiguous countries in the Third World (e.g. the Iran–Iraq War, 1980–88) and independence struggles (e.g. Eritrean guerrilla activities in Ethiopia after 1972) are now the most common kinds of conflicts. Wars have become less global and more local since 1945.

The second development is that wars are increasingly confined to certain parts of the world. Regions of 'stable peace' (Boulding, 1978) are growing to include now Western Europe, North America, Australasia and, perhaps in the near future, South America. These are defined as areas where the states have not fought each other and have no plans to attack their neighbours. Among the NATO states and the neutrals in Europe, only the Greek–Turkish dyadic relationship seems capable of degenerating into conflict. It remains to be seen if the renewed state-making activities of the former Second World (Soviet Union and its allies) will, by and large, be peaceful and, if not, how far spillover effects will ripple outward. War over the breakup of Yugoslavia has resulted in peace-making efforts from the outside (European Community and the United Nations) and renewed tensions with neighbours (e.g., with Greece over the fate of Macedonia). The dissolution of the former Soviet Union is accompanied presently by the renewal of internal (e.g. Nagorno-Karabakh) and external (e.g. Rumanian irredentism in Moldavia) grievances that may spark wider wars.

War, like deadly infectious diseases, famine and political repression, is increasingly a Third World phenomenon. Because state-making and the construction of civil society in the countries of the Third World happened only after the end of the European colonial era, the process is still far from complete and the stresses of competing ethnic, religious and linguistic nationalisms are compounded by the consequences of economic deprivation and declining life-chances. From the perspective of Boulder or Amsterdam, it is easy to think of war as a distant phenomenon, even if the US or The Netherlands send troops into battle, as both did in the 1991 Gulf War. From the perspective of Kabul, Asmara, Mogadishu, Basra, Nablus or Jaffna, the image of peace looks very different.

The changing nature of war in the twentieth century

As well as changing the temporal and spatial pattern, wars have also changed in other ways in the twentieth century. War-making and state-making are closely linked in the evolving world-system. One influential interpretation sees the development of the modern state as the natural outcome of the growth of the state apparatus, intertwined with and servicing the needs of the state war machine. War determined the state in two ways. First, the borders between states were the outcomes (ceasefire lines, in many cases) of wars between them, and second, preparation for war in the form of tax revenues created the internal structures of the states. Around 1800, just after the French Revolution, states stopped paying mercenary armies and began to raise their own forces through conscription, as well as continuing to encourage volunteers. The consequences of this parallel development of the war arsenals and state apparatuses in Europe made states more similar than they had previously been.

European states started in very different positions as a function of the distribution of concentrated capital and coercion. They changed as the intersection of capital and coercion altered. But military competition eventually drove them all in the same general direction. It underlay both the creation and the ultimate predominance of the national state. In the process, Europeans created a state-system that dominated the entire world. (Tilly, 1990, 190–91.)

The model of the national state became universal.

In another article, Tilly (1985) has explicitly defined the four state roles related to war. First, states engage in war-making, which is defined as the elimination or neutralizing of their own rivals outside their territories. A recent example would be the US attempt to push Saddam Hussein from power in Iraq. Second, states engage in state-making, defined as the elimination or neutralizing of rivals within the state. A recent example is the Nigerian government's success in quashing the secessionist Ibo independence movement in Biafra (S.E. Nigeria) in the late 1960s. A third state role is protection, defined as the elimination or neutralizing of enemies of client states or movements. A recent example is the US support for the Salvadorean government in its civil war against left-wing FMLN guerrillas. Lastly, states engage in extraction, which is the means to carry out the first three activities. As well as taxes, governments extract a great deal from civilian populations in order to engage in war-making.

Income and other taxes appeared widely after the turn of the century as the costs of war preparation escalated rapidly. Governments also floated war loans, allowed and encouraged women to work in factories and other locations outside the home, persuaded civilians to forfeit or postpone consumer goods, acquired control of all areas of social and economic life (including stringent espionage provisions and penalties and severe punishment for war opposition), accepted pressure from unions to share decision-making and to bargain with the state and employers, and in general extracted a lot more from the civilian population (Howard, 1975). Unlike wars up to 1900, with the possible exception of the American Civil War (1861–5), the typical war was no longer a distant affair affecting only the families of servicemen, but was increasingly a *totaler Krieg* (total war), involving a complete dedication of the state apparatus to its operation and a massive commitment of civilian resources and efforts to its pursuit. Schaefer (1989, 4) calculates that in World War I, there were twenty soldiers killed for every civilian; in World War II, the ratio was 1 : 1, and in the Korean War the ratio was one soldier to five civilians.

Until the Napoleonic wars (1797–1815), wars were relatively sudden affairs, that is, they were not preceded by a long build-up of tension or a racheting-upwards of war preparations. In the subsequent century, wars became increasingly a matter of competing technologies so that, for the first time, we see the phenomenon of arms races. The last decade of the nineteenth century was marked by the intense German–British naval race to see who could produce the greatest amount of battleship tonnage. The naval race was not the cause, but the consequence of the larger imperial struggle and the resulting perceptions of state policy-makers that has not changed much since classical Greek days. We can substitute any of the modern Great Power rivalries in the Thucydides statement that 'What made war inevitable was the growth of Sparta power and the fear that this caused in Athens' (quoted in Howard, 1983, 21). In an increasingly competitive international environment, leaders feel that they have more to gain from going to war than remaining at peace.

Another development of the late nineteenth century was the beginning of research in weapons of mass destruction. In the 1880s, the internal combustion engine was invented and the first aeroplane flight occurred in 1903. Concurrent research and development of chemicals was part of the industrial upsurge of the global growth phase after 1890. Within twenty years, chemicals, planes and the internal combustion engine had been developed as weapons on the World War I battlefields in the form of gas, bombers and fighters, and tanks. Within another twenty-five years, biological and nuclear weapons had been added to the weapons menu, to be accompanied by further development of even more sophisticated and destructive tanks, chemical weapons (now forbidden under the Geneva Convention), rockets and planes. The combined effect was not only to produce greater battlefield casualties, but also to render civilian areas part of the front line, though they were geographically distant from the battlelines. Coventry, Hiroshima and Dresden, though the best known, were only three of many cities severely damaged or destroyed by the aerial bombardment (Hewitt, 1983).

Wars, wars everywhere: is not a theory there?

From many perspectives war may appear to be irrational, and the frequent occurrence of war may be puzzling. Unfortunately, the simple fact is that coercion works (Tilly, 1990, 70). A decision to go to war rarely happens without an evolving and escalating background of inter-state conflict. Though state decision-makers may miscalculate about the opponent's reaction, as Saddam Hussein undoubtedly did about the intentions of regional and global leaders in summer 1990, the war decision is a bid to gain advantage as a result of rational calculation. To use Bueno de Mesquita's (1981) language, governments act when they calculate a favourable 'net utility' as the benefits of action are greater than the costs of inaction. One can, of course, quarrel with their metric of costs and benefits. To view World War II as the actions of a mad leader, Adolf Hitler, is to ignore both the global stage of long imperial competition between Germany and the other Great Powers, as well as Hitler's own careful calculations about the reactions of his opponents.

Idealist and realist views

Since the end of World War I, there have been countless academic and popular attempts to try to understand why wars happen, why specific disputes escalate to war, how many troops and civilians died, under what conditions they died, and how wars can be grouped and classified. International relations as a discipline owes its origins to the search for clues for answers to these questions. The horrors of trench warfare, the mobilization of society to fight 'total war', the suffering of refugees and civilians previously relatively immune from the direct effects of war, the rush to develop more destructive weapons, including some of mass destruction, all helped to generate an international consensus that a 'great war' like that of 1914–18 must not be allowed to happen again. The Geneva Convention (1925) outlawed the use of poison gas and regulated the treatment of prisoners-of-war and of refugees, while the League of Nations (1920) tried to offer, for the first time in world history, a permanent international mediative assembly in which issues of inter-state dispute and territorial claims could be arbitrated.

Early works of international relations had a strong tinge of moralism and idealism, and pursued the notion of diplomacy as the forum of conflict resolution while supporting international principles of self-determination, justice and disarmament. While scholars, jurists and historians worked to create a new academic discipline, one that had as its guiding principle the avoidance of war, writers, artists and others expressed strong feelings of alienation in the decade immediately after the Versailles Treaty (1919). T. S. Eliot's *Waste Land* (1922) expresses the shared feelings of hopelessness after such a tragic decade, and the cumulative writings of the (anti)war poets and novelists, especially Erich Maria Remarque's *All Quiet on the Western Front* (1928) kept the horror of the war in the minds of the public.

The idealist authors of international relations texts of the inter-war years by-passed the two writers, Thucydides and von Clausewitz, whose works became the basis of the new 'realist' school of international relations after the Second World War. Idealists continued to place their faith in international institutions to prevent war occurrences, but most scholars drifted back to a belief in a balance of power system, despite the fact that such a system had failed in 1914. The global failure to respond to the expansionism of Nazi Germany, Fascist Italy and Imperial Japan helped to rekindle a belief in the efficacy of power politics. Political geographers of the inter-war period were uniformly of the realist school of thinking, despite the strength of the idealist vision in related disciplines. Realism, the dominant school of international relations of the post-1945 period, views the world order as essentially anarchic, with each individual country competing to improve its position and striving to maximize its self-interest. Self-interest is measured by increasing power status and domination of other states; power is the central concept of the realist world vision. In the oft-cited passage by Hans Morgenthau (1960, 27), 'International politics, like all politics, is a struggle for power. Whatever the ultimate aims of international politics, power is always the immediate aim.' Order of a sort is maintained by large states so that states of weaker power recognize that it is in their own self-interest to follow the lead of stronger states.

Since the century-old balance of power system collapsed into global war in 1914 in the face of the Great Power competition, it is no wonder that many international relations researchers believe that peace is most likely under a dominant superpower that can organize the world-system and keep competition in check. Thompson (1988, xxi) is one of many (Modelski, 1987b, is another) who believe that hegemonic powers can maintain world peace through their world ordering capabilities. In his examination of the past 500 years of war, Thompson notes that the most peaceful periods are after a global war which had led to a concentration of military and economic power in one state. The most recent example of this concentration is in the United States after 1945 when, using Modelski and Thompson's (1988) seapower index, the hegemon accounted for almost 100 per cent of global power. (Most other power indices show the US at about 50 per cent of global power in the late 1940s: O'Loughlin, 1993.) Later, as the global power dissipates and other challengers gain in strength, the opportunities for war increase in an atmosphere of power uncertainty and shifting alliances. In the multi-polar world, the number of possible global conflicts is magnified dramatically so that a shift from a bi-polar world (the US–Soviet order, 1945–90) to a five-power multi-polar one would raise the number of possible conflicts from one to ten. It is no surprise that Mearsheimer (1990) can make the case that 'we will soon miss the Cold War'.

In the realist vision, wars are the result of the power competition and the failure of the checks and balances inherent in the power hierarchy. But not all states behave as expected. Instead, we can cite numerous instances where the stronger state did not pursue the logical (power-political) course of action to achieve its obvious objectives. The US decision not to engage in full-scale war with North Vietnam is only one recent example. It is possible to conclude now, after a half-century of realism, that this approach offers no definitive account of why wars happen. There is no single explanation of war, partly because they are different kinds of wars (Howard, 1983, 13). Over the past half-millennium, wars have become intimately linked to the development of the inter-state system and the processes of state-formation that occur inside that system. These processes, though comparable, differ in the states at certain points in time. Echoing Jean-Jacques Rousseau, if we had no states, we would have no wars, to which one could counter with the famous riposte of Thomas Hobbes that we would also be very unlikely to have peace (Howard, 1983, 11). Inside the state, the violence of civil war easily spills over into the arena of international relations and can often only be contained by a certain amount of coercion. But coercion in itself gives rise to renewed aggression that again may be linked to international violent conflict. As Tilly (1990, 225) aptly states the dilemma: 'Destroy the state, and create Lebanon. Fortify it, and create Korea. Until other forms displace the national state, neither alternative will do.'

Taking the episodes of a fully mature hegemonic state as a borderline case, war cannot be excluded in a state-system as an ultimate resort due to its partly unpredictable character. That is not to say that war is totally inevitable. Under conditions of hegemony, the leading power sets many of the terms and limits for other state-makers. The absence of hegemony gives state-makers much more free rein. What policies, peaceful or war-like, that a hegemonic power

and ordinary state-makers will follow depends on their economic and political interests and on the network of relations that already link them. Between a lot of pairs of countries, the idea of war to sort out differences is absent because there are no relations, therefore no differences. In other pairs, like the dyads in a stable peaceful area, the thought of using conflict to resolve a dispute does not cross the minds of policy-makers. War has apparently become banned from the considered repertoire of mutual actions as relations have become harmonious and/or adequate devices have developed to sort out differences. In other pairs, the power differences are so preponderant that use of conflict is ruled out by one side and unnecessary on the other. The ensuing types of peace are just as different as these varying types of conflict situations.

Singer (1981), in trying to summarize and evaluate the immense literature on the causes of war, concluded that trying to see which countries were most war-like was an intellectual cul-de-sac. He argued that the causes of wars should not be sought in the domestic characteristics of states, but should be pursued at the level of the international system. In other words, we should seek explanations of war that are system-level, rather than trying to account for war on the basis of country characteristics. Dessler (1992) notes that we have hundreds of 'correlational' studies of war that indicate which factors are related to war behaviour, but no overarching causal theory because of the fragmented nature of war. We share the view that the best approach to the study of war is to examine the evolving structural conditions of the international system. Unlike geographers of the past, who have usually been wedded to a narrow geopolitical perspective that, as noted earlier, shared the realist world vision, we see war as a product of the world-system structure and its constitutive relations that play themselves out under the impetus of a set of temporally- and spatially-specific conditions.

It is worth repeating that most geographic writings about war, as well as about international politics, have been deliberately subjective. Geopolitical writers have tried to evaluate the impact of the changing global distribution of power for their home country and have usually proceeded to advocate a geographically-determined foreign policy strategy, often based on control of specific territories (O'Loughlin and Heske, 1991). While most geographers have now abandoned the geopolitical (realist) approach, it can still be found in the strategic studies literature, where references to the legacy of Sir Halford J. Mackinder, Alfred T. Mahan and Nicholas Spykman are commonplace. We try not only to understand the time–space distributions of different kinds of wars; we also explore in this chapter why certain time-periods and certain parts of the globe have been peaceful. We believe that the world-system perspective allows us to account for these uneven patterns, though in a non-complete manner. There are some wars whose causes lie in the very specific relations between two or more states, usually in a locally-bounded context and isolated from major world developments. In these circumstances, control of borders, irredentist issues and territorial claims are paramount.

Historical-structural approaches to the study of war

Like most international relations researchers, we do not provide a detailed interpretation of the specific sequence of events leading to individual wars. Instead, we are more concerned with the larger picture of how states move to the war preparation stage, as a result of their international position and state self-interest. Obviously, not all hostile situations devolve to war; the Berlin crisis of 1948 illustrates that. But if the international and national circumstances are ripe, it takes little to initiate the irresistible sequence of events that end in war. We see the interactions of the Great Powers against the international background of imperial competition of forty years as the explanation of World War I, rather than the spark that set off the conflagration – the assassination by a Serbian nationalist of the successor to the Austro-Hungarian throne, Archduke Ferdinand, in Sarajevo on 28 June 1914. Like most world systems theorists, we interpret World War II as the extension of the Great Power competition because the First World War did not resolve who would be the successor to the British world

leadership of the nineteenth century. World War II was not just 'Hitler's War' (as the *Economist* likes to state), but was the logical outcome of the unresolved hegemonic battle and Germany's continued bid for world leadership against other global pretenders – Russia, United States, United Kingdom, France and Japan.

We have stated repeatedly that we share the historical-structural view of world politics and that this perspective determines our explanation of conflict. We now need to elaborate on the elements of this view and why we believe that it is valuable in the attempt to understand the variable temporal and spatial occurrence of war. We have referred to the world-system, believing that there is a shared economic-political one. From a reading of the literature, it would be easy to come to the conclusion that there are two overlapping but separate world-systems, one economic and one political. Modelski's (1983, 1987b, c, d, 1990) long-cycles approach is based fundamentally on the changing order of world power, measured by the global reach of states. It does not stray far from the realist emphasis on power in world politics and it defines the cycles by the global wars that begin and end each one (Thompson, 1988; Modelski and Thompson, 1988). Like realists, long-cycle theorists see states engaged in the business of maximizing their own power resources and using their resources to organize the world-system to their advantage. (For a review of the slippery notion of power in world politics and a comparison of many different power indices, see Stoll and Ward, 1989.)

For Modelski and his followers, state power is measured best by the ratio of world shipping tonnage; following the classic geopolitical theorist, Alfred T. Mahan (1890), Modelski believes that seapowers (in contemporary terms, states able to project their power over vast distances through both air and water) always prevail in global contests. The explanation for this is provided by Chase-Dunn (1989, 161), who notes that island countries did not have to provide for the immense costs of protecting land frontiers and thus were able both to keep the costs of government down and to devote more capital to the tasks of economic development, leading in turn to the ability to give more state resources to seapower. Seapower was then used to promote the country's interests overseas as well as to protect the investments of capitalists from the seapower state. This perspective of the importance of trade route protection is shared by Wallerstein (1979, 1984, 1991) and his co-authors. A review of the outcomes of global wars over the past five centuries supports this argument. Global wars, accounting for four-fifths of all battle-deaths since 1500 and lasting on average twenty-seven years (Rasler and Thompson, 1989, 15), are crucial. From each (see the list earlier in this chapter), a new system leader emerges with new international responsibilities and overhead burdens.

Wallerstein's (1979, 1984, 1991) perspective on the world-system is both different and similar to the long-cycle approach. He shares with, Modelski the central idea of power cycles in both political and economic terms, the concept of global leaders, the power of global leaders to shape world orders, the reorganizing effects of global conflict, the intimate links between economic and political developments, the timing of the beginning of the modern world-system and the importance of a multi-disciplinary perspective on world politics (Taylor, 1989). Wallerstein and Modelski also have similar sequences of global leadership, both in timing and actors. Why, despite their different operating definitions and foci, should the sequences be the same? The answer lies in the intersection of the focus on seapower and the maritime strategy of accumulation through trade. Control of trade routes is all-important, and avoidance of direct overhead costs, such as land-based forces in garrisons to protect a colonial possession (Thompson and Zuk, 1983), are also significant. Shipping is the cheapest form of transport; control of its routes, and expertise and leadership in its provision are important comparative advantages in the race for global leadership (Chase-Dunn, 1989).

Unlike Modelski's vague references to the accumulation of seapower, Wallerstein has a clear central driving force for his vision of how world politics operates. Since about 1500, capitalism has diffused from Europe to all corners of the earth, bringing with it the interests of the states whence the capitalists came. Incorporation of Africa, North and South America, Asia and Australasia into the capitalist world-economy went hand-in-glove with the colonization of these regions by the respective European states. As the capitalist race for resources, markets and control intensified in the late nineteenth century, so too did the colonial conflicts, leading eventually to an agreement by the Great Powers to divide the spoils at the Congress of Berlin in 1884–5.

As in the nineteenth century, so too in the twentieth century global political and economic relations are intimately connected. Overwhelming economic power translates into global political and military leadership. The term hegemony is usually used to express this double concentration in one state and its acceptance by others. For Wallerstein, there have only been three brief periods of hegemony and, unlike Modelski, his examination of the process by which hegemony is attained leads him to define it carefully. Based on an early productive advantage of a new set of products, in turn the result of research and development of a new technology such as chemicals after 1890, a state translates its lead into higher global trade shares and later, using its accumulated capital, into a dominating position in the international financial markets. Through its accompanying political and military might and economic strength, a hegemon is able to organize the global institutions to its wishes, thereby ensuring its continued success. Thus, The Netherlands was able to establish global hegemony from 1625 to 1672, the United Kingdom was hegemonic between 1815 and 1873, and the United States had a period of hegemony from 1945 to 1967 (Wallerstein, 1984).

Rosenau (1982) has presented a very useful summary of approaches to the examination of world politics. State-centric theory is equated with the realist position; this we find deficient because it ignores non-state actors; it emphasizes military power; its historical horizons are limited; it denies the order in inter-state relations, and its methods of analysis are too bound by time and place. We also reject both the purely multi-centric (Modelski) and the purely global-centric (Wallerstein) theories. They share many features. As already noted, separate analyses of the evolution of world politics have generated similar conclusions on periodicity and hegemonic (global leader) succession. In both approaches, there is no consistent top-level system dominance; interdependence among states and regions reigns; power concentration is variable; wars are not examined *in vacuo*; history is important, and world order is possible, usually regulated by a hegemon.

The two theories differ in many respects, too. They have different emphases on the importance of non-state, especially economic, actors. They differ on why order happens. (The global-centric/world-system theory approach relies on the exchange of economic goods as its organizing device.) They differ on the central question for examination (for multi-centric theorists it is global change and transformation, while for world-systems theorists it is global inequality). They differ on the unit of analysis, which is variable: usually it is the state for multi-centric theorists, and the global division into core and periphery for the global-centric theorists. As noted earlier, the theories differ significantly in the relative importance of military and economic strength for global power status (Thompson, 1988).

Wars and the world-economy

Recently there has been a *rapprochement* between the long-cycle and world-system theorists, producing what Goldstein (1988) has called a 'hybrid' theory. The hybrid theory is based on an integrated political-economic logic since 'in the capitalist world-economy, system dynamics are produced by a single logic in which capitalist production interacts with the processes of geo-politics, state-formation, class-formation and nation-building' (Chase-Dunn, 1989, 154). Contrary to intuition, capitalism is promoted by a decentralized and competitive state-system rather than a global empire, which would suffer political and normative constraints on resources allocation and the danger of monopolies. So we get an arrangement in which each state promotes the interests of its own capitalists against the competing interests of other capitalist states. States will also go beyond simple promotion to providing military protection of installations and trade routes; to forcing local leaders to accept imports and open markets; to naked territorial

aggression in order to acquire important raw material, and to reciprocal actions when their own capitalists are attacked. Like Chase-Dunn (1989) and Bergesen (1985), we believe that separating political and economic developments makes little sense when trying to understand the evolution of world power and the occurrence of wars. While not all wars have both economic and political causes, wars with global implications do.

It is no coincidence that war and economic cycles are linked. The evidence is clearly presented in Goldstein (1988), who takes the view that wars are more common in times of global prosperity since states have more resources in those periods to devote to military activities. More than that, for Goldstein, wars play a crucial role in relation to economic growth. The argument is that economic growth generates national surpluses which allow wars to be pursued, as in the Tilly argument earlier. But the costs of war drain these surpluses, and by moving the domestic civilian economy away from consumer goods, war disrupts economic growth, leading to a downturn not only in the war-making states, but in other parts of the world-system because of the integrated nature of the world-economy. Obviously, the larger the number of states involved and the bigger their economic output, the greater the ripple effects of war in the world-economy.

The Goldstein thesis runs counter to the best-known relation of war and the economy, that of Lenin (1939, first published 1917). Lenin believed that major power wars (or imperialist wars, as he called them) would be more likely in times of global downturn since then the major powers would each be suffering from economic stagnation. States will be forced to expand in the increasingly competitive world environment, and as they do they come into conflict with other states engaged in the same behaviour, leading eventually to war over control of resources. As stated by John Maynard Keynes (1936, 381), 'wars have several causes . . . [Above all] are the economic causes of war, namely, the pressure of population and the competitive struggle for markets'. In modern parlance, this imperialist struggle is known as 'lateral pressure theory', so named after its use for an understanding of the interactions between the Great Powers after

1870, leading up to the First World War (Choucri and North, 1975). The analysis is very complicated with many causal arrows linking the domestic and international variables that constitute the building blocks of the theory. It was calibrated using a set of simultaneous equations.

At its core, lateral pressure theory relies on three key indicators – population size, level of technology and domestic resource availability. It argues that changes in these conditions will determine a country's international behaviour. As population increases and technology advances, the country will eventually exceed its domestic resource capacity and look to foreign sources for both raw materials and markets. This expansion abroad is lateral pressure, and if other states are simultaneously engaged in it, the conditions for conflict are ripe as the limited global resources are demanded by competitive states. States marshall their own economic, political and military power to promote their interests, thereby leading to similar reactions by the competitors and eventual arms, resources, population and territorial races. This is precisely what happened in the late nineteenth century as the race was on to see who would be most successful in establishing markets and who would gain most from the second wave of colony-grabbing. Though the central race pitted Germany against Britain, the other Great Powers (the US, France, Russia, Austro-Hungary and Japan) were also engaged in it (Choucri and North, 1975).

In this chapter, we reject the choice of either the multi-centric or the global-centric approaches. We appreciate the integration of the economic and the political/military conditions promoted by the world-systems approach, while at the same time we find that the emphasis on the importance of global wars for the reorganization of the global system in long-cycle theory to be an important framework for our understanding of war. But what we find lacking in both models is an appreciation of any geographic contexts. Both models are high-level approaches, that is, they do not really care much about the nuances of time and place. What distinguishes political geography from international relations is an appreciation of the complexity of the earth and its diverse regions. As we have tried to argue in our own empirical research (O'Loughlin and van der Wusten, 1990), we need to integrate the specifics of regional and local developments with the structural-historical theories that we have examined above. For example, local wars do not play a prominent role in either long-cycle or world-system theories, yet they are the most common form of war in the late twentieth century. Consequently, we need an approach that allows the incorporation of local social, economic and political processes with broader global-level theories. The convergence of political geography and historical-structural approaches allows such a development.

Great power developments in the twentieth century

The twentieth century began in about 1890. This seemingly inaccurate statement defines the twentieth century, not by dates, but by global political and economic developments. By many indicators, the world order changed about a decade before the end of the century. Probably most significant of these developments, though scarcely recognized at the time, was the beginning of the end of the European-dominated world. Since about 1500, the accepted start of the modern era, world politics had been the playground of the large European states, Portugal, Spain, the United Kingdom, The Netherlands, France, Germany and Italy. By 1890, only France, Germany and the United Kingdom were left on the world stage, to be joined there within a short time by extra-European powers, the United States, Russia (after 1917, the Soviet Union) and Japan. The world had finally become global at the level of political relations and was well on its way to economic integration too, as the world-economy continued its seemingly inexorable incorporation of external areas.

Regardless of the precise definition of power, all common measures include economic size (Stoll and Ward, 1989). Though Modelski and Thompson (1988) use naval strength as their index of Great Power status, in their other writings they accept that military strength is based, in turn, on economic output. Wallerstein (1984) has always equated hegemony with economic domination as translated into trade shares, producti-

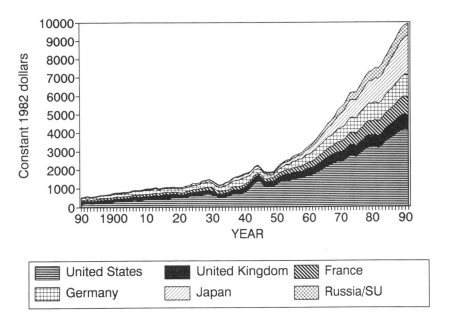

Figure 2.1 *Growth in the economies of the six Great Powers, 1890–1991. Values are the Gross National Products in 1982 US dollars. Sources for the data are Maddison (1982), Mitchell (1975, 1982), Leisner (1991) and recent World Bank tables. Calculated by the authors.*

vity leads, new technological products and financial dominance. For him and his followers like Chase-Dunn (1989), the role of military power is to ensure continued economic leadership and access to resources and markets. The reason why economic size is important to the story of war is that in explaining the most destructive of all wars, global conflicts, we need to account for the rise and fall of the system leaders and the consequent struggle for leadership (Thompson, 1988, 112).

We can examine the development of Great Power fortunes by examining their comparative economic output. Obviously, some countries will have military power disproportionate to their economic output (large as in the case of the Soviet Union from 1945 to 1991, and small in the case of Japan from 1960 to 1991). But the trends in Figure 2.1 indicate a lot about the fortunes of the Great Powers over the past century. The Gross National Product data, the standard measure of economic output, were taken from Leisner (1991), Mitchell (1975, 1982), Maddison (1982) and recent World Bank tables. This set of six countries is the usual one examined by

Wallerstein (1984), Levy (1980), Modelski and Thompson (1988), Thompson (1988) and Goldstein (1988). Our decision on the composition of the list of the twentieth century's Great Powers leaves out China. This is especially debatable for the last half of the twentieth century after this old imperial power had transformed itself through revolution into a new kind of polity grafted on to a Marxist-Leninist creed. Apart from the severe problems of collecting reliable economic data for China in the early years, it is quite clear that China on the particular GNP indicator is still not a Great Power.

All six Great Powers have seen substantial economic growth over the past century. Using 1982 US dollars as the comparative unit of measure (this procedure of currency comparability is very difficult and riddled with inconsistencies), the total Gross National Products (GNPs) of the six countries has increased twenty-fold since 1890 (Figure 2.1). Economic growth was steady for all countries to 1945 and really exploded after the Second World War. Obviously, there have been times of economic downturn (the 1930s recession, part of the

Kondratieff IIIB phase, is clearly visible on the graph, as are the hiccups of the early 1970s and early 1980s) but the overall trend is continued strong growth in all the Great Powers.

However, Figure 2.1 hides the relative fortunes of the individual countries. For that picture, we need to turn to the relative proportions of the total Great Power economic output (Figures 2.2–2.7). As the leading power of the nineteenth century, the United Kingdom had lost considerable ground and was reduced to 22 per cent of the Great Power output by 1890 (see Figure 2.2). It has been noted by many historians, among them Modelski (1988) and Kennedy (1987), that the UK share of world power was never very large during its period of global dominance and that it was only in the military and territorial indicators that the country had overwhelming strength. Since 1890, the UK has experienced relative decline, and its ratio of Great Power economic strength is now below 10 per cent. The past 100 years has been a century of slow and steady decline for the nineteenth-century hegemon.

The causes of this trend of British decline has been the subject of numerous speculations, and the best known and most hotly debated remains Kennedy's (1987) 'imperial overstretch' thesis. In a detailed historical account, he shows that the costs of empire severely stretched British resources and how the global wars of this century finally drove the country to the verge of economic collapse. It took United States monies to keep it afloat after 1945 and US embrace of British territorial commitments to make a gradual withdrawal from empire and its associated military expenditures possible in exchange for British support of US global positions (Taylor, 1990). Thompson and Zuk (1986) clearly show the switch in British military expenditures away from the Navy to the Army in the middle of the nineteenth century as the country became bogged down in more and more colonial wars and commitments. Like Kennedy (1987), they attribute the declining fortunes of the UK to a wrong-headed national decision to pursue territorial ambitions and their attendant military expenditures, rather than pursuing policies dedicated to rebuilding the domestic industrial infrastructure. They warn the US against a similar historical fate.

The country with a fate most similar to the United Kingdom is that of France (Figure 2.3). It has also experienced slow steady relative decline

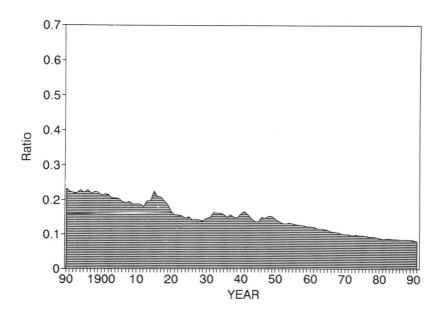

Figure 2.2 *United Kingdom proportion of the total Gross National Product of the six Great Powers, 1890–1991. Sources as in Figure 2.1. Calculated by the authors.*

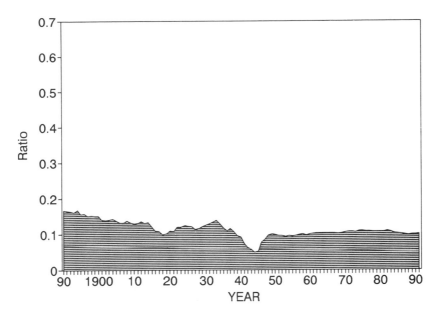

Figure 2.3 *French proportion of the total Gross National Product of the six Great Powers, 1890–1991. Sources as in Figure 2.1. Calculated by the authors.*

since 1890, though its ratio was not as high as that of the UK at the beginning and remained higher at the end. Both of these European states had been at the heart of world politics since the early seventeenth century, and their struggles for leadership in the global wars of Louis XIV (1688–1713) and of Napoleon (1792–1815) are clearly part of the historical power competition

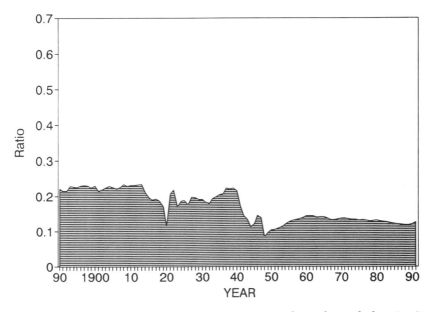

Figure 2.4 *German proportion of the total Gross National Product of the six Great Powers, 1890–1991. Sources as in Figure 2.1. Calculated by the authors.*

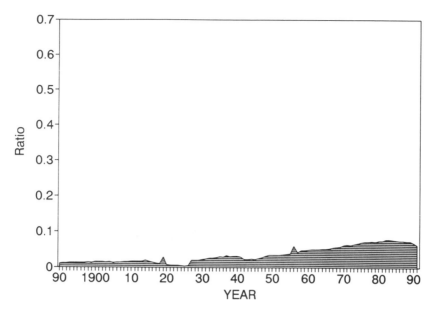

Figure 2.5 *Russian/USSR proportion of the total Gross National Product of the six Great Powers, 1890–1991. Sources as in Figure 2.1. Calculated by the authors.*

(Modelski, 1987b). Both states also led the nineteenth-century race for colonies and overseas possessions and both had the difficult experience of decolonization after World War II.

By the beginning of the twentieth century, France had been eclipsed by a neighbour as the European challenger to the United Kingdom; Germany's rapid population, economic and

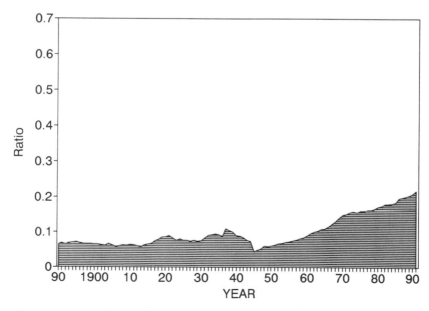

Figure 2.6 *Japanese proportion of the total Gross National Product of the six Great Powers, 1890–1991. Sources as in Figure 2.1. Calculated by the authors.*

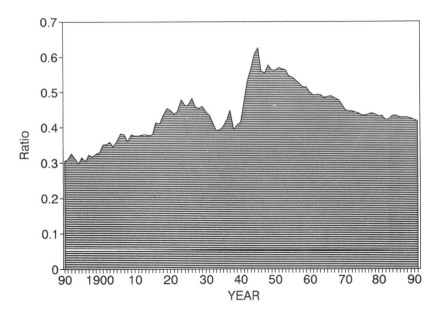

Figure 2.7 *United States proportion of the total Gross National Product of the six Great Powers, 1890–1991. Sources as in Figure 2.1. Calculated by the authors.*

military growth fundamentally changed the European balance of power. By 1914, France was firmly entrenched in the Dual Alliance with Russia. The larger question of with whom the UK would ally had also been settled as the threat from Germany to British leadership in Europe and increasingly around the world pushed the UK towards the Franco-Russian alliance. The stage was set for the first act of the global wars of the twentieth century.

In contrast to the fortunes of the other two European states, Germany maintained its relative position to the time of the Second World War (Figure 2.4). At the beginning of the century, its GNP figures made Germany the largest European economic power and that position was maintained to 1945. Like all the other five states, Germany experienced sharp recession between the wars but recovered to the point that its share of the Great Power GNP was the same in 1940 as in 1890. While power was diffusing from Europe, Germany, for a while, managed to hold its own. Like Japan, the other decimated Great Power of World War II, Germany experienced a sharp contraction between 1942 and 1950 but returned to a middle-level Great Power ratio by 1960. Its economic share has hardly changed

since then. Clearly, Germany had the economic resources by 1890 to challenge any other single Great Power for global leadership but, unlike previous hegemonic successions, the declining power, Britain, managed to remain part of the victorious coalition that led to a changed leadership under US hegemony. The twentieth century has been shaped fundamentally by the failed German challenge to British leadership and the emergence of US hegemony from the debris of the power transition.

Russia has always been a peripheral European power. The nineteenth century was punctuated with conflict against other European states, especially Britain, about local spheres of influence and territorial possessions on the margins of the growing Tsarist empire. Serfdom had been abolished in Russia only in 1861, and though the empire was enormous in size and population, its economic output was disproportionately small as a result of its traditional agricultural character. In 1890, the Russian share of Great Power economic strength was tiny (see Figure 2.5), and it was not till Stalin's rapid industrialization policies of the 1920s and 1930s that the Soviet Union began to experience relative growth. However, it should be noted from Figure 2.5 that the Soviet

Union has never crossed the 10 per cent share threshold. Though it is notoriously difficult to acquire accurate data for the Soviet Union, these figures are a fair reflection of the general economic status of that huge country. Since World War II, there was a very slow growth in the Soviet economic share but that turned sharply downward in the late 1980s. As Haass (1988) has noted, the Kennedy thesis of 'imperial overstretch' is more accurately applied to the Russian/Soviet state than to any western nation, since the costs of internal and external conflict drove military spending to the highest ratio of government spending of any Great Power and precluded the necessary investment in basic infrastructure. Russian/Soviet power in twentieth-century world affairs rested on a large population and a large resource and territorial base, combined with an enormous military machine and an anti-systemic ideology in the form of communism after 1917.

Japan is a twentieth-century power. With the exception of the years immediately after World War II, Japan has seen continued steady economic expansion and growing power shares for the past 100 years. In 1890, Japan's share was smaller than all except for Russia (whom it defeated in 1905), but by 1970 its share was larger than all except for the United States (Figure 2.6). Like Germany, with whom it formed an anti-Western (hegemonic) alliance in the 1930s, Japan's share of total Great Power GNP remained fairly constant from 1890 to 1942, but unlike Germany and every other Great Power its share has grown fast since 1950. In the aftermath of World War II, Japan's experience was the antithesis of the Soviet Union as it used its forced (by the Western alliance) demilitarization and restricted expenditures to redevelop its economic base and pursue an export-led strategy of growth. The results have been so spectacular that Japan, despite its relatively small armed forces and limited theatre of operations in the western Pacific is considered in the late twentieth century as a world power and capable of challenging the US for the global leadership role (Goldstein and Rapkin, 1991).

Finally, we turn our attention to the second global leader in the period under consideration. Since about 1940, the US has been the global hegemon, though this status did not become obvious till the global conflict that finally enshrined that role was over in 1945. As noted above by Peter Taylor in Chapter 1, the twentieth century is the tale of two hegemons, the declining British one to 1940 and the triumphant American one since then. We can clearly see the basis for later US hegemony in the GNP shares in 1890 which was already the largest at about 30 per cent. This ratio rose dramatically to over 60 per cent by 1948, achieving a share that had never before been approached by any single power (Figure 2.7). Of course, this enormous ratio was a temporary aberration as a result of the destruction in the war of the economies of the other Great Powers, but it is worth noting that, though US relative decline has occurred, the current share is about the same as it was during the years of rapid US growth in the 1920s (Elliott, 1991). American fears of decline are vastly exaggerated, like Mark Twain's death (Russett, 1985; Strange, 1987; Nye 1990; Nau, 1990); these fears are artificially generated by a fake comparison to the immediate post-war years and in the proper long-term perspective it can be seen from Figure 2.7 that the US has managed to retain its large historical proportion of Great Power GNP.

As we have noted earlier, the past 100 years is best conceived as two periods, the first half-century (1890–1945) corresponding to the waning years of British hegemony, and the years since 1945 as the apex of American hegemony. In Modelski's (1987b, c) long-cycle model, these past 100 years cover four phases of power transition, each phase lasting about a generation or twenty-five years. A global leader emerges from twenty-five years of war that determines the succession, and then the leader ensures its lead by organizing the peace terms and building an alliance system that promotes its global interests and enshrines its position. Following this period of unchallenged leadership, the leader begins to see a reduction of its lead over other Great Powers and its leadership is no longer unquestioned, especially if the leader is promoting a specific international policy. This is the phase of de-legitimation as other states offer alternative world visions and begin to build up their economic and military forces as part of the demand to share in global political power. The final twenty-five years of the long cycle is the period of power de-concentration as the total world power becomes shared increasingly by the aspirant and the existing leaders.

The trends in the graphs (Figures 2.2 to 2.7) indicate the validity of Modelski's phases. In the phase from 1890 to 1914, we can clearly see the growth of the power of the US, Germany, Japan and, to a lesser extent, of Russia as power deconcentrates from the erstwhile nineteenth-century leader, the UK. These global trends continued in the next phase, that of world war, from 1914 to 1945, so that by the end of this phase the United States was clearly the global hegemon as Germany, France, Japan and the United Kingdom declined rapidly in strength. The succeeding quarter-century, 1945–70, was the heyday of American power and global dominance, a development recognized by all commentators on the development of world power. The cycle is completed in the past quarter-century as American leadership of both the Western (anti-Soviet) alliance and the world community has come increasingly into question. One clear indicator of the trend is the increasing isolation of the US in the UN General Assembly, a body that seemed to be completely dominated by the US in the 1950s and 1960s. Within the Western alliance, France (in the 1960s), then increasingly Germany in the 1980s and later some smaller countries no longer followed the US lead unquestioningly. The American conduct of the Vietnam War and the US pursuit of the second Cold War against the Soviet Union in 1979–85 were important developments in de-legitimizing the US leadership. In the phase of de-concentration, the slippage of power by the global leader is not yet certain, but the signs are appearing. The final decision on American relative decline is not yet in, but there is no doubt that many individual indicators of global status show a decline or flattening from the US-dominated post-World War II years (O'Loughlin, 1993).

World order has been an ephemeral concept since the demise of the European colonial empires after the Second World War. Until then the European states maintained internal order in most of the Third World through repression. After 1945, new regional conflict formations developed over certain issues, such as the Arab-Israeli conflict in the Middle East; in East Africa over the extent of the Ethiopian state; in South Africa over apartheid and the international position of the white regime; in South Asia over regional dominance between India and Pakistan, and in South East Asia over the successor regime to the French colonial administrations (Väyrynen, 1987). Some, like Cohen (1982), have thought that at such regional levels, powers might come to the fore that would play a local hegemonic role under the tutelage of a global power such as the US. Such plans with the Shah's Iran, with Nigeria, Brazil and India have thus far been wishful thinking. The regime of these countries proved too weak, their regional dominance too quixotic, and outside intervention too dominant and conflicting.

Fragile underpinnings of peace

In Modelski's (1983) view, global conflict ensued for the first time at the end of the fifteenth century because it was only then that political power could be projected at the world scale through the use of naval forces. New technology in shipbuilding and armaments gave the Europeans the winning edge as they groped around the world. Portugal (according to Modelski, 1983) was the first predominant power with global reach. This precarious capability to control distant shores became ever more robust over time, but it remained in the hands of only a few powers. A higher military profile further and further from home was, of course, in most cases directly caused by visions of profit. As a result, the capitalist world-system developed from its European base into a truly global system. Over time, the major powers got increasingly involved in overseas adventures and relationships. Most states, however, hardly ventured outside their own immediate surroundings, militarily or otherwise. This is still the case. However, their economies have increasingly been incorporated in a capitalist world-system with global reach.

This section focuses on some of the factors that are conducive to peace or tend to stabilize peace. Peace is a multifaceted phenomenon. It is taken for granted that peace can primarily be considered as a negative – the absence of violence. Violence refers to the set of human actions that inflict bodily harm or at least severely limit the freedom of movement of the individual. Violence in these instances is direct and physical. The concept can be widened in various directions. One of these revolves around the notion of structural violence, the set of limitations installed by systemic factors that preclude humans from realizing their full potential and life span. Physical structural violence is the lack of food or medical care on account of the allocations of the economic system that shortens people's life spans. In this chapter, we focus our attention on direct physical violence (called behavioural violence), though it is evident that other versions of the violence concept can usefully be introduced in normative considerations. There are close relations between behavioural and structural violence, especially in the poorest regions of the world. Recent conflicts in the Horn of Africa have exacerbated the damage caused by the recurrent famines.

Apart from the universally shared idea of absence of violence, peace has varying connotations in different parts of the world related to the outlooks that people have on life as a whole. Different cultures have different social cosmologies (Galtung 1985). Whereas Buddhism and Confucianism relate peace primarily to inner personal harmony, the Christian and Islamic world, as well as the Japanese one, share a deeply-held conviction that peace is a social construct exclusive to their respective worlds, part of which is an effort to exclude outsiders. In addition, the Christian world knows another heritage of peace as social construct, universally inclusive and therefore applicable to all parts of the world. These different connotations obviously do not facilitate the search for common ground as far as positive peace policies are concerned. Here we stick to the notion of peace as, at least, the absence of violence (Boulding 1977; Galtung, 1987), knowing the full peace is based on a harmony of interests and voluntary co-operation.

Peace in the twentieth century knows two different stages as we have seen. The first half of the century was overshadowed by the fifth global war coming in two stretches, 1914–18 and 1939–45. That war was primarily fought between the major powers in the inter-state system, despite the role of smaller powers in their outbreaks, sucking in others as it developed (Fox, 1959). The second half saw a flurry of state-making efforts where European political dominance gave way to European withdrawal, the distribution of power within and between the newly-established states, and the positioning of old and new major powers in the various regions gave rise to a host of often serious wars.

Peace, we state, can be maintained by isolation through physical inaccessibility, or through dissociation where the potential for contact is not realized. Once relations are maintained states stay in contact, with the overt desire to prevent conflicts from deteriorating into war and to enhance co-operation. We start from the certainly naive assumption that this is indeed contributing to peace by itself, and briefly describe the evolving web of international relations of the twentieth century. The question then becomes under what conditions, and in what way, this set of international links effectively helps engender peace and prevent war.

First, it may be that the existing set of international relations simply works as the transmission belt of the hegemonic power facilitating its preponderance in international affairs. In this perspective, the structure of international relations contributes to the hegemonic power that, through the network, maintains its order. Consequently, it deters other parties from instituting violent conflict and does not have to wage war itself because other means to get its preferred outcomes are normally available. Additionally, it has been argued that even if the internal power-base of the actual hegemon declines, the institutions it has formed to ascertain its hegemony may prolong its effective operation because these institutions reflect the values and preferences of the hegemonic power (Keohane, 1984). The question, of course, is how long this will continue. In the long run, change seems to be predicated on the change in degree of hegemony in the system of international relations.

Second, the international relations that states maintain suggests that they act as 'satisfiers', that is to say, states are reluctant to change the status quo. Their independence is their most precious asset, so any change in the status quo may jeopardize their future chances of survival, as the end result of changes is impossible to predict in this multi-actor arena. Consequently, they try to maintain the existing structure of their international relations. Extra-cooperative gestures are compensated by slightly conflictive acts and the other way round. Increasing intensity is followed by a decrease, etc. It is clear that such views may well account for stability, but they cannot account for change in any direction.

Third, some see the possibility of a dynamic of co-operation to be built into the network of international relations. If states start to co-operate on some technical issue where they have a shared interest, this may extend to other policy sectors. This spillover will, if it develops, engender wholesale pacification of mutual relations and the disappearance of the repertoire of threats and military actions that has so often undermined the evolution towards a peaceful world. This functionalist vision (Mitrany, 1976) has inspired a number of efforts of international co-operation. Basically, the strengths of co-operation should be able to overcome the strains of conflict. In this way, a security community of countries could come about where the thought of violence to sort out differences would have vanished (Deutsch, 1957). Hegemonic preservation of peace is compatible with either status quo maintenance by a set of satisfiers, or a functionalist process of increasing co-operation. These last two mechanisms do not go together at the same place at the same time, but they may well simultaneously occur at different places in the world-system.

Isolation implies absence of violence, a situation of minimal peace. Sheer distance in this way pacifies, for it incapacitates individuals and groups to get involved with each other (O'Sullivan, 1986). A reluctance to connect to the outside world in the sector of security policy often has the same background. Such policies of passive neutrality have long been followed by smaller European states even if they could not or did not want to disconnect from the world-economy. The Irish Republic has long been tempted in that direction, at least from the 1930s to the 1960s, and in some respects, up to the present. This is very different from the active policy of neutrality in which states try to mediate between opposing camps or to persuade parties that the current lines of conflict should be redrawn. An example is Sweden in the Cold War period, 1945–90.

Once states have fought, withdrawal or dissociation often accompany the initial endings of violent conflict. Agreements on ceasefire lines that separate fighting forces are necessary steps to end hostilities. Historically, they have often become permanent borders. The present Dutch–Belgian border follows a ceasefire line between Spanish and Dutch troops in 1609 that interrupted the so-called 'war of liberation' of The Netherlands. Even if they do not end as agreed international borders, ceasefire lines can become semi-permanent with political consequences. An example is the 'Green Line' separating Israel from the West Bank territories, the ceasefire line interrupting the Arab–Israeli conflict in 1948. In 1967, Israel occupied the West Bank in another round of hostilities, but the Green Line still serves as an important division between the kinds of territories that the Israeli state administers.

Demilitarized zones extending the idea of a ceasefire line and creating more space between opposing armed forces were agreed after World War I for the Rhineland, and in Korea at the end of the early 1950s war there. A variation on that idea is the withdrawal of weapon systems from a certain area, like proposed nuclear-free zones along either side of the former Iron Curtain. These are dissociative policies with highly symbolic overtones because many of the weapons systems concerned have such reach that their geographical withdrawal does not really diminish the threats they provoke.

States may try to disentangle themselves from a web of international links in order not to get involved in conflict, violent or otherwise. Third World states have been encouraged to get rid of their links with the capitalist world. Those who did could hardly maintain their isolation (Holsti, 1982). The Soviet Union after 1945, accompanied by a number of client states, temporarily tried to cut as many tangible links as possible with the capitalist developed world, but to no avail in the long run. Even while seeking isolation, ongoing rivalry between the two Cold War camps mutually focused a lot of attention on the other side. During the 1970s and 1980s, core western states, particularly the US, shunned those arenas of international negotiation where their viewpoint tended to get overruled consistently. Various UN institutions (e.g., the United Nations Economic, Social and Cultural Organization, UNESCO and the International Labour Office, ILO) have been temporarily boycotted by some of their most powerful members. This dissociative policy has been seen as the expression of a more permanent situation where interrelations, particularly on a multilateral basis, between North and South could be relinquished if the South was not more compliant to the North (Krasner, 1985). However, it is difficult for major powers to be permanently absent from active networks of international relations; it remains in their long-term interest to belong.

In the twentieth century, distance and isolation still count for much more than global village imagery would suggest. Even if instant reports crowd the television screens and can be heard on radios everywhere, tangible relations are still limited and constrained by distance or induced by propinquity. This is clear from any study of the pattern of world trade flows, but also from the geographical distribution of state participation in bilateral and multilateral diplomatic networks and from war involvement.

The regionalized, or at least distance-dependent, character of a major part of the web of international relations is even more obvious in the second half of the twentieth century than for its initial decades. The end of European colonialism in the first period after the last global war finished a global system directed from Western Europe, with some adjuncts elsewhere through political dependency relations. Politically at least, regions in different parts of the world are reasserting themselves, though major core powers are occasionally significant players in the regional arenas of action (e.g. the role of the US and even Britain in the Middle East for decades). Since 1945, global networks have been maintained as a multilateral political arena around the UN and its specialized agencies, as well as a series of webs of bilateral relations centred around the major powers, the United States, and until the 1990s around the Soviet Union.

Only very few countries can use military force everywhere in the world. The British Navy, still paramount at the beginning of the century, could operate only with great difficulty in the Malvinas/Falklands War of 1982, significantly aided by the quiet assistance of the Americans in distant waters. Most countries could not operate so far from home at the outset of the century, and could not at the end either. Major powers with worldwide interests are rare. Some powers are more intent and better able to maintain a worldwide network of relations in one sector than in the other. The defeated parties, Germany and Japan, of the last global war have increasingly reasserted their global positions as major powers based on trading interests. By contrast, the Soviet Union made its presence felt worldwide from 1945 to 1990 as an ideological protector of anti-systemic forces with a missionary policy, plus massive arms sales and a small amount of development aid.

Isolation and the use of dissociation as a policy is, of course, not universally applicable nor desirable. If strains develop between states that cannot avoid each other's presence, existing relations may limit the chances of escalating conflict. At the end of the twentieth century, direct bilateral diplomatic relations between the countries of the inter-state system are maintained in only one-third of all possible dyads (pairs of countries). These contacts are indeed highly dependent on distance, but also on the power positions of the two countries in the dyad. Units at the apex of the power hierarchies maintain close contacts everywhere or, in any case, keep track of the actions of all others; weak powers relate primarily to their more immediate surroundings.

In the finite globe of the twentieth century, major powers tend to be universally involved; former major powers as a rule find it difficult to disentangle from their overseas relations. This is not only true for the larger colonial powers, the United Kingdom and France, but is even true for former minor powers with overseas interests. Relations between The Netherlands and Indonesia and between Belgium and Zaïre have remained special and sensitive. Some states may still try to be left alone (one of the most persevering is Myanmar, formerly Burma), but they cannot ignore their neighbouring region or even distant Great Powers. Particularly during the period of superpower rivalry during the Cold War period, major powers, especially the US, prompted small ones to action and consequently immersed them in conflicts. A good example is the southern part of the string of alliances that the US wove around the Soviet Union in the 1950s in the form of CENTO and SEATO. In the 1970s and 1980s, the superpowers found themselves increasingly sucked into conflicts of primarily regional relevance.

Inter-state relations vary widely in intensity. Once the relations become intense, they can become so co-operative that the distinctions between separate states start to disappear. This has been happening in Western Europe since 1945 in a long-winding process of fits and starts that has gradually taken in ever more countries, now up to twelve in the European Community (EC) and likely to expand to about twenty by the year 2000. It has now come to the stage where, in some countries such as The Netherlands, serious discussion takes place about the future need for a separate minister of foreign affairs, since so many of the country's foreign interests are now concentrated within the EC. The EC is beginning to have its own foreign relations (e.g. a diplomatic representative in Washington), and much of each country's interests in Brussels is handled directly by technical sector ministries.

Intense relations may also be conflictual, occasionally deteriorating into war. European integration is partly an answer to this having happened twice between France and Germany in the two phases of the twentieth-century global war following an earlier encounter in 1870. The relations between the post-war superpowers was also intensely conflictual, though long frozen into a Cold War. The question then becomes how the strengths of intense interrelationships inducing co-operative behaviour absorb, neutralize or cope with the strains that necessarily arise from time to time (Boulding, 1978) or fail to do so, thereby pushing the parties towards sustained conflict.

These strengths are located in the structural and cultural features of relation patterns, crossing state boundaries that merge in institutions or regimes. They give rise to associative peace conditions eventually supported or instigated by policy. These are by no means purely relations between state apparatuses. The strengths of these relations pointing towards co-operation either derive directly from their nature (e.g. more technical than political) and pattern (e.g. criss-cross rather than polar), or they are mainly instigated by attributes of the various states and the people that inhabit them (e.g. democratic, rich).

Taking the relations themselves as the basis for stable peace conditions, some major changes in the inter-state system should be stressed. After the long shakeout of the European state-system that took place between 1500 and 1700 (see the war data in Levy, 1983), the internal scope of state involvement grew. State boundary-crossing relations became more intense as states wanted to know what was happening in other places. Subsequently, more regularized ways to maintain contact were installed in the post-Napoleonic system of international relations on a bilateral as well as a multilateral base. The Austrian secretary at the Congress of Vienna, Friedrich von Gentz, codified the notion of a balance of power that was to be the working device of regular Congresses where occasional disputes could be resolved. The diplomatic machinery of the 'Concert of Europe', based on an assumption of a system of collective security, slowly evolved up to the beginning of the twentieth century. Although no permanent secretariat developed, international Congresses were frequently held, various sides became simultaneously involved in the preparation of these conferences, and it was agreed that a single record of the proceedings would be produced. After Congresses in The Hague around 1900, the beginnings of international, judicial institutionalization and a permanent machinery for mediation and arbitration were installed in an International Court sitting in a newly-built Peace Palace at The Hague (Netherlands) from funds supplied by the American millionaire, Andrew Carnegie.

In the course of the nineteenth century, various international relations were facilitated, executed or controlled by newly-instituted international organizations. A World Health Organization was created to control the spread of contagious diseases, while the International Postal Union facilitated the increasing international mail flows. Agreements about time and weights and measures were made internationally to ease traffic and trade, as was the production of comparable statistics and the sharing of weather information. This happened so thoroughly in the long run that American-supplied world weather information could not be withheld from the Iraqis when the Gulf War broke out in January 1991. At the same time, in the field of security policy, major states, increasingly and ever more permanently, allied themselves to others in order to empower their positions. This reflected the erosion of the Concert of Europe understandings in that field and further contributed to their demise. These agreements were primarily meant as an insurance against potential aggression, but they in fact triggered similar reactions from others. A counterproductive way to solve the power–security dilemma, they eventually contributed to the Great War of 1914–18.

The League of Nations was a further effort to build strength in a peaceful international network. The League was hesitantly conceived as a compromise between President Woodrow Wilson's drive to oust the existing kind of state-system altogether, the French reluctance ever again to be confronted with German aggression, and the British preference for flexible adaptations of the regulations of the preceding era. In the end the US Senate blocked American membership. Britain and France, the European victors of the Great War, became the principal managers of the League. As a result, the League lacked the leadership of the ascending hegemon, the US, while the UK was no longer in a position to provide unequivocal guidance. The League installed a permanent secretariat and the kernel of an international civil service. The talents nurtured in this environment were not wasted, despite the League's ineffectiveness. One talented administrator was Jean Monnet, who was to become the major instigator of the European Community. The League had a Council mainly consisting of the major powers, and an Assembly readily accessible to all internationally-recognized states. In its Headquarters at the end of its existence, located in another newly-built Palais (des Nations) in Geneva, Council and Assembly regularly convened. The League took under its wing most of the pre-Great War international institutions and organizations. It also entered the field of economic co-operation based on the reparations questions that spoiled the post-Versailles world order, and it executed technical co-operation programmes, mainly in China.

Only slowly and reluctantly did the League move on disarmament, considered to be the heart of an international security policy. When the Disarmament Conference called by the League of Nations finally convened in 1932, it was too late. The Germans, forced to disarm immediately after 1918 with the understanding that other powers would follow suit, were on the verge of rebuilding the Reichswehr over the Allied-imposed limit of 100,000 troops. Even then, the conference merely proposed a freeze on current armament levels (arms limitation and control instead of disarmament), clearly putting the Germans at a disadvantage. Outside the League framework, naval disarmament agreements between the major naval powers had already been made in the early 1920s, perpetuating the dominant position of the US and Britain at lower cost to these countries.

In the course of the 1930s, the multilateral League of Nations machinery declined, particularly in the field of security policy as the mechanisms to enforce collective security (trade embargoes, common intervention, etc.) clearly were not effective. The functional organizations maintained their operations, but the League lost its aura as a peace promoter. It had, of course, never been granted much leeway even by its initial main protagonists. Just before global war resumed in 1939, the Bruce Commission recommended a general overhaul of the organization. This report was a major stepping-stone to the United Nations launched by the victorious powers in San Francisco in 1945.

The UN was part of the set of organizational devices instituted in order to build the framework of an American-dominated global order of sovereign states, inspired by liberalism and free trade (IMF, World Bank, GATT, etc.). Although truly global reach for these institutions perhaps will only be achieved at the end of the twentieth century, they were essential for the preservation of the capitalist world-system, and their impact reached beyond their formal membership. Only the United Nations was a really global organization of independent states from the outset, though some states were late in joining (East and West Germany, North and South Korea, and the People's Republic of China) because of post-war geopolitical divides. The organization became the main venue for initial international recognition of new states. As a result, it more faithfully reflected the trends and contradictions in the field of international relations than other early post-war institutions.

The United Nations showed some major changes from the League of Nations. There was a clearer separation of roles between the Council (now called the Security Council) and the (General) Assembly. The secretary's position was strengthened and the idea of an international civil service further elaborated. The Secretary-General of the UN was typically recruited from the smaller countries (and eventually from different continents), not from the major powers, as was customary in the League of Nations. Recruitment from the smaller countries had been tried in 1919 but had failed. The UN was firmly tied to US participation even by its location in New York, though Geneva and later Vienna were maintained as secondary headquarters. Many of the League staff served again in the UN. New functional organs were instituted (e.g. UNICEF, UNESCO), though many of the older League of Nation institutions became part of the UN family.

The functioning of the UN was impeded on a number of occasions by the East–West bi-polar divide that soon overshadowed the incentives for global multilateral co-operation. The earlier lessons of multilateral inter-state co-operation were now very effectively used in western-dominated (IMF, World Bank), purely western (OECD) or West European (EC) bodies that were only reluctantly opened up to new members. The UN was at the same time thoroughly transformed through the massive influx of former colonies; the final stage of a temporary withdrawal process by colonial powers was already cautiously started under League of Nation auspices.

Twice in its history, the UN was the vehicle for a coalition of states that guaranteed the status quo in the inter-state system by massive armed intervention against military attack – in Korea in the early 1950s and in the Gulf War, 1990–91. There were procedural differences in the two cases: uniting for a peace resolution in the General Assembly to circumvent Soviet veto power in the Security Council in the Korean case, versus a generally-supported Security Council resolution in the Iraq case. In both cases, a US-led coalition under the auspices of the UN managed in the end successfully to reinstate the status quo ante. In these two instances a rough approximation of a collective security system was in operation, but this is not to imply that both operations were totally successful. The division of Korea in two mutually sealed-off camps has now existed for four decades and it has proven very difficult to begin the process of mutual *rapprochement*. Encouraging steps are being taken finally in 1991–92. Similarly, the outcome of the 1991 Gulf War does not, by itself, help to solve the intricate problems that gave rise to Iraq's assault on Kuwait in August 1990. Structural conditions for associative peace in that area have not improved since that time.

In addition to the two UN military actions in Korea and Kuwait, the United Nations has undertaken more than a dozen peace-keeping operations. These tend to freeze often-precarious situations where strengths of a peaceful settlement are still far away and the strains of the situation may easily erupt again. Under those circumstances, freezing clearly is not sufficient for long-term peace, but warring parties typically accept the status quo. The longer the UN is present, the more they grow accustomed to its valuable contribution in maintaining peace, such as the case in Cyprus. In most interventions of this type, ceasefires have been maintained between parties related to separate states. The trend of UN intervention in civil wars rests on the agreement of governments unable to maintain order, advocated by outsiders and replacing intervention by neighbours. Boundaries have become more and more sacrosanct in the inter-state system, but not the regimes that are supposed to administer the fixed territories.

Massive armed intervention under UN auspices has been very rare and the military provisions of the UN Charter have hardly been used. The small-scale peace-keeping operations have gradually been elaborated on an *ad hoc* basis and they seem to occur more and more frequently. The major powers (the US, the former USSR, China) have only seen fit to use the UN machinery in questions of security policy in a small number of instances. On the side, they have pursued major military interventions on their own often with disastrous results, for the US in Vietnam and for the ex-USSR in Afghanistan. The same holds for the increasing interventions by weak neighbours in civil wars in the Third World.

The other institutions of the UN family have played enabling global roles, sometimes sorting out differences (the Court of Justice) and sometimes assisting in development aims (UNCTAD, UNIDO). However, states have remained paramount in regulating the use they have made of the UN presence. Occasionally, powerful states have threatened to leave or have tried to force changes in organizations like UNESCO and the ILO. Slow payment of dues, or failure to pay, is a well-developed tactic to demonstrate disagreement with the organization's thrust. This reliance on the coffers of the member countries, of course, provides more leverage for high-dues-paying members.

The web of international relations at the global level, now mediated by the UN-related organizations, has incrementally become more important. It has provided some strengths to a still-precarious inter-state system, allowing the maintenance of a semblance of Pax Britannica first and Pax Americana later in the century. The strengths of the organizations on their own are insufficient to guarantee an associative form of peace in all parts of the world. In a world-system perspective, a hegemon is needed to back up such strengths that these links can provide.

In the years leading to the Great War of 1914–18, the peace-providing framework of the incipient global institutions was too weak compared to the strengths of two opposing camps of major European states. The British hegemon finally handed in its role as balancer of the Concert of Europe by unequivocally allying itself to one side, threatened as it was by the German challenge. During the inter-war years, the UK and France lacked the means and were not prepared to use the League of Nations as the vehicle to organize an associative peace order – particularly in Europe, but also in the rest of the world, such as for Japanese–Chinese relations. The US, the impending hegemon, still held itself aloof from political leadership at the global level. In an even more interconnected post-war world, the US,

through the UN and other institutions, has tried to impose an American-dominated world order. This was fiercely resisted by the communist camp, as well as by neutral and nonaligned states. American influence has nonetheless been pervasive during the second half of the twentieth century. A large number of regional-level and intra-state violent conflicts have generally been contained at that level, but they have occasionally been extremely bloody through the general availability of very lethal weaponry, as in the Iran–Iraq war of 1980–1988. Harff and Gurr (1989) provide an overview of the worst cases of genocide.

Forms of hegemony are also at stake where segments of the world-system are exclusively dominated by one overpowering state whose authority is hardly in doubt. In terms of distribution of power resources, the Soviet bloc was the best example. In the West the parallel is not so much NATO, as it is the panregional US plus Latin America. Such blocs have been part of the post-war security structure. Their primary reason was, of course, the conflict between East and West. One of their main functions was to provide international pacification by means of a bloc leader. Members following their lead were rightly called bloc-provinces (Weede 1975), as they had relinquished part of their sovereign power, particularly in nuclear defence policy. These hegemonic segments were perhaps more peaceful than they might have been otherwise, though this is difficult to prove. The Soviet Union interfered in its own camp in order to preserve its hegemony, legitimized by the so-called Brezhnev Doctrine. By contrast, the Americans had great difficulty in thwarting the Greek–Turkish conflict that erupted over Cyprus in 1974. Unconvincing efforts have been made to devise American policies aiming at the erection of pseudo-hegemons in other parts of the world. Under the ultimate guidance of the US, these regional powers were supposed to provide order, stability and peace. In the late 1970s, countries like the Shah's Iran, Brazil and Nigeria had been cast in that role, but they have hardly been able to perform (Väyrynen, 1987; Cohen, 1982).

During the inter-war years, Germans particularly advocated the idea of Panregions

(Haushofer, 1931), where one state would be paramount and the patterns of relations would be ordered according to its preferences based on an overall Pan-ideology, unimpeded by other major powers who would look after their own Pan-regions (O'Loughlin and van der Wusten, 1990). This predecessor of post-war policies was more or less put into practice by the Germans and the Japanese, but they were unable to achieve an enduring success because of their defeat in war. It is only after the demise of the Soviet empire and the weakening of American hegemony that developments in this direction can again be detected. A shared uneasiness in South East Asia within ASEAN about growing Japanese economic preponderance, as well as reluctance within Europe to grant Germans unequivocal leadership of the EC, are based on similar historical experience. The outcome of a combination of structural change and workings of memory is not easy to predict.

Despite superficial impressions to the contrary, the basic pattern of international relations changes only very slowly. The number of countries that drastically reorder their international relations in a short period is low. China has made a number of sudden sharp turns, from inward- to outward-oriented, and from Soviet- to American-oriented toward an independent position in a triangular relation (Holsti, 1982; Goldstein and Freeman, 1990). Cuba made one of the most spectacular changes in foreign policy profile around 1960 after the victory of Fidel Castro over the American-oriented former leader, Batista.

Countries seem very often careful not to upset the status quo, satisfied as they are with their own survival. Event analysis shows how they try to balance slight gestures in one direction with similar actions in the opposite direction; they support another country in one instance and, as a function of that gesture, deny it a favour in the next instance. This is a useful demonstration of the gradual character of most changes in the realm of international affairs. It also helps explain that countries most of the time do not rush easily to war, but it does not provide clues where the brakes should be sought if countries nevertheless tend to go to war, and neither does it account for the road towards stable peace. The roads to war

or to a more harmoniously connected set of countries can only be explored by other means.

Functionalism has, for a long time, seemed a road towards more peaceful international relations. It was initially proposed at the world level, but it has most successfully been applied in the case of European integration. Functionalism maintains that peaceful relations in one area will spill over to other arenas of action and may finally pacify the dyadic relationship as a whole, leading to ever-higher forms of integration. Functionalism as a predictive theory has known its ups and downs with the evolution of the European Community. After initial enthusiasm in the 1950s and 1960s, it has seen a phoenix-like rebirth after the renewed vigour of the EC in the later 1980s. There is no similar example of functionalist movement towards integration and a security community elsewhere in the world. If the mechanism is to work in reducing state sovereignty and building international organizations, there seem to be extra conditions that have to be met. In Deutsch's (1957) collection of historical cases, where pairs of states in conflict in the long run became co-operative partners (e.g. Canada and the US), one of the basic necessary conditions turned out to be a set of shared basic values. The apparent rule of international relations that democracies do not fight each other (Lake, 1992) is a main point in this argument; it by no means excludes democracies' involvement in warfare generally, but it points in a specific direction as regards the kinds of values that should be shared.

A set of middle-sized countries in the core of the world capitalist system share a high degree of economic development that has made interdependence with other countries a natural and inevitable state of affairs. Such interdependence should not be perceived as too asymmetrical in benefits. Liberal democracies are so far the most successful political constructions to cope with changing and challenging economic conditions (Fukuyama, 1992). Laqueur (1980) has argued the case of a general diffusion of peace feelings through major parts of the developed world starting even long before the last global war. He baptizes the process as 'hollanditis', a reference to the country where its earliest practitioners can be found, though a reference to most of

Scandinavia would have been just as accurate. In these countries, there is a growing acceptance of the notion that fighting has become too costly or impossible. It calls for the act of withdrawing the use of violence from the repertoire of foreign relations, and the idea might now be spreading to other countries. It is strongly motivated by the realization that the arms technology available for modern war is so abhorrent, and if fully used would result in no military force being able to fight. Keegan (1976) commented that though the exercises were dutifully performed by the respective military forces in the Central European theatre, battle there was effectively impossible because no army, let alone civilians, would be able to sustain it for any length of time.

Summarizing this section on the fragile underpinnings of peace, we can conclude that the network of international relations has grown incrementally during the twentieth century. This is mainly a function of state interests and societal development and does not by itself contribute to the direction of either peace or war. Distance and isolation are still important in accounting for the absence of violent conflict. But lack of communication in a conflictual relationship may further enhance the chances of war. A factor limiting the chances of extreme outcomes in all directions is the uncertainty about the final result of drastic changes. In most cases, countries have a preference for non-war. If a hegemonic position in the world-system is in doubt (the power transition hypothesis), or if the polity of a neighbouring country is destabilized, states may prefer drastic changes and go to war. Deterioration of British hegemony and the resulting global uncertainty was apparently a major factor in the outbreak of global war in 1914–45. But this leaves unexplained the low degree of other warfare during the British hegemonic period. Maximal American hegemony after 1945 contributed to the long peace (Gaddis, 1987), even if it was sometimes disguised as a Cold War. It did not, however, prevent the bloody state-making wars in the Third World during the same years.

Functionalist relationships between core societies that are increasingly interconnected has probably reached a state in Western Europe where they have become an effective means to absorb the strains that arise in the various bi-lateral relationships. Populations in these countries are wary of war, and their democratic politics now translate such preferences accurately into actual policies. In much of the rest of the world, states are still trying to gain sufficient stability. Where that stability has been acquired, the maintenance of the status quo in international affairs has become a major, though rarely-expressed, policy goal. As state societies, such as South Korea move up the development ladder, their populations become massively and directly involved in transnational relations, and a basis for a functionalist spiralling process of cooperation comes into place. Such a developmentalist view of regional state-systems assumes a gradually widening core in the world capitalist system, as is happening in East Asia. But it underestimates the unsuccessful rivalries, exclusions and outright failures that we see in many other parts of the world, especially in Africa.

Major powers have become involved in local and regional conflict formations, as they have seen their interests threatened or on account of rivalry with other major powers. This generalization applies to the Balkans at the beginning of the twentieth century, as it does at its end (Nijman, 1991). At the end of the century, the collapse of the communist world has added a formidable series of state-formation problems to the existing long list.

Once states depart from a non-war status quo and launch war, strong factors work in the direction of the expansion of violence. Some states cannot avoid involvement; others are afraid to miss an opportunity to affect final outcomes. Particularly in Europe, such diffusion has historically been strong (Faber, Houweling and Siccama, 1984). But escalation in the direction of increased co-operation may also be contagious. Again, Europe provides a good example. The EC provoked the establishment of EFTA (European Free Trade Agreement), and has gradually taken in members of that organization. The remaining membership and the newly-established regimes in East-Central Europe now compete for early entry. Just as wars have traditionally spread, peace may unleash similar forces of diffusion if conditions are right.

Temporal and spatial distribution of war

As well as the power trends, we can see the global war period that marks the beginning and the end of each of Modelski's long cycles in the temporal distribution of war over the past century (Figure 2.8). Using the indicator of battle-deaths (a common measure, but one that significantly understates the total destruction of life in war), we can identify the two peaks of World Wars I and II at 8.56 million and 15.29 million battle-deaths respectively. As we have indicated earlier, world wars differ fundamentally from local and regional conflicts in their scale, both temporal and spatial extent, and outcomes for the world order. The next most destructive conflict, that of Korea in the early 1950s, caused only one-fifth as many casualties as the First World War and about one-ninth as many as the Second World War. And, as we have noted, the two world wars are seen as two halves of the global conflict – the one that would eventually determine the global successor to British hegemony.

What is also interesting about the world wars is that neither the waning global power (the United Kingdom) nor the eventual victor (the United States) suffered the largest casualties in either conflict. They had respectively 1.18 million and 0.53 million battle-deaths in the two wars, compared with totals of 1.56 million in France, over 1 million in Japan, 5.3 million in Germany and 9.2 million in Russia/Soviet Union (Small and Singer, 1982, 89–91). These numbers reflect, in part, the locations of the major battles on the continent of Europe, and they illustrate the argument that the main staging-ground of world power competition will be Eastern Europe because that is where the seapowers interact with the landpowers (Mackinder, 1904; Brzezinski, 1986; Gray, 1988). While the seapowers (the UK and the US) escaped relatively lightly from the global conflict, it was the landpowers (Germany and Russia/Soviet Union) that suffered most. The battle-death casualty total for these two European states in the 1914–18 and 1939–45 conflict was 5 million more than all the other twenty-six participants in both world wars combined.

A second important feature of Figure 2.8 is the absence of significant violence in many years. Wars are not continuous, but show strong evidence of spiking and cyclical behaviour and, as noted earlier and shown on Figures 2.9 and 2.10, they are increasingly confined to certain regional

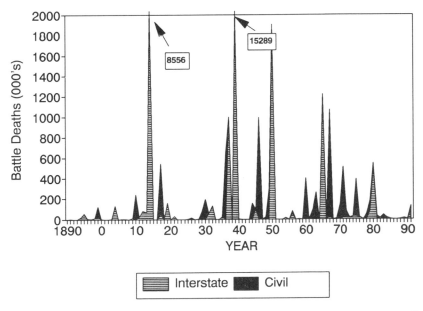

Figure 2.8 *Yearly distribution of battle-deaths in war 1890–1991. Sources are Small and Singer (1982) and SIPRI yearbooks 1980–1991. Calculated by the authors.*

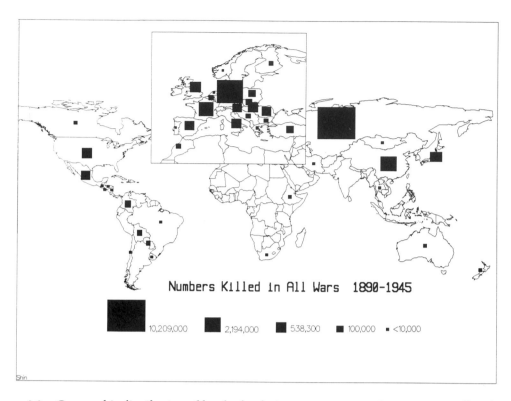

Figure 2.9 *Geographic distribution of battle-deaths in war 1890–1945. Sources are Small and Singer (1982) and SIPRI yearbooks 1980–1991. Calculated by the authors.*

locations. A third feature is the different trends in civil and inter-state wars. Increasingly over time, inter-state wars are becoming relatively infrequent as civil conflicts abound. Since 1890, there have been ten inter-state war-years with more than 100,000 battle-deaths, while there have been fifteen war-years with civil war deaths exceeding that figure (Figure 2.8). Three of the ten inter-state peaks have occurred since 1945, while nine of the fifteen civil war peaks are found in the past half-century. We will note many times in this chapter that local wars, including civil wars, have become the stereotypical conflicts of the late twentieth century. Local wars can be enormously destructive, as were the Chinese civil war of the 1940s, the Korean War of the 1950s, the Vietnam War of the 1960s and 1970s, and the Iran–Iraq war of the 1980s.

Another way to classify conflicts is not by their size and extent, but by their causes. A very useful classification is provided in Holsti (1991).

In a major study of conflict in the world order since 1648, he groups the issue-areas of conflict and the (sometimes) resulting wars into six categories, namely, territorial conflict, ideological conflict, nation-state creation, economic conflicts, human sympathy (ethnicity, religion, etc.) and other causes. Though his time-periods do not correspond to the ones used in this chapter, we can re-order his data for the periods 1918–45 and 1945–89 and include the data for the nineteenth century for comparison. The resulting figures are shown in Table 2.1.

Looking at the trends in the causes of international conflicts and of war since the seventeenth century, Holsti (1991, 321) concludes that

relatively abstract issues – self-determination, principles of political philosophy and ideology and sympathy for kin – have become increasingly important as sources of war while concrete issues such as territory and wealth have declined. One explanation for this pattern might lie in the ability of governments to create legal and other conflict-avoiding regimes for concrete-type issues, while for abstract issues regulation is difficult to develop.

Looking at the past nearly 200 years, it is possible to arrive at the same conclusions. While there seems to be little trend in the causes of international issues, there seems to be a growing tendency for wars to be at least partly caused by 'human sympathy' (ethnicity, religion and kinship). This category constitutes the largest set of war causes and is closely related to the ideological and territorial categories. A typical example of this kind of conflict is the series of Arab–Israeli wars since 1948, as Arab states support Palestinian aspirations for an independent state. What is surprising is the relative infrequency of war caused by economic reasons or by strategic considerations, subsumed under the 'Other' category in Table 2.1. State-making, kinship and political ideology account for most wars, since such disputes usually have competing territorial

claims that provide a basis for the ensuring conflict.

Another thematic classification of conflicts is to use the scheme developed in Johnston, O'Loughlin and Taylor (1987). In their study, these authors provide an historical-structuralist account of violence, including structural and personal as well as behavioural violence (the latter type being the focus of this chapter). They first identify two territorially-based dominant–dominated relations – that between core and periphery (North and South in the world-economy), and that of class relations. They accept the world-system argument that war is closely related to the competition engendered by the global struggle for economic control and resources, waged by capitalist enterprises but assisted and abetted by state sponsors. From the resulting class and core–periphery relations, eight types of violence can be identified. They are listed in Table 2.2 with their associated numbers

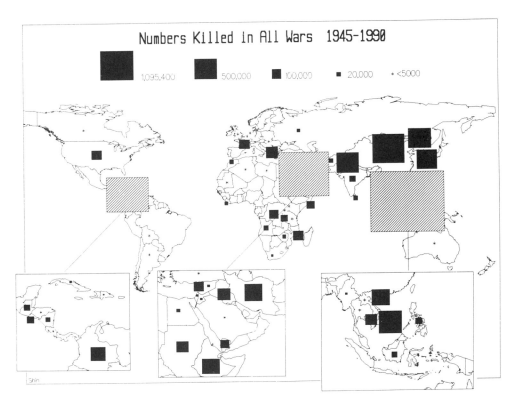

Figure 2.10 *Geographic distribution of battle-deaths in war 1945–1990. Sources are Small and Singer (1982) and SIPRI yearbooks 1980–1991. Calculated by the authors.*

Table 2.1 *Causes of international conflicts and wars, 1816–1989*

Cause	International issues			Wars		
	1816–1914	*1918–45*	*1946–89*	*1816–1914*	*1918–45*	*1946–89*
Territorial	18	23	18	53	77	52
Ideological	10	16	13	71	67	70
Nation-state	37	4	17	53	33	52
Economics	5	6	5	25	20	30
Human sympathy	10	23	12	32	78	97
Other	11	33	23	35	47	41

Source: Holsti (1991). The numbers are percentages of total issues per period for the international issues columns and frequency of issue as percent of wars. Numbers may sum to more than 100 per cent because of the multiple causes of wars.

of battle-deaths since 1890, divided in turn into two geopolitical epochs – the British long cycle to 1945 and the American long cycle since then.

The figures on battle-deaths, taken from Small and Singer (1982) and updated using the SIPRI yearbooks, show that over 37 million have died in conflicts since 1890. This is an underestimate since both sources have a minimum size cutoff for the conflict. So, for example, the conflict in Northern Ireland is not included in Small and Singer's counting, though over 2,500 have died there in the past two decades. And, as noted

earlier, these huge numbers do not count most civilian casualties. Most of the deaths occurred before 1945 as a result of the global conflict of 1914–18 and 1939–45. This conflict is included in the 'imperialist' war category in Table 2.2. The world wars involved mostly a deadly struggle for accession to the post-British hegemony, but because of the alliance patterns, all sorts of local and territorial disputes became entangled in the central contest for hegemony. Other examples of imperialist conflicts have been the Japanese invasion of China, the US involvements in Vietnam

Table 2.2 *Distribution of battle-death by type of conflict, 1890–1990*

	Frontier	Imperialist	Colonial	Neo-colonial	Collabora-tionist	Resistance	State resistance	Territorial	Totals
1890–1990									
Number of wars	11	11	11	8	3	96	0	27	167
Numbers killed	119,003	27,428,367	1,314,100	1,520,926	10,000	5,459,460	0	1,200,069	37,051,925
Average	10,818	2,493,488	119,464	190,116	3,333	56,869	0	44,447	
1890–1945									
Number of wars	11	8	7	0	3	25	0	11	65
Numbers killed	119,003	24,209,225	1,161,400	0	10,000	2,327,810	0	530,000	28,357,438
Average	10,818	3,026,153	165,914	0	3,333	93,112	0	48,181	
1946–1990									
Number of wars	0	3	4	8	0	71	0	16	102
Numbers killed	0	3,219,142	152,700	1,520,926	0	3,131,650	0	670,069	8,694,487
Average	0	1,073,047	38,175	190,116	0	44,108	0	41,879	

and Korea, and the Russian/Soviet Union invasion of Afghanistan. The largest most destructive wars have been between the Great Powers and also happened as they have attempted to extend their territorial sphere of influence. In the eleven imperialist wars since 1890, the average battle-death figure has been about 2.5 million, with significantly fewer dying in post-1945 wars because the Great Powers have avoided direct conflict with each other.

Another major cause of war has been 'resistance' violence. In this category are wars that would usually be classified as civil wars, involving majority–minority conflicts based on ethnicity, religion, ideology and governmental control. Pure class warfare is the exception, as most Third World conflicts (where this type of conflict is most common) involve a potent mix of class and ethnic issues. Again, it is noteworthy that this kind of conflict has come to predominate recently, with 71 per cent of the conflicts of the past half-century of this type. Most of these have been of the state-making kind, in Africa, Latin America and Asia. During the colonial period up to 1960, civil conflict in the colonial possessions of the European powers would usually have been classified as 'colonial' or 'frontier'. The major change in the nature of war over the past century has been the switch from a few very destructive imperialist contests to many local, often civil, wars of less individual destruction – but cumulatively they have resulted in over 5 million deaths.

Frontier and colonial violence are quite difficult to separate. Frontier violence is involved in extending the world-economy into previously peripheral areas, such as the contemporary incorporation of the Amazonian Basin. But all significant frontier violence occurred before World War I as the European powers (and the United States at war against Native Americans) pushed their influence into hitherto non-capitalist regions. These wars have been relatively few in number and small in destruction. Colonial conflicts, defined as wars that happened when the peoples of the periphery tried to shake off the imperialist core control, have also been more common early in this century. India's freedom struggle against the UK is a good example and was quite destructive, as well as resulting in

further violence over territorial issues between the newly-independent states.

Neo-colonial violence, by definition, involves core states and technically independent regimes who depend on core allies for survival. The core power may work to keep a friendly ally in power (as in the case of US, French and Belgian military and economic support for President Mobutu of Zaïre), undermine an opponent in their sphere of influence (as in US attempts to destabilize the Castro regime of Cuba) or provide internal support for political opponents (such as US support for the Contras in Nicaragua, 1981–90). Collaborationist violence is quite similar. It consists of state repression, usually by a conservative pro-capitalist regime, against internal opposition. The internal class conflicts of Latin American countries offer the best examples, but these have been small-scale conflicts by world standards of violence.

It proved impossible to find a war of the past 100 years that met the 'state resistance' definition of peripheral countries attacking core interests. Clearly state-sponsored terrorism by anti-West regimes, like Libya and Iran, would qualify, but the scale of this kind of violence is too small to be included in the usual compendia. The final violence type, territorial, is fairly evenly distributed between the two half-centuries in both the number of occurrences and degree of destruction. It is usually the result of dissatisfaction of one or more parties with the outcome of previous conflict and the resulting ceasefire or border lines. The Iran–Iraq war of 1980–88 is a good example of the legacy of territorial conflict and, of course, the potential for this kind of violence is huge because of the large number of unresolved border disputes and the involvement of outside powers supporting local allies.

The geographic distribution of battle-deaths in wars since 1890 clearly indicates how war has changed. In Figure 2.9, showing the distribution of deaths from 1890 to 1945, it is obvious that Europe was the locus of significant conflict. Russia/Soviet Union and Germany accounted for most deaths, but almost all the European states experienced significant casualties in the imperialist wars of 1914–18 and 1939–45; the conflict has been called a *europäischer Bürgerkrieg* (European civil war). Other large casualty figures in Europe were related to the Spanish Civil War of the 1930s and the ongoing Balkan state-making conflicts before World War I. In contrast to Europe, significant numbers of deaths are found in only a few other countries – China, Japan, Mexico, the United States, Bolivia, Columbia and Paraguay. In the Latin American countries, the main cause of deaths in wars has been the result of civil strife, usually over governmental control, rather than regional separatism. In China, the large number (over 2 million) is both a result of Japanese and other imperialist invasions as well as civil strife. And, of course, Japan and the United States were intimately involved on opposite sides in the biggest imperialist contest of all, as well as fighting local wars to produce desired regional dominations.

The second map of the distribution of conflict since 1945 (Figure 2.10) is drastically different from the map of the 1890–1945 deaths. The relationship between the two distributions is close to perfect negative correlation. With the exceptions of the French and the Greeks, Europeans have not experienced major war destruction since 1945. France's battle-deaths have come about mostly as a result of decolonization conflicts in Indo-China and in Algeria, while the large figure for Greece is the result of the short nasty civil war there in the immediate aftermath of World War II between communists and conservatives. The real victims of recent conflict are in the Third World, especially in the countries involved in recent imperialist wars in Vietnam (1963–75) and Korea (1951–3). Other major conflicts have involved India and Pakistan (four times), Iran and Iraq, and the extended Arab–Israeli conflict, all involving territorial delimitations. A third kind of destructive post-1945 conflict is civil war, which has been especially damaging in El Salvador, Colombia, Sudan, Mozambique, Zaïre, Ethiopia, Lebanon, Yemen, Cambodia, Somalia and Sri Lanka. Though the overall numbers of people killed are down from the peaks of the early twentieth century, the displacement of conflict to the Third World should not close our eyes to the continued devastation that wars of various types continues to wreak there.

War cycles and economic cycles in the twentieth-century

Until this point, the chapter has focused on relations between Great Powers as the main reason why wars of the twentieth century have been so destructive. We have looked at trends in the developments of Great Powers, at the scope and geographic distribution of conflict since 1890, and we have discussed the theories that try to explain why hegemony switches from one Great Power to another. We have noted that over time wars are becoming more local, that is, involving only local powers with less and less extra-regional activity, while the global institutions designed to reduce inter-state conflicts have grown in size and power. However, the time-frame of forty-five years may be too short to state that these constitute a definite trend. As we shall see in the section below on local wars, there was plenty of Great Power involvement in local wars in a time of supposed global peace, 1945–91. But the fact remains that, for more and more countries, wars are distant affairs as the zone of Stable Peace grows.

In the discussion of the intimate links between war and hegemonic change, it was noted that there is a dispute between those (following Lenin) who hold that wars are more likely in a time of global recession as states try to hold on to declining shares and expand into other states' areas of influence, and those who hold (like Goldstein, 1988, 1991) that states are more likely to engage in major wars when their economies will allow it, after a period of economic growth, general prosperity and government revenue surplus. In an effort to clarify this dispute, Goldstein (1988, 244–57; 1991) has engaged in a careful time-series analysis of 500 years of war and economic data. He first shows convincingly that there are waves or cycles in the various economic series of price, innovation, production and investment data. Using Granger causality, a procedure to determine which series leads (whether the economic trends lead those of war, or vice versa), he concludes that 'the severity of great power war correlates strongly with the long wave, leading prices by about one to five years' (Goldstein, 1988, 257). He defines the war/economic cycle sequence as upturn in production, upturn in capital investment, downturn in innovation, upturn in war severity and upturn in prices and downturn in wages.

The results of Goldstein's long-run analysis are very important for our chapter. We have consistently argued a historical-structural approach to the examination of the political geography of war. In our view, war is not a random event, nor is it determined by the unexpected actions of unpredictable leaders, though obviously the precise timing of the start of war will be decided by the actions of individuals. If war were random, there would be no evidence for trends, cycles or regularity since the beginning of the modern world-system about 500 years ago. Instead, we have both strong evidence for war cycles (Thompson, 1988), economic cycles (Goldstein, 1988) and war related to economic cycles (Goldstein, 1988). Recent work by Thompson (1992) indicates that long cycles can be identified all the way back to 1500 in the European countries, and that the pre-1800 cycles had more frequent fluctuations than later cycles because of the more varied pattern of war. In a sense, war and economic developments are mutually reinforcing.

There are certain periods during the 100-year-long cycle (or the approximately equal hegemonic cycle, as Hopkins and Wallerstein (1982) call it) that global war is more likely. This happens towards the end of the power deconcentration phase as challengers reduce the power gap between themselves and the erstwhile hegemon. The correspondence between the long-cycle (Modelski/Thompson) and world-systems (Wallerstein) perspectives is fairly clear. Over the past century, these have been successively (with the corresponding Kondratieff phases): Kondratieff IIIA 1890–1914 power deconcentration (US aspiring hegemony); Kondratieff IIIB 1914–45 global war (hegemonic victory); Kondratieff IVA 1945–70 US, world power (hegemonic maturity), and Kondratieff IVB 1970–2000? power de-legitimation (declining hegemony) (see Table 1.1). From the world-systems perspective, the past century has been the 'American Century' of a hegemonic rise and relative decline.

The world of 1890 differed a lot from the world of 1990. The most obvious difference was the increase of the number of members of the global system from about forty to about 160, mostly due to the independence of about 100 countries from the former European empires. But hidden beyond the obvious political-cartographic changes lie some major power, military and political developments. We have already noted the challenge to Britain by Germany after 1890 as both countries turned their attentions back to the European scene after the frantic race for colonies of the previous twenty years. The First World War was preceded by a succession of nine crises that propelled Europe to the brink of all-out conflict (Quigley, 1966, 218). By 1918, sheer exhaustion on the part of all global powers centrally involved in the 1914 crisis produced uncertainty in the hegemonic-succession race, though the extra-European powers that entered the war later – Japan and the United States – emerged with enhanced prestige in 1918. Though the US failed to take on the global role expected from its relative power and standing in the isolationist years of 1919–41, Japan's ambitions in East Asia and the Pacific were unaltered. Therefore, when the second round of global war began, it was still centred on Europe but now had an important Pacific dimension as Japan and the US vied for regional dominance.

The years since 1945 have been termed the 'Long Peace' (Gaddis, 1987). The term is only accurate if we consider peace between the Great Powers. Clearly, as can be seen from Figure 2.10, wars were common among regional and local powers as well as between small and large powers. US hegemony was unusual because unlike the previous hegemon, Britain, it was never global. Instead, about one-fifth of the countries of the world were removed from the American orbit by Soviet anti-systemic power and the allied arrangements in the waning days of World War II (Taylor, 1990). Recent post-revisionist writings on the Cold War (Cox, 1990; Kaldor, 1990; Wallerstein, 1991) view it as having served the needs of both superpowers in a sort of mutual complementarity between the capitalist and communist systems. Both sides needed the other to justify and regulate relations within its bloc. By raising the communist threat, the US managed to keep erstwhile and future competitors (Japan and Western Europe) in US-controlled economic, financial, military and political alliances, while the Soviet Union was allowed free rein in its sphere of influence. But after 1970, with the beginning of the global downturn, the superpowers engaged in an all-out race for friends and allies in the Third World (Nijman, 1991). The result was a sharp escalation in the levels of arms sales and in the resulting scale of destruction in the subsequent conflicts. We can conclude that our 100-year time-span shows the same development as Kennedy (1987, 536–7) notes: that 'the history of international affairs over the past five centuries has all too frequently been a history of warfare, or at least of preparation for warfare – both of which consume resources which societies might use for other "goods", whether public or private.'

Global wars

World Wars I and II overshadow all other fighting during the first half of the twentieth century in terms of geographic scope, deaths and political consequences. They have been accurately lumped together as the last round in the recurrent bloody leadership selection game in the long cycle analysis of war. In this view, there was one long-drawn-out challenge, particularly by Germany and later by Japan, to the established distribution of influence. In fact, it is far from clear that these countries aspired to become alternative hegemonic powers. Even in his wilder fantasies, Hitler remained primarily though not exclusively interested in Europe, while the Japanese confined their bid for enhanced power primarily to East and South East Asia. But the two countries challenged the status quo at important points (e.g., naval supremacy, restoration of the former European position after 1918, division of influence in China), and refused to be accommodated easily by the declining (UK) and ascending (US) hegemonic powers.

The conventional view of the outbreak of World War I is very much in line with a structuralist perspective. Various parties, particularly in the European state-system, were tied to each other by 1914 in such ways that any little disturbance might have tipped the scales towards war. Various earlier crises had just been con-

tained during the preceding years. The murder of the successor to the Austro-Hungarian throne and his wife in Sarajevo finally surpassed the capacity of the European state-system to accommodate its political differences. Contemporaries were less surprised by the outbreak of war itself than by its length and severity once it had begun. There was a strong feeling in August 1914 that it would all be over by Christmas. The First World War was a classic tragedy in two senses. The mechanics of its outbreak were apparently beyond human control. Once underway, these mechanics resulted in an unmitigated disaster and had consequences far beyond the Balkans, since the second round of fighting in 1939–45 was a direct result of the unresolved global succession of the Great War (Taylor, 1961).

The inter-war years were indeed an intermission of an unresolved dispute. To paraphrase Clausewitz, World War II was a continuation of the First World War with similar means. There is a different view that considers World War II to have been essentially Hitler's war. With different leadership, Germany might have refrained from challenging the status quo in a violent manner. Major powers were willing to make considerable concessions to German grievances, as the events before World War II showed. The München agreement 1938 showed the extremes to which western leaders were willing to go, allowing the Germans to fulfil their Czechoslovak demands. Thus, it was Hitler who pushed the state-system over the brink towards war, and the Japanese assault on the US fleet in Pearl Harbor and elsewhere followed in 1941 (Hinsley, 1987).

These arguments are based on counterfactuals and it is not easy to choose between them. With respect to world-system analysis, two issues are particularly important. The end of World War I did not in fact result in the global political leadership of the US that could have occurred, based on its economic position at the time. On the contrary, the US took a detached, though not an isolationist, view of the European state-system. Europe still remained the political centre of the world, despite the ascending position of new arenas at the western shore of the Pacific. World War II ended with the US in a position of preponderant power, but by then the US was more willing to play the role of political leader that its power position warranted. At the same time its major ideological challenger, the USSR, had acquired a strong and widely-acclaimed position at world level. This occurred through its own effort and through lavish Allied assistance, despite incredible failures of judgement, wasteful forced development, army and other purges, and the underestimating of the chances of a German offensive in 1941. Europe became a vital but dependent arena in the rivalry between the preponderant power and its junior partner in superpower status. The US was never fully able to acquire a hegemonic position in its legitimation, since worldwide American influence was denied by Stalin and his successor Soviet leadership from the outset (Cox, 1990, Kaldor, 1990).

World Wars I and II were fought between geographically-similar rival camps, but the contenders differed to a considerable extent. In wars of this magnitude, the coalition with the greatest resources eventually wins. Both wars were concentrated in Europe, though the Japanese challenge produced a major Asian twist to the global struggle. The Second World War had more direct effects on individuals far from the battlefield as war became more indiscriminately the affair of societies as a whole. For the US, however, the actual war was still remote. It was mediated through images, pictures and reports and the experience of loved ones elsewhere at risk. In London during the Blitz, in much of Germany during a major portion of the war period, and also in Japan during the last stages of the war, massive air strikes of civilian targets and arms manufacturing plants brought tangible evidence of war to the domestic front (Hewitt, 1983). Even this destruction paled into insignificance compared to the human suffering along the gigantic fronts in Eastern Europe and the associated killings by both sides.

War as collective human action became incredibly massive, purely grim and overpowered by hardware and its technology. World War I had massive killings of those who filled the trenches as its icon, particularly as these troops were forced into the offensive. As Quigley (1966) notes, World War I essentially ended in stalemate because military defences were able to withstand any level of assault. The British in 1917 at Passchendaele (southern Belgium) launched five

tons of artillery shells per yard of the 11-mile battle-front; this incredible firepower did not produce a breakthrough, but instead the British suffered over 400,000 casualties in the ensuing assault. By 1945, offensive weapons again had the upper hand with the atomic bombing of Hiroshima and Nagasaki. Only the United States of the six Great Powers emerged from the second round of global fighting with its power enhanced and its global position strengthened. The result was confirmation of Kennedy's (1987, xxiii) generalization that 'there is a very strong correlation between the eventual outcome of the *major coalition wars* [his italics] for European or global mastery, and the amount of productive resources mobilized by each side . . . A lengthy grinding war eventually turns into a test of the relative capacities of each coalition.'

World War II is epitomized by huge tank battles, scenes of slaughter of infantry and tank crews alike in West Russia and North Africa, plus naval warfare on an unprecedented scale with the marines, ship crews and pilots on the carriers at risk, all topped off with the atomic bombs on Japanese cities. These were the ultimate incidents of organized warfare where any human scale was definitively superseded. This trend at the upper levels of technological advancement has continued since 1945. Military historians (Keegan, 1976; van Crefeld, 1991) argue that humans can no longer sustain the strains of this type of warfare for any length of time. This applies to the military machine itself. The scale of this type of warfare makes other parts of society more and more ephemeral.

World War I ended in an Allied victory, after the Russian part of the victorious alliance had collapsed as a result of the tensions of the war. The Versailles Treaty, after a series of negotiations in and around Paris, was vindictive to the defeated, particularly Germany. The new Weimar regime there inherited the country's international position created by the war, and this was a factor in its eventual replacement by the Nazis. This vindictiveness was mainly the result of French effort, since that country felt most immediately threatened by a strong Germany. Versailles undoubtedly helped prepare the ground for another round of war in another generation, in spite of public opinion at large in western Europe being shocked out of any complacency in this regard by the vivid memory of the 1914–18 horrors.

World War II, contrary to the tradition in global war, did not result in an overall settlement. There was a separate, speedy treaty with Japan. Peace treaties were negotiated with most of the minor European powers at Hitler's side in a co-ordinated way soon after the war. Peace with Austria was finally made after Stalin had died, during a lull in East–West tension in 1955. The victorious former allies could not, however, agree on the peace terms for Germany, as they had become inextricably locked in their own mutual disputes. Germany was eventually divided in two states after 1948. Its western part became a very successful economic power in the post-war bi-polar world. Germany's remaining dependency in military and security policy was less and less proportionate to its economic power and dominant position in the EC after 1970.

After Gorbachev had effectively renounced the claims of the Soviet Union on the hegemony of Eastern Europe in 1989, but before the actual demise of that superpower, the borders between East and West Germany were reopened, the removal of the Berlin Wall being the core symbol of change. A strong dynamic towards German unity ensued within the two halves of the divided country. The West German government crested the waves at the same time, encouraging further progress and taking a decisive lead to procure international recognition of the reunification process. The so-called 2 (East and West Germany) + 4 the USSR, the UK, France and the US) talks ended in agreement on German unification, as well as the end of occupational rights that had long been notional in many respects. The results of the 2 + 4 talks were then accepted in a solemn all-European gathering (and including the Americans and Canadians) in Paris in November 1990 (Leurdijk, 1991). A few weeks later the first all-German elections occurred and a formerly-defeated power finally occupied the centre-stage and directed the principal negotiations for the ultimate peace settlement of the fifth global war. The challenge of the Soviet Union, fruit of the first stage of the war in 1918, was abolished by self-defeat in the final stages of the endgame of the global war of the twentieth century.

Contemporaries called World War I 'The Great War', expressing its extraordinary character quite unlike anything during the nineteenth century, which had been very peaceful in terms of major-power war. In its aftermath, the war was damned by winners and defeated powers alike. The inter-war years saw a very strong, widespread, anti-war sentiment, while at the same time sizeable population groups were seized by extreme and radical-right political forces that capitalized on the mood of anomie and dislocation induced first by the Great War and, later, by the Great Depression.

World War II is remembered quite differently in different countries. Terkel (1984) called his oral history of the period, 'The Good War'. Within the western populations on the Allied side there was a widespread reluctance to go to war, but at the same time a widely-shared conviction that the Allies represented the right side. France was torn between broad support for the collaborationist Vichy regime and minority resistance divided into a communist and a non-communist wing. The communists had only wholeheartedly entered the resistance in 1941 but then put up a fierce fight. After two generations, French public opinion is still confused over the war record of that country. So are the Japanese, who only reluctantly dwell upon the wartime atrocities of their army in China and South East Asia, while at the same time denouncing the use of nuclear arms and mourning their compatriots who died in Hiroshima and Nagasaki. The most insistent, widespread and deeply emotional debate on the war has been going on in West Germany – not in the former East Germany, where any continuity with the Nazi past was denied. Memories surfaced again in the rightly famous *Historikerstreit*, an unusually fierce argument among historians over the Hitler regime's extermination of the Jews, and its comparability with and causation by Stalin's murderous repression in the Soviet Union (Augstein, 1987; Bullock, 1992).

These memories are not merely fictions of the mind. The facts of the global war of 1914–45 are still visible in the landscape, in the age–sex structure of the populations most involved in the wars, and in innocent social customs, like pub closing-hours in Britain. Smaller age cohorts of those born during the war years and extremely unequal numbers of males and females for cohorts who were then in their twenties, particularly in both Germanies and in Russia/ USSR, still show up on population maps of Europe (Decroly and Vanlaer, 1991, 104–6). The twentieth-century global war had further consequences. Its second stage in 1939–45 irreparably scarred Eastern Europe, owing in particular to the deliberate mass murder of Jews and others by the Germans and their allies, the forced boundary changes (Poland and adjoining areas), and the enforced or induced mass migrations, particularly of Germans. In the rest of the world, it sparked the end of western colonialism. The colonial masters had been shown to be weak in war, particularly in South East Asia. Within western societies, both stages of the global war spurred wider social and political participation. After 1914–18, participation was widened by an extension of suffrage to women and to others previously denied electoral rights, while the 1939–45 stage instituted the welfare state in western countries. The consequences of the terrible conflict were not only in shaping the post-war geopolitical hierarchy, but in affecting deeply the internal conditions in participating and spectating countries.

Local wars

As we have noted, the political structure of the contemporary world-system is essentially a product of the European experiences. Tilly (1990) shows how the creation and development of the European state-system was strongly influenced by war-making. The central argument was that European states started off from different positions of wealth, military power and foreign interests, but that the military and related territorial competition between them drove them all in the same general direction – the ultimate dominance of the national state. Rasler and Thompson (1989) develop this argument to show the intimate relationship between state-making, war-making and the effect of cumulative national debts. As the European powers de-colonized in the 1950s and 1960s, they left little accumulated capital behind them, but instead 'bequeathed to their successor states military forces drawn from

and modelled on the repressive forces they had previously established to maintain their own local administrations' (Tilly, 1990, 199). In a study of African state-making after 1960, Kirby and Ward (1991) show strong evidence for the central relationships of Tilly's European state-formation model, in which state formation depends on the interplay of capital accumulation and political consolidation, urbanization plays a key role, capital accumulation is reduced under coercive regimes, and the overall process has distinctive geographic forms. Strong states with large militaries became the norm in the post-colonial Third World.

There is one major difference between European and Third World state formation. In Europe since 1500, large armies were formed to protect national territory from attack and to launch aggression on neighbouring and colonial peoples. Since about 1900, however, an increasing proportion of wars have been within states in the form of civil strife. Recomputing Small and Singer's (1982) data on war deaths, Tilly (1990, 201) shows how since 1959, the numbers killed per year in civil wars has exceeded by a substantial margin the numbers killed in inter-state disputes or extra-systemic conflicts (usually colonial wars). Even more clear-cut is the finding that civil wars constituted 6.6 per cent of wars in the period 1937–47, 15.9 per cent in 1948–58, 59.4 per cent in 1959–69 and 89.0 per cent between 1970 and 1980. Tilly (1990, 203) is able to conclude that: 'The continued rise of war couples with a fixation on international boundaries. With a few significant exceptions, military conquests across borders has ended, states have ceased fighting each other over disputed territories, and border forces have shifted their efforts from defense against direct attack toward control of infiltration. Armies . . . concentrate increasingly on repression of civilian populations.'

Another element of the ways in which local inter-state and civil wars are developing is the disparity in the level of attention that they receive from outside powers. Since 1945, Great Powers have been involved directly in civil conflicts (the US in Korea and Vietnam, Russia/ Soviet Union in Afghanistan) as well as supporting different sides in both civil (US and Russia/ Soviet Union in the Angolan and Nicaraguan conflicts) and local inter-state wars, (the tangled Israeli/Arab conflicts since 1967). Other civil and local inter-state wars have mostly been free of outside direct intervention, such as the Nigerian civil war in the late 1960s, the Pakistani/Indian wars, the Iran/Iraq war of the 1980s and the ongoing conflicts in Sri Lanka, Peru and the former multi-national states of Yugoslavia and the Soviet Union.

What distinguishes the level of commitment by the outside powers is the perceived importance of the country in their overall strategic plans (O'Loughlin, 1989). Thus, the former Soviet Union was most committed to maintaining stability in the form of pro-Soviet regimes on the borders of that country (Czechoslovakia, Hungary and Afghanistan), while the US was much more actively involved in Central American conflicts than one would have expected from the size and importance of the countries in the region. Of the approximately 350 military interventions by the superpowers in the period 1946–85, most happened in the same circular belt of states on the margins of the Eurasian landmass, in the so-called Rimland. The large majority of the interventions were carried out by the US as part of the Containment strategy (O'-Loughlin, 1987). A serious upsurge in the level of commitment by both sides occurred after the Brezhnev Doctrine of supporting local communist movements was promulgated by the Soviet Union in the 1960s.

The superpowers have not been the instigators of local conflicts. Instead, they typically became involved either when help was requested by one of the warring parties, or by offering assistance to a perceived ally. These assistance offers often backfired as changing fortunes pushed allies out of power and installed former enemies. However, the misfortunes of US and Soviet former allies and the subsequent switching of sides in the Cold War in Iran, Cuba, Nicaragua, Egypt, Somalia, Ethiopia, Afghanistan and China did not deter the superpowers from further strenuous efforts on behalf of client states. The end of the Cold War in 1989–91 was marked by and caused by the withdrawal of the former Soviet Union from Third World commitments, especially in Cuba, Vietnam, Ethiopia, Angola, Afghanistan and Nicaragua.

We have already classified wars according to different criteria. It is important to note that most wars have multiple causes – usually some combination of ethnic, ideological, economic and territorial strife. Territorial disputes can be identified as a root cause of many contemporary conflicts (Diehl and Goertz, 1991). The best known are claims to the same territory by competing national groups, as in Palestine between Israelis and Arabs and in Sri Lanka between Singhalese and Tamils. Goertz and Diehl (1992) have counted sixty-seven current territorial conflicts, mostly in the Third World, and since 1890, they can identify 487 territorial conflicts. In trying to account for the large number of disputes of this nature, Goertz and Diehl (1992) attribute it to both the intrinsic (the resources in the disputed region) and relational (the geographic positioning of the disputed region in relation to the warring parties) importance of territory. An example of the former is the oilfields of Kuwait, and of the latter, the strategically-placed Golan Heights between Syria and Israel. In looking at the temporal distribution of territorial changes since 1815, we identify two peaks, one around 1870–90 corresponding to European colonial aggrandizement, and another about 1960 as the Third World states gained independence from the colonial powers.

There are three types of dispute that disproportionately generate territorial conflicts (Goertz and Diehl, 1992). State-formation conflicts happen as the population of a region is fighting a war of independence as well as (frequently subsequent) internal wars in a state as competing nationalisms fight for regional autonomy. An example would be the Israeli fight for the creation of the Jewish state as well as the post-independence conflict with the Arab residents of Palestine and the adjoining Arab states. Africa, especially, has been wracked with this form of conflict since the independence wave of the 1960s. A second type is inter-state conflict between established members of the international system, usually over a region along their joint border. The classic example of such a dispute is the Iran–Iraq war over the positioning of their common boundary. A third type of territorial conflict is called 'recurring conflict' by Goertz and Diehl (1992) in order to focus attention, not so much on past wars, but on the likelihood of future ones. As Erich Weede (1973, 87) notes, 'the history of war and peace is largely identical with the history of territorial changes as results of wars and causes of the next war'. Partition, frequently seen as a (temporary) geographical solution to vexing claims, rarely produces permanent peace but only a inter-bellum breathing-space (Waterman, 1984). The long history of recurring war in Western Europe, and most recently in the Middle East, between the same participants or their successor states should give pause to any thoughts of a general era of global peace at the end of the late twentieth century. Looking at the war record since 1815, almost 40 per cent of territorial changes have been followed by another conflict between the same parties within a generation of the earlier dispute (Goertz and Diehl, 1992, 109).

The term 'local wars' seems to suggest limited and bounded conflicts, but there is growing evidence that wars have a tendency to spread. Siverson and Starr (1991) use what is called an 'opportunity and willingness framework' to examine the evidence for the diffusion of conflict and the reasons why it should happen. Opportunity refers to the geographic and political environments in which the conflict occurs, and willingness to the choices that must be made by other states to get involved or to remain on the sidelines. Of the environmental factors, geographic proximity in the form of sharing borders has been shown to be very important at both the global and regional scales. In Africa, O'Loughlin and Anselin (1991) have identified regional conflict formations consisting of countries intimately connected to each other's disputes, and have shown that a state's geographic position is an important factor in the level of its conflict involvement, in addition to the usual predictors of military expenditures and size. Inter-state alliance behaviour combines with the scale of geographic proximity to affect the rate of infectious diffusion of conflict. However, a clear distinction should be made between civil and international conflict here. There is much less propensity for civil wars to diffuse in a manner that involves neighbouring states than is the case for inter-state disputes (Anselin and O'Loughlin, 1992). One of the informal rules of international

relations ('the enemy of my enemy is my friend') determines who allies with whom and who is likely to join an ongoing conflict. In the North East African conflict-formation of the 1970s and 1980s, Egypt, Ethiopia, Sudan, Libya and Somalia developed a staggering array of support links and alliances between each other and the internal opponents of the regimes in power.

If we know that wars are likely to draw a crowd, the question remains as to where wars occur in the first place. We have consistently argued in this chapter that war is very unlikely to occur across international boundaries in the zones of Stable Peace in Western Europe, North America and Australasia. It is possible to anticipate that the next countries to be incorporated into this expanding zone will be the newly industrializing countries of East Asia and perhaps the countries of South America. Because there seems to be a correlation between the wealth of a region, its level and strength of democracy, and the absence of international conflict, we can expect that stable peace will proceed as part and parcel of a process of democratization, which seems also to have the character of geographic diffusion (Starr, 1991). Increasingly, the locations of wars are the locations of poverty. There is a very complicated web of connections between the relative poverty of a country, its rate of militarization, its dependence on outside benefactors and military allies, and its rate of internal and external conflict, so that very poor countries like Ethiopia, Cambodia, El Salvador and Sri Lanka are the locations of current conflicts in the late twentieth century.

One of the most enduring geographic notions in international politics is that of a 'crush zone' or a 'shatterbelt' (O'Sullivan, 1986). This is a region not controlled by any of the Great Powers, but slotted into the interstices of their spheres of influence. Eastern Europe, between Germany and Russia/Soviet Union, was an early example, and in the past twenty years three contemporary shatterbelts have been identified in the Middle East, South East Asia and sub-Saharan Africa (Cohen, 1982, 1991). Shatterbelts are complex regions of many states, nations, cultures and economic activities whose resources and locations attract the attention of outside Great Powers trying to bring them into their orbits. The expectations deriving from internal cultural, political and economic complexity combined with external involvement is that the shatterbelts will be the locations of disproportionate amounts of conflict.

Using the Correlates of War dataset once again (Small and Singer, 1982), van der Wusten and Nierop (1990) find that indeed shatterbelts have the highest levels of violence of both the civil and inter-state kinds, as well as militarized disputes. Looking at post-1945 interventions by extra-regional powers, the authors also conclude that intervention is a shatterbelt phenomenon, but that most of the intervening powers are not Great Powers as might be expected from the concept, but instead are smaller powers. Recent examples of such kinds of intervenors are Saudi Arabia in the Horn of Africa wars, Iran in the Afghan conflict, India in the Sri Lankan civil war, Libya in the Chadian civil war, Zaïre in the Angolan conflict, and Tanzania in the Ugandan civil war. In all of these cases, the intervention was in a bordering country. There are very good reasons why small states intervene in the affairs of neighbours and fight with them. As Boulding (1962) and O'Sullivan (1986) show, the 'loss of strength gradient' operates even in the post-war years of mass communication and ease of transport. Constraints in the ability to move large amounts of troops and *matériel* affect all but a tiny handful of the members of the international system and confine the interventions to immediate or near-immediate neighbours. Of course, events and developments in these states will be of most immediate concern to the intervenor.

In trying to anticipate the development of local wars, we must be cognizant of some important parallel contemporary trends. First, there is a sorting out of the world's regions. As we noted, for reasons of capability, most states in the international system interact intensively only with their neighbours and with a couple of very strong extra-regional powers. The maps in Anselin and O'Loughlin (1992) illustrate this situation for a variety of African states. This local interaction is also economic. Maps of groupings of states based on their shared international memberships show increasing clarity of regionalization in the world-system (Nierop, 1989), and trade regions are also well defined, though now it is increasingly

according to the trade shares of a major economic power – Germany, Japan or the United States (Nierop and de Vos, 1988). A second development is that the regions are dynamic in nature. A clear example is the dramatic changes in formerly homogeneous Eastern Europe after 1989. Economic, cultural and geopolitical regions will not necessarily be coincidental.

Until now the international community has been unwilling to intervene as a world body in civil conflicts. There are written rules in the charters of the Organization of African Unity and the United Nations that intervention by other member states acting collectively would only be allowed with the consent of the government of the country under scrutiny. Borders left by the departing European colonial powers were considered inviolate and civil wars to be an internal matter, though individual members of the international community had no hesitation in intervention. Since so many of these Third World conflicts became struggles for pawns on the chessboard of superpower competition, local conflicts developed global implications. Regional organizations have sent peace-keeping forces into neighbouring states to maintain order, such as the Arab League forces in Lebanon in 1975 and West African forces in Liberia in 1990. Since 1989 and the end of the Cold War, the United Nations has stationed peace-keeping forces in conflicts along the Iraq–Kuwait border and along the Iran–Iraq border, as well as in countries with major civil conflicts (Afghanistan, El Salvador, Yugoslavia, Namibia, Cambodia). These operations are in addition to ongoing peace-keeping ones in Cyprus, Lebanon, and along the Israeli–Syrian and Israeli–Egyptian borders. How long a united world organization will retain the willingness to send and pay for peace-keeping forces in the face of growing conflict in formerly peaceful areas of the ex-Soviet Union remains to be determined.

Conclusion

It is easy to be optimistic about world order and general peace in 1992. The causes of war seem to be confined to ethno-nationalist disputes, though these are globally distributed from Northern

Ireland to Bosnia-Herzegovina to Armenia–Georgia to Israel–Palestinians to Kurds–Iraq to Kashmir to South Africa to Sudan and to Sri Lanka, among others. The Iraqi incursion into Kuwait in August 1990 appears more and more anachronistic as the event recedes into the mist of history. International borders are unlikely to shift as frequently as they have in the past as a result of war; instead, we can expect more internal partitioning of states along ethnic lines, as we have seen in the former states of Yugoslavia and the Soviet Union. The United Nations is wary of over-committing its peace-keeping resources as more and more protagonists accept the mediating efforts of the world body. The International Court of Justice in The Hague (Netherlands) is demonstrating a renewed authority, as states appeal to it to redress perceived grievances with neighbours and other states rather than resorting to force of arms. The future looks bright indeed.

To try to understand why such a happy prospect of general peace lies in store for us, Francis Fukuyama (1992) has offered a view of historical evolution predicated on the Hegelian notion of progress. History in this view is not the series of events that comprise the stuff of textbooks and courses. Instead, history is a single, coherent evolutionary process towards human progress. Progress, for Fukuyama, is not just economic improvement and increased life spans, but it is a gradual shift towards the central principles of liberal democracy, especially equality, freedom, individual human rights and liberties, and the construction of institutions that guarantee those principles. 1989, the year of the revolutions in the former Eastern Europe, marked the end of an era in this view. The near-total rejection of communism by the populations of the six countries put paid to any notion of a continued global struggle between liberal democracy and communism. Communism was only the last of the rival ideologies to fall before the march of liberal democracy, according to Fukuyama. In this century alone, the ideologies of Fascism and hereditary monarchy have suffered a similar fate. At the end of the twentieth century, Fukuyama sees a remarkable global consensus on the value of liberal democracy, spanning the globe across widely different regions and cultures. In terms of world politics, this development means that

'liberal democracy replaces an irrational desire to be recognized as greater than others with a rational desire to be recognized as equal. A world made up of liberal democracies, then, should have less incentive for war, since all nations would recognize one another's legitimacy' (Fukuyama, 1992, xx).

The Fukuyama belief about the peaceful intentions of democratic states fits neatly into one of the few general principles that seem to govern international relations. In their review of the discipline, Wayman and Singer (1991, 248) conclude that 'at least three dyadic relationships appear strong enough that some scholars have argued that they are deterministic: democratic dyads do not fight each other; overwhelming preponderance guarantees peace, and power transitions between the leading powers lead to war.' Some have even argued that the only law of international relations is that democracies do not fight each other. Lake (1992) found that rule to hold, and further that democratic states are likely to prevail in wars with autocratic states because they have greater resources in security (they spend less in internal protection), enjoy greater societal support for their policies and tend to form counter-coalitions against expansionist autocracies. The Gulf War of 1991 is a perfect example of the argument. Even more deterministic is Modelski's (1990) argument that democratic states will generally be more successful in international politics since their economies prosper in free-enterprise operations. In 'global endgames', democratic and more open societies will prevail. Consequently, he trumpets Immanuel Kant's claim of two centuries ago that democracy is the road to world peace. Since 'the United States is best equipped for forwarding the processes of democracy and the agenda of democracy' (Modelski, 1990, 249), the US will have two terms of global leadership.

In a world of only one ideology, according to Fukuyama (1992, 11), we can expect conflicts to be less intense. He attributes the enormous devastation of the Russian and Chinese revolutions, as well as the Second World War, to a magnified form of brutality not seen since the European wars of religion of the sixteenth century. Not just territory and resources were at stake – matters on which opponents might compromise – but value systems and ways of life of entire populations were endangered. In the terms that we used earlier in this chapter, we can expect 'stable peace' to grow as regions of peaceful and democratic states coalesce in the Fukuyama world vision.

And yet and yet, uncertainties abound about a peaceful world in development. First, the process of state-building is by no means over, as potential conflicts of two types continue to loom on the horizon. There are still forty-three dependent territories in the world (Goertz and Diehl, 1992, 58), and although not all of them have active independence movements nor any clear desire for a change in their political status, there remains conflict potential of three types – internal, external with the controlling power, and external with neighbours. Puerto Rico is an example of the first, New Caledonia of the second, and the Malvinas/Falkland Islands is of the third kind. A second type of state-building conflict is the continuing struggle for ethno-territorial autonomy by dozens of groups around the world. In early 1992, over thirty active conflicts of this type could be identified. The nation-state, that nineteenth-century model of the ideal political arrangement, is on the one hand still very much alive as more and more are created, but on the other hand the state-system itself is being undermined by global economic forces and by a gradual waning of sovereignty as a result of the growth of international organizations, especially in Europe. Peering into a crystal ball, we can equally see a world of 250 separate and often antagonistic states by the middle of the next century, or alternatively a world of fewer than a dozen multi-country groups, within which state roles are confined to local services.

A second reason why the world may not resemble Fukuyama's peaceful order is the continued presence of a counter-ideology. Fukuyama dismisses religion as a threat to the materialist beliefs on which western liberal democracy is based. Though he is scathing in his comments on the emptiness of the materialist philosophy, he nevertheless believes that it will triumph. Islamic fundamentalism as a counter-ideology to the western world-view is limited in Fukuyama's vision to certain regions, and any contest between 'Jihad and McWorld' will be decidedly in favour of the latter (Barber, 1992). This view is not shared by the world-system theorists Taylor (1992) and Wallerstein (1991). In their perspective, there has increasingly been a rejection of the European universalistic culture by the Third World. Taylor views Saddam Hussein's popularity among the populations of some Islamic countries during the Gulf War as a symbol of this rejection, despite the support of the governments of these countries for the US-led coalition. Wallerstein evaluates the 'geocultural' rejection by parts of the Third World of the Europe-based universalistic ideology first propagated in the nineteenth century as an 'anti-systemic movement'. This movement rejects western notions of how economies and polities should be structured and increasingly challenges the Hegelian definitions of progress as championed by Fukuyama.

There are three possible outcomes in a world with two ideologies, western liberal democracy aligned with capitalism and an anti-systemic Third World (mostly Islamic-based) rejection of that model. Through the power of mass communications, multinational corporations and the aggressive promotion of Western (mostly US) cultural values and products, most observers believe that McWorld will triumph, so that in 100 years Bangkok will resemble Boston, Birmingham and Berlin. The emptiness of the western materialist vision, similar to Oscar Wilde's cynic who knows the price of everything and the value of nothing, may be its Achilles' heel. A powerful ideological appeal based on emotion and heartfelt beliefs may turn back the seemingly-irreversible tide of western cultural, economic and political beliefs. More likely, though, is a co-existence of the two orders, like the bi-polar world that existed between capitalism and communism between 1945 and 1990. Whether this co-existence will be competitive, conflictual, isolated or mutually respectful is still unclear.

And what of the future for the Great Powers? As we have seen, by far the most destructive conflicts are those with two or more Great Powers. The United States has certainly declined in relative strength from its apogee in the late 1940s, but after the demise of the Soviet Union and in the absence of any obvious military or political challenger in the late twentieth century, this country is the only Great Power according to the traditional definitions. O'Loughlin (1992) envisions as the most likely global scenario a world in which the US will continue to pursue its political interests unilaterally at the cost of inter-national co-operation. By contrast, Goldstein and Rapkin (1991) think that slow hegemonic decline for the US combined with Great Power co-operation under lingering US leadership, is the most plausible scenario for the future. They do not believe that the historical scenarios will be repeated in the form of US hegemonic revival, rapid hegemonic decline leading to anarchy or hegemonic war, or a peaceful transition to Japanese hegemony. Recognizing the costs of hegemony, the US in this most plausible scenario will be willing to share global leadership with other Great Powers of similar cultural, political and economic beliefs.

Clearly, hegemonic war of the military kind fought periodically in the past half-millennium would devastate the globe, including any poss-ible victor. As a result, Modelski (1987a) antici-pates that any future global war is not likely to be a military one, but one fought with political and economic weapons, including tariffs on goods, subsidies, boycotts, etc. Goldstein (1988) antici-pated a global war at the end of the current economic-political cycle in about 2030, but be-cause of the global sea-changes since 1989, he presents a much less grim picture in a later scenario-building article (Goldstein and Rapkin, 1991). Based on power transition from a hege-mon to a challenger – the core of both the long cycle and world-system schools – global war could be expected again in about 2030–50. In that case, the world has about forty years to build the mechanisms that would undermine the historical global war institution as a way of reorganizing world leadership and power transitions (Thomp-son, 1988).

The method of historical parallels is the basis for the prediction of global war in the next century. A typical example is Kahler's (1979) analogy between the world in the late 1970s and the world before 1914, with the obvious impli-cation that the precarious balance of power and cautious relations between the rival blocs could break down and devolve to war, just as happened in the summer of 1914. But the dramatic changes in Eastern Europe since 1989 should have put paid to any notion of historical inevitability. The jury will be out for a long time on the 'end of history' thesis of Fukuyama. And there are always new issues and battlegrounds, including access to global resources, especially oil. This critical resource was at the core of the American mobilization of the Western bloc against the Iraqi occupation of Kuwait in 1990 – the first and almost certainly not the last post-Cold War crisis. Not only was the twentieth century the 'century of oil', but the twenty-first century appears to be too (Yergin, 1991).

It would be nice to end this long chapter on a cheerful tone. Those of us in the West could easily close our eyes to the structural violence endemic in much of the Third World and to the contemporary ethno-territorial conflicts in all continents. These appear anachronistic in our world of plenty, freedom and tolerance. Yet the levels of suffering in many parts of the world demand our constant concern, and not just when the television pictures of some atrociously vio-lent act or devastating famine command our brief attention. For previous generations, long periods of general peace, such as in the early nineteenth century, were at the time considered a break with a past of endemic violence degenerating period-ically into all-out conflict. The 'war to end all wars' of 1914–18 only marked the beginning of a much more devastating conflict. It is to be hoped that this era of global peace will deepen and widen to encompass all parts of the world so that the admonition of Siegfried Sassoon, a survivor of the World War I battlefields, will enter our collective conscience and produce its desired result for peoples everywhere.

> You smug-faced crowds with kindling eye,
> Who cheer when soldier lads march by.
> Sneak home and pray you'll never know
> The hell where youth and laughter go.
> Siegfried Sassoon (1886–1967)
> *Suicide in the Trenches* (1919)

Acknowledgements

Work on this chapter by John O'Loughlin was supported by a US National Science Foundation grant on the spatial analysis of international conflict and co-operation. The authors would like to thank students at the University of Colorado for comments that improved the content substantially. The encouragement and patience of Peter Taylor and Iain Stevenson helped a lot at critical junctures.

References

Agnew, J. A., 1983, An excess of 'national exceptionalism': towards a new political geography of American foreign policy, *Political Geography Quarterly*, 2(2): 151–66.

Anselin, L. and O'Loughlin, J., 1992, Geography of international conflict and cooperation: spatial dependence and regional context in Africa. In M. D. Ward (ed.), *The new geopolitics*, Gordon and Breach, New York, pp. 39–75.

Augstein, R. (ed.), 1987, *Die Dokumentation der Kontroverse um die Einzigartigkeit der national-sozialistischen Judenvernichtung*, Piper, Munich.

Barber, B. R., 1992, Jihad vs. McWorld, *Atlantic Monthly*, March, 53–63.

Bergesen, A., 1985, Cycles of war in the reproduction of the world economy. In P. M. Johnson and W. R. Thompson, eds., *Rhythms in politics and economics*, Praeger, New York, pp. 313–31.

Boulding, K. E., 1962, *Conflict and defense*, Harper and Row, New York.

Boulding, K. E., 1977, Twelve friendly quarrels with Johan Galtung, *Journal of Peace Research*, 14(1): 75–86.

Boulding, K. E., 1978, *Stable peace*, University of Texas Press, Austin, TX.

Brzezinski, Z., 1986, *Game plan: a geostrategic framework for the conduct of the US–Soviet contest*, Atlantic Monthly Press, Boston.

Bueno de Mesquita, B., 1981, *The war trap*, Yale University Press, New Haven, CT.

Bullock, A., 1992, *Hitler and Stalin: parallel lives*, Alfred Knopf, New York.

Buzan, B., 1983, *People, states and fear: the national security dilemma in international relations*, Wheatsheaf Books, Brighton.

Chase-Dunn, C., 1989, *Global capitalism: structures of the world economy*, Basil Blackwell, Oxford.

Choucri, N. and R. C. North, 1987, *Nations in conflict: national growth and international violence*, W. H. Freeman, San Francisco.

Cohen, S. B., 1982, A new map of geopolitical equilibrium: a developmental approach, *Political Geography Quarterly*, 1(3): 223–42.

Cohen, S. B., 1991, The emerging world map of peace.

In N. Kliot and S. Waterman, eds., *The political geography of conflict and peace*, Belhaven Press, London, pp. 18–36.

Cox, M., 1990, From the Truman Doctrine to the second superpower detente: the rise and fall of the Cold War, *Journal of Peace Research*, 27(1): 25–41.

Decoly, M. and Vanlaer, J., 1991, *Atlas de la population européen*, Editions Université de Bruxelles, Brussels.

Dessler, D., 1991, Beyond correlations: toward a causal theory of war, *International Studies Quarterly*, 35(3): 337–55.

Deutsch, K. W., 1957, *Political community in the North Atlantic area: international organisation in the light of historical experience*, Princeton University Press, Princeton, NJ.

Diehl, P. F. and G. Goertz, 1991, Interstate conflict over exchanges of homeland territory, 1816–1980, *Political Geography Quarterly* 10(4): 342–55.

Elliott, M., 1991, America: a better yesterday, *Economist*, October 25.

Faber, J., H. Houweling and J. Siccama, 1984, Diffusion of war: some theoretical considerations and empirical evidence, *Journal of Peace Research*, 21(4): 278–88.

Fox, A.B., 1959, *The power of small states: diplomacy in World War II*, University of Chicago Press, Chicago.

Fukuyama, F., 1992, *The end of history and the last man*, Free Press, New York.

Gaddis, J. L., 1982, *Strategies of containment: a critical appraisal of postwar American national security policy*, Oxford University Press, New York.

Gaddis, J. L., 1987, *The long peace: inquiries into the history of the Cold War*, Oxford University Press, New York.

Galtung, J., 1985, Global conflict formations: present developments and future directions. In P. Wallensteen, J. Galtung and C. Portales, eds., *Global militarization*, Westview Press, Boulder, CO., pp. 23–74.

Galtung, J., 1987, Only one friendly quarrel with Kenneth Boulding, *Journal of Peace Research*, 24(2): 199–203.

Goertz, G. and Diehl, P. F., 1992, *Territorial changes and international conflict*, Routledge, New York.

Goldstein, J. S., 1988, *Long cycles: war and prosperity in the modern age*, Yale University Press, New Haven, CT.

Goldstein, J. S., 1991, A war-economy theory of the long wave. In N. Thygesen, K. Velupillai and S. Zambelli, eds., *Business cycles: theories, evidence and analysis*, Macmillan, London, pp. 303–25.

Goldstein, J. S. and J. R. Freeman, 1990, *Three-way*

street: strategic reciprocity in world politics, University of Chicago Press, Chicago.

Goldstein, J. S. and D. P. Rapkin, 1991, After insularity: hegemony and the future world order, *Futures*, 23(9): 935–59.

Gray, C. S., 1988, *The geopolitics of superpower*, University Press of Kentucky, Lexington, KY.

Haass, R., 1988, The use (and mainly misuse) of history, *Orbis*, 32(3): 411–19.

Harff, B. and T. R. Gurr, 1989, Victims of the state: politicides and group repression since 1945, *International Review of Victimology*, 1(1): 23–41.

Haushofer, K., 1931, *Geopolitik der Panideen*, Zentral Verlag, Berlin.

Heske, H., 1988, *Und morgen die ganze Welt: Erdkundeunterricht im Nationalsozialismus*, Focus Verlag, Giessen.

Hewitt, K., 1983, Place annihilation: area bombing and the fate of urban places, *Annals of the Association of American Geographers*, 73(2): 257–84.

Hinsley, F. M., 1987, Peace and war in modern times. In R. Väyrynen, C. Schmidt and D. Senghaas (eds.), *The quest for peace: transcending collective violence and war among societies, cultures and states*, Sage, London, pp. 65–79.

Hobsbawn, E. J., 1987, *The age of empire. 1875–1914*, Penguin, New York.

Hoffman, G. W., 1982, Nineteenth-century roots of American power relations, *Political Geography Quarterly*, 1(4): 279–92.

Holsti, K. J., 1982, *Why nations realign: foreign policy restructuring in the postwar world*, Allen and Unwin, London.

Holsti, K. J., 1986, The horseman of the apocalypse: at the gate, detoured or retreating?, *International Studies Quarterly* 30(4): 355–72.

Holsti, K. J., 1991, *Peace and war: armed conflicts and the international order, 1648–1989*, Cambridge University Press, New York.

Hopkins, T. K., and Wallerstein, I., 1982, Cyclical rhythms and secular trends in the capitalist world-economy: some premises, hypotheses and questions. In T. K. Hopkins and I. Wallerstein (eds.), *World-systems analysis: theory and methodology*, Sage Publications, Beverly Hills, CA., pp. 104–20.

Howard, M., 1975, *War in European history*, Oxford University Press, New York

Howard, M., 1983, *The causes of war and other essays*, Harvard University Press, Cambridge, MA.

Johnston, R. J., O'Loughlin, J. and P. J. Taylor, 1987, The geography of violence and premature death: a world-systems approach. In R. Väyrynen, D. Senghaas and C. Schmidt, eds., *The quest for peace*,

Sage and International Social Science Council, Beverly Hills, CA., pp. 241–59.

Kahler, M., 1979, Rumors of war: the 1914 analogy, *Foreign Affairs*, 47(4); 374–96.

Kaldor, M., 1990, *The imaginary war: understanding the East–West conflict*, Basil Blackwell, Oxford.

Keegan, J., 1976, *The face of battle: a study of Agincourt, Waterloo and the Somme*, Penguin, Harmondsworth, Middx.

Kelman, H. C., 1991, A behavioral science perspective on the study of war and peace. In R. Jessor, ed., *Perspectives on behavioral science: the Colorado lectures*, Westview Press, Boulder, CO., pp. 245–75.

Keohane, R., 1984, *After hegemony: cooperation and discord in the world political economy*, Princeton University Press, Princeton, NJ.

Kennedy, P. M., 1987, *The rise and fall of the great powers: economic change and military conflict 1500–2000*, Random House, New York.

Kennedy, P. M., 1987, *The rise and decline of the great powers*, Random House, New York.

Keynes, J. M., 1936, *The general theory of employment, interest and money*, Harcourt, Brace Jovanovich, New York.

Kirby, A. M. and Ward, M. D., 1991, Modernity and the process of state formation: an examination of 20th-century Africa, *International Interactions*, 17(1): 113–26.

Klingberg, F. L., 1979, Cyclical trends in American foreign policy moods and their policy implications. In C. W. Kegley and P. McGowan, eds., *Challenges to America: United States' foreign policy in the 1980s*, Sage Publications, Beverly Hills, CA, pp. 37–55.

Krasner, S. D., 1985, *Structural conflict: the Third World against global liberalism*, University of California Press, Berkeley, CA.

Lake, D. A., 1992, Powerful pacifists: democratic states and war, *American Political Science Review*, 86(1): 24–37.

Laqueur, W. Z., 1980, *The political psychology of appeasement: Finlandization and other unpopular essays*, Transaction Books, New Brunswick, NJ.

Leisner, T., 1989, *One hundred years of historical statistics*, Facts on File, New York.

Lenin, V. I., 1939, *Imperialism: the highest stage of capitalism*, International Publishers, New York.

Leurdijk, J. N., 1991, The allied powers and the peace with Germany after the Second World War. Paper presented to the Conference on the Great Peace Congresses, Utrecht.

Levy, J. S., 1983, *War in the modern great power system, 1495–1975*, University Press of Kentucky, Lexington, KY.

Mackinder, H. J., 1904, The geographical pivot of history, *Geographical Journal* 23(4): 421–37.

Maddison, A., 1982, *Phases of capitalist development*, Oxford University Press, New York.

Mahan, A. T., 1890, *The influence of sea-power on history, 1660–1783*, Little Brown, Boston.

Mandel, E., 1980, *Long waves of capitalist development*, Cambridge University Press, New York.

Mearsheimer, J. J., 1990, Why we will soon miss the Cold War, *Atlantic Monthly*, August, 35–50.

Mensch, G., 1979, *Stalemate in technology: innovations overcome the depression*, Ballinger Publishing Company, Cambridge, MA.

Mitchell, B., 1975, *European historical statistics, 1750–1970*, Columbia University Press, New York.

Mitchell, B., 1982, *International historical statistics: Asia and Africa*, New York University Press, New York.

Mitrany, D., 1976, *The functional theory of politics*, St. Martins Press, London.

Modelski, G., 1983, Long cycles of world leadership. In W. R. Thompson, ed., *Contending approaches to world system analysis*, Sage Publications, Beverly Hills, CA., pp. 115–39.

Modelski, G., 1987a, A global politics scenario for the year 2016. In G. Modelski (ed.), *Exploring long cycles*, Lynne Rienner, Boulder, CO, pp. 218–48.

Modelski, G., ed., 1987b, *Exploring long cycles*, Lynne Rienner, Boulder, CO.

Modelski, G., 1987c, *Long cycles in world politics*, University of Washington Press, Seattle, WA.

Modelski, G., 1987d, The study of long cycles. In G. Modelski, ed., *Exploring long cycles*, Lynne Rienner, Boulder, CO., pp. 1–15.

Modelski, G., 1990, Global leadership: end game scenarios. In D. P. Rapkin, ed., *World leadership and hegemony*, Lynne Rienner, Boulder, CO, pp. 241–56.

Modelski, G. and Thompson, W. R., 1988, *Seapower in global politics, 1494–1993*, University of Washington Press, Seattle, WA.

Morgenthau, H., 1960, *Politics among nations: the struggle for power and peace*, 3rd ed., Knopf, New York.

Nau, H., 1990, *The myth of America's decline*, Oxford University Press, New York.

Nierop, T., 1989, Macro-regions and the global institutional network, *Political Geography Quarterly*, 8(1): 43–65.

Nierop, T. and de Vos, S., 1988, Of shrinking empires and changing roles: world trade patterns in the post-war period, *Tijdschrift voor Economische en Sociale Geografie*, 79(5): 343–64.

Nijman, J., 1991, The dynamics of superpower spheres of influence: US and Soviet military activities, 1948–78, *International Interactions*, 17(1): 63–91.

Nye, J. S., 1990, *Bound to lead: the changing nature of American power*, Basic Books, New York.

O'Loughlin, J., 1987, Superpower competition and the militarization of the Third World, *Journal of Geography*, 86(6): 269–75.

O'Loughlin, J., 1989, World-power competition and local conflicts in the Third World. In R. J. Johnston and P. J. Taylor, eds., *A world in crisis: geographical perspectives*, Basil Blackwell, Oxford, pp. 289–332.

O'Loughlin, J., 1992, Ten scenarios for a new world-order, *Professional Geographer*, 44(1): 22–9.

O'Loughlin, J., 1993, Fact or fiction: the evidence for the thesis of US relative decline, 1946–91. In C. H. Williams, ed., *The political geography of the new world order*, Belhaven Press, London.

O'Loughlin, J. and Anselin, L., 1991, Bringing geography back to the study of international relations: spatial dependence and regional context, 1966–1978, *International Interactions*, 17(1): 29–61.

O'Loughlin, J. and H. Heske, 1991, From 'Geopolitik' to 'Geopolitique': converting a discipline for war to a discipline for peace. In N. Kliot and S. Waterman, eds., *The political geography of peace and war*, Belhaven Press, London, pp. 37–59.

O'Loughlin, J. and van der Wusten, H., 1990, The political geography of panregions, *Geographical Review*, 80(1): 1–20.

O'Sullivan, P., 1986, *Geopolitics*, St. Martin's Press, New York.

O'Tuathail, G., 1992, Putting Mackinder in his place: material transformations and myth, *Political Geography*, 11(1), 100–18.

Quigley, C., 1966, *Tragedy and hope: a history of the world in our time*, Macmillan, New York.

Rasler, K. and Thompson, W. R., 1989, *Global war and state-making processes*, Unwin Hyman, Boston, MA.

Rasler, K. and Thompson, W. R., 1991, Relative decline and the overconsumption–underinvestment hypothesis, *International Studies Quarterly*, 35(3): 273–94.

Rosenau, J. N., 1982, Order and disorder in the study of world politics. In R. Maghroori and B. Ramberg, eds., *Globalism versus realism: international relations' third debate*, Westview Press, Boulder, CO., pp. 1–7.

Russett, B. M., 1985, The mysterious case of vanishing hegemony, or, is Mark Twain really dead?, *International Organization*, 39(2): 207–32.

Schaefer, R. K., 1989, War in the world-system. In R. K. Schaefer, ed., *War in the world-system*,

Greenwood Press, Westport, CT, pp. 1–19.

Singer, J. D., 1981, Accounting for international war: the state of the discipline, *Journal of Peace Research*, 18(1): 1–18.

SIPRI Yearbooks, various years, Stockholm International Peace Research Institute.

Siverson, R. M. and H. Starr, 1991, *The diffusion of war*, University of Michigan Press, Ann Arbor, MI.

Sloan, G. R., 1988, *Geopolitics in United States strategic policy: 1890–1987*, Wheatsheaf Books, Brighton.

Small, M. and Singer, J. D., 1982, *Resort to arms: international and civil wars, 1816–1980*, Sage, Beverly Hills, CA.

Starr, H., 1991, Democratic dominoes: diffusion approaches to the spread of democracy in the international system, *Journal of Conflict Resolution*, 35(2): 356–81.

Stoll, R. J. and M. D. Ward, eds., 1989, *Power in world politics*, Lynne Rienner, Boulder, CO.

Strange, S., 1987, The persistent myth of lost hegemony, *International Organization*, 41(4): 551–74.

Taylor, A. J. P., 1961, *The origins of the Second World War*, Premier Books, New York.

Taylor, P. J., 1989, The world-systems project. In R. J. Johnston and P. J. Taylor, eds., *A world in crisis: geographical perspectives*, Basil Blackwell, Oxford, pp. 333–54.

Taylor, P. J., 1990, *Britain and the Cold War: 1945 as geopolitical transition*, Pinter, London.

Taylor, P. J., 1992, Tribulations of transition, *Professional Geographer*, 44(1): 10–12.

Terkel, S., 1984, *The good war: an oral history of World War Two*, Pantheon Books, New York.

Thompson, W. R., ed., 1983, *Contending approaches to world system analysis*, Sage, Beverly Hills, CA.

Thompson, W. R., 1988, *On global war: historical-structural approaches to world politics*, University of South Carolina Press, Columbia, SC.

Thompson, W. R. and G. Zuk, 1986, World power and the strategic trap of territorial commitments, *International Studies Quarterly*, 30(3): 249–68.

Tilly, C., 1985, Warmaking and statemaking as organized crime. In P. Evans, D. Rueschemeyer and T. Skocpol, eds., *Bringing the state back in*,

Cambridge University Press, New York, pp. 169–91.

Tilly, C., 1990, *Coercion, conflict and the European states, AD 990–1990*, Basil Blackwell, Oxford.

van Crefeld, M., 1991, *On future war*, Brassey's, London.

van der Wusten, H., 1985, The geography of conflict since 1945. In D. Pepper and A. Jenkins, eds., *The geography of peace and war*, Basil Blackwell, Oxford, pp. 13–28.

van der Wusten, H. and T. Nierop, 1990, Functions, roles and form in international politics, *Political Geography Quarterly*, 9(3): 213–31.

van Duijn, J. J., 1983, *The long wave in economic life*, Allen and Unwin, Boston, MA.

Väyrynen, R., 1987, Global power dynamics and collective violence. In R. Väyrynen, D. Senghaas and C. Schmidt, eds., *The quest for peace*, Sage and International Social Science Council, Beverly Hills, CA., pp. 80–96.

Yergin, D., 1991, *The prize: the epic quests for oil, money, and power*, Simon and Schuster, New York.

Wallerstein, I., 1979, *The capitalist world-economy*, Cambridge University Press, New York.

Wallerstein, I., 1984, *The politics of the world-economy*, Cambridge University Press, New York.

Wallerstein, I., 1991, *Geopolitics and geoculture: essays on the changing world-system*, Cambridge University Press, New York.

Waterman, S., 1984, Partition: a problem in political geography. In P. J. Taylor and R. J. Johnston, eds, *Political geography: recent advances and future directions*, Croom Helm, London, pp. 98–116.

Wayman, F. L. and J. D. Singer, 1990, Evolution and directions for improvement in the Correlates of War project methodologies. In J. D. Singer and P. F. Diehl, eds., *Measuring the correlates of war*, University of Michigan Press, Ann Arbor, MI., pp. 247–67.

Weede, E., 1973, Nation–environment relations as determinants of hostilities among nations, *Peace Science Society (International) Papers*, 20(1): 67–90.

Weede, E., 1975, *Weltpolitik und Kriegsursachen im 20. Jahrhundert: eine quantitativ-empirische Studie*, Oldenbourg Verlag, München.

Williams, W. A., 1980, *Empire as a way of life*, Oxford University Press, New York.

3

The rise and decline of the corporate-welfare state: a comparative analysis in global context

R. J. Johnston

Whereas most of the chapters in this book focus on the inter-state system during the twentieth century, attention here is directed to the contemporaneous intra-state situation. Within its own territory, the state has become an increasingly important institution, extending its power and influence very substantially – from the 'night-watchman state to the nanny state'. In the last two decades, that power and influence have been subject to criticism and the focus of moves to reduce the state's role.

The goal here is to understand these changes. This is done, first, by exploring the nature of the state through an examination of relevant theory and, secondly, by appreciating the context of state operations through an outline of general trends in the world-economy during the twentieth century, with particular emphasis on Kondratieff waves. Most of the chapter concentrates on the capitalist world, and especially the countries at the core of the world-economy.

Even in a chapter of this length, the treatment of such a large topic calls for generalization. Over-generalization is characteristic of much late-twentieth-century political geography writing, however (Driver, 1991). To counter that possibility it is particularly important to distinguish between the *necessary roles of the state within a mode of production* and the *actual performance of those roles in particular circumstances*.[1] Detailed case study material on New Zealand, West Germany, the United States and the United Kingdom provides empirical depth to illustrate the latter point.

The state in operation: some quantitative indicators

There are many ways of indexing the importance of the state within a society, such as the size (relative and absolute) of its workforce, its dependent population and its budget. Whichever is used, the almost certain conclusion will be Rose's (1984, 1) that:

Government is big in itself, big in its claims upon society's resources and big in its impact upon society. By every conventional measure, government looms large in the life of every Western nation today; governments differ from nation to nation only in their degree of bigness. Because government is dynamic not static, it can grow bigger still. Whether or not big government is regarded as a big problem, it is certainly a major issue of politics in the 1980s.

Rose used five indices of government size: the number of laws passed; the raising of taxes; the number and work of public employees (see also Rose, 1985); the multifarious organization of the state apparatus; and the range of government programmes. Comparative data illustrating the size and growth of governments in Western Europe since the Second World War have been published by Lane and Ersson (1991). Figures 3.1–3.3 plot three of their indices for the sixteen countries. Figure 3.1 shows the percentage of GDP collected as taxes by central and local governments combined. In Sweden, just over half of the entire national GDP was collected through taxation in 1985, an almost exact doubling over the thirty years: the percentage exceeded 40 in another six countries and was below 30 in only one – Spain (where it had doubled in the previous two decades). Financing government activity absorbs a major part of every country's annual national income, and that part has increased in each case during the four decades, with only seven examples of the percentage declining over a particular five-year period.

Governments raise income in order to purchase goods and services and to make transfer payments to citizens. Figure 3.2 shows the percentage of GDP in each of the sixteen countries expended by governments (central and local combined) on final consumption, involving public administration, defence, health and education. It exceeded 20 per cent of GDP in four cases in 1985 and was below 15 in only two – Spain and Switzerland. The percentage increased substantially in every country over the 35-year period – though interestingly there was a decline between 1980 and 1985 in exactly half of the cases. (Seven of those eight spent above the median percentage in 1980.) Finally, Figure 3.3 shows the percentage of GDP redistributed by governments via social security transfers (which include social security benefits, social assistance grants, unfunded pension and welfare benefits for employees, and transfers to private, non-profit-making institutions which serve house-

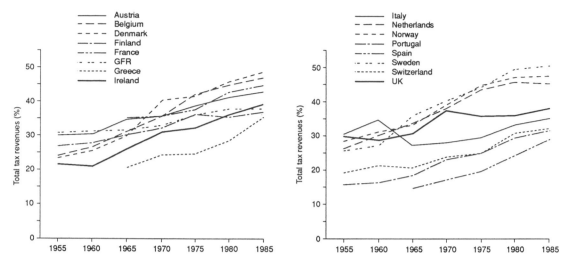

Figure 3.1 *The size of government in sixteen West European countries, 1955–85, as shown by the percentage of GDP taken by government in taxation.*
Source: data from Lane and Ersson (1991, Table 10.3).

holds, such as voluntary agencies). Over 10 per cent of GDP was thus distributed in all sixteen countries by 1985, with the percentage at least doubling since 1950 in all but the Federal Republic of Germany (West Germany). The rate of growth declined over the period in many of the countries, however.

The scale of government activity indicated by Figures 3.1–3.3 is matched by the size of the public sector workforce, as illustrated by Rose's (1985) data for France, West Germany, Italy,

Sweden, the UK and the USA. In the early 1980s, the five European countries had between 27 and 39 per cent of their non-agricultural workforce employed by government: the US figure was 19 per cent. All five European countries had experienced a substantial increase over the previous three decades, ranging from 5 per cent in the UK through 11 per cent in West Germany, 13 per cent in Italy and 15 per cent in France to 23 per cent in Sweden: the USA was an exception, with a growth of only 1.3 per cent. In all six cases

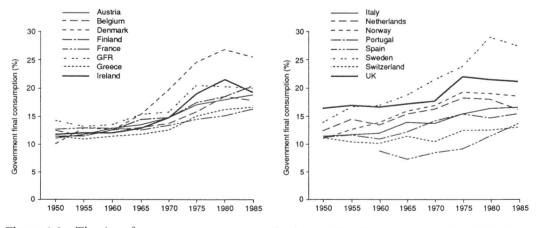

Figure 3.2 *The size of government as consumer in sixteen West European countries, 1950–85, as shown by the percentage of GDP expended by government on final consumption.*
Source: data from Lane and Ersson (1991, Table 10.5).

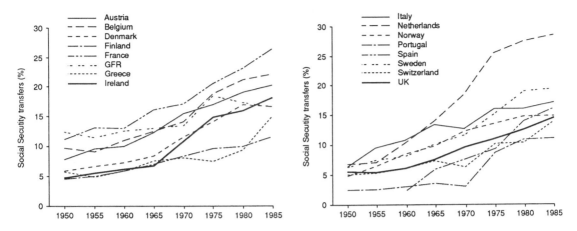

Figure 3.3 *The size of government as a redistributive agent in sixteen West European countries, 1950–85, as shown by the percentage of GDP disbursed as social security payments. Source: data from Lane and Ersson (1991, Table 10.9).*

public sector employment grew more rapidly than that in the private sector, which means that 'government is disproportionately responsible for the creation of new jobs in response to an increase in the size of the labour force' (Rose, 1985, 12).

These comparative data illustrate the scale of the task essayed here: the state dominates the labour markets of core countries in the capitalist world, and its domination has increased during the twentieth century. (Rose (1985, 9) estimates, for example, that between 1914 and 1951 British public sector employment more than tripled as a percentage of the workforce, whereas France and Italy experienced a doubling and the USA an eightfold increase.) They also indicate a slowing-down in the rate of growth of government, however – a point highlighted by Lane and Ersson (1991, 348). Rose (1984, 6) notes a similar trend with his chosen indicators: the rate at which laws are being added to the statute books is not increasing; the increase in taxation is more a result of inflation than an expansion of the government's share of GDP; 'the number of government organizations does *not* multiply endlessly'; and most growth has not been in new programmes, but rather in the large and well-established welfare state policies providing 'pensions, health care and education for tens of millions of people'.

The state is very big as a proportion of total activity, in the core countries of the capitalist world-economy at least. It has become much bigger during the twentieth century, but the rate of growth has slowed in recent years. Understanding these trends is the goal here.

Why the state?

In their introduction to theories of the state, Dunleavy and O'Leary (1987, 1–2) argue that:

Some form of government is intrinsic to human society, because a society which is totally uncontrolled, unguided and unregulated is a contradiction in terms. By contrast, the state – defined organizationally – is not intrinsic to human society. There have been, and still are according to anthropologists, some state-less societies, such as segmentary tribal systems or small, isolated bands, in which rules and decisions are made collectively, or through implicit negotiation, with no specialization of government in the hands of one set of persons.

Government, then, is a function – a means of making and implementing rules. The state is a particular context for government.

The state form of government is distinguished, they claim, by five characteristics (Dunleavy and O'Leary, 1987, 2):

1) It is recognizably separate from the rest of society, comprising a *public sector* which is distinct from the *private*.

2) It is sovereign within its territory, being the

119

ultimate legal authority and having the mono-poly of force with which to impose that authority.

3) Its sovereignty applies equally to all indi-viduals within its territory, and is distinguish-able from the influence wielded transitorily by individuals holding positions within the state apparatus.

4) Its personnel are recruited and trained as managers within a bureaucratic system.

5) It has the capacity to raise revenue from the population of its territory through taxation.

But why has this particular governmental form evolved, so that it now embraces virtually the entire population of the world?

Within the *capitalist mode of production* the necessity for the state has been portrayed by Harvey (1985) in a major theoretical essay (which also has implications for many of the geopolitical arguments developed elsewhere in this book). The mode of production brings together three separate factors – land, labour and capital – for the creation of commodities that can be sold at a profit and thus yield wealth to those involved: this wealth is almost always unequally distributed, with the owners of land and capital benefiting greatly, at the expense of the owners of labour.

Two of the three factors of production differ from the other: land is fixed, but capital and labour are mobile. For production to occur, 'capital and labour power must be brought together at a particular point in space' (Harvey, 1985, 145). What he calls a 'structured coherence' is thus created – a regional space 'within which production and consumption, supply and demand (for commodities and labour power), production and realisation, class struggle and accumulation, culture and life style, hang together'. The nature of a place reflects this coherence.

The creation of a structured coherence is somewhat paradoxical, however, because it both provides the context for wealth-creation, and yet inhibits the very mobility of capital and labour which is central to capitalism's success (as demonstrated in another Harvey essay (1982)). As Harvey (1985, 146) expresses it: 'The ability of both capital and labour power to move at

short order and low cost from place to place depends upon the creation of fixed, secure, and largely immobile social and physical infra-structures.' A built environment and a society are necessary to attract virtually any capitalist investment. However, by making that fixed in-vestment, with workers making their fixed invest-ments in places also, the paradoxical situation is created in which: 'A portion of the total capital and labour power has to be immobilised in space, frozen in place, in order to facilitate greater liberty of movement for the remainder.' A major tension is thus created. Capitalists may wish to shift their investment to realize larger profits elsewhere, and workers may wish to shift theirs to realize higher wages. But that shift may mean losing part of the invested value, which they cannot take with them (Harvey, 1985, 150–51): 'All economic agents (individuals, organisations, institutions) make decisions on the circulation of their capital or the deployment of their labour power in a context marked by a deep tension between cutting and running to wherever the rate of remuneration is highest, or staying put, stick-ing with past commitments and recouping values already embodied.' The state is central to the decision whether 'to cut and run' or 'to stay'.

Once investments in a place have been made, and a structured coherence created, there is a need to protect them and thus to promote favourable conditions for the continued crea-tion of wealth through profit-making activities. Such protection is provided by *regional class alliances* – 'loosely bounded within a territory and usually (though not exclusively or uniquely) organised through the state' (Harvey, 1985, 151):

Land and property owners, developers and builders . . . Those sectors of production which cannot easily move (by virtue of the fixed capital they employ or other spatial constraints [such as a fixed resource like a mineral deposit or a tract of highly fertile agricultural land]) will tend to support an alliance and be tempted or forced to buy local labour peace and skills through compromises over wages and work conditions. Fac-tions of labour that have through struggle or out of scarcity managed to create islands of privilege within a sea of exploitation will also just as surely rally to the cause of the alliance to preserve their gains.

Part of the structured coherence therefore involves most if not all of the parties in a place supporting an institution – almost certainly the state – which promotes their collective interests against those of people in other places. Thus the capitalist world-economy becomes a mosaic of regional class alliances, of separate states acting to advance the interests of local capital and labour and so sustain, if not enhance, relative wealth and income there. (Within states, too, 'subsidiary regional class alliances' may well develop to represent the interests of people in particular parts of the state's territory: see Cox (1989).)

The state is not a creation of capitalist regional class alliances, however: it predates them, and provides a viable shell within which they can operate. It is particularly attractive as such a shell, however, because (Harvey, 1985, 152):

The state is different . . . from other agents in a variety of respects. First, territory and the integrity of territory is the objective of its personnel to a degree uncharacteristic of other agents. Second, by virtue of its authority, it can give firmer shape and cohesion to regional class alliances through the institutions of law, governance, political participation and negotiation, repression and military might. Third, it can impose relatively firm boundaries on otherwise porous and unstable geographical edges. Finally, by virtue of its powers to tax and to control monetary and fiscal policy, it can actively promote and sustain that structured regional coherence to production and consumption to which capitalism in any case tends and undertake infrastructural investments that individual capitalists could not tackle.

If the state, or a comparable institution with the five characteristics identified by Dunleavy and O'Leary, did not exist prior to the emergence of capitalism, it would have been necessary for capitalism to invent it.

Harvey's essay indicates why the state is necessary to capitalism, but does not account for the prior existence of a territorial institution within which sovereign power is exercised by those who gain control of the state apparatus. That is too large a topic for the present chapter (Mann, 1986, offers the needed wider perspective). The concern here is with the roles that the state, as regional alliance, is called upon to play in order to protect and enhance the interests of those living within its territory (though little is

said about the geopolitical role, which is more than adequately covered elsewhere in this book).

The three roles of the capitalist state

Observation of modern capitalist states readily indicates the great breadth and depth of the power and influence which they exercise. Mann terms this *infrastructural power*, which he defines as 'the capacity of the state to actually penetrate civil society, and to implement logistically political decisions throughout the realm' (Mann, 1984, 189).

[Its] powers are now immense. The state can assess and tax our income and wealth at source, without our consent or that of our neighbours or kin (which states before about 1850 were *never* able to do); it stores and can recall immediately a massive amount of information about all of us; it can enforce its will within the day almost anywhere in its domains; its influence on the overall economy is enormous; it even directly provides the subsistence of most of us (in state employment, in pensions, in family allowances, etc.). The state penetrates everyday life more than did any historical state. . . . there is no hiding place from the infrastructural reach of the modern state.

This is done with the implicit consent of the population concerned,[2] and in some countries – the liberal democracies – their explicit consent, since those in charge of the state apparatus (usually termed the government) are directly accountable to the majority of the territory's residents through the electoral process.

Understanding why people give this consent to the exercise of state infrastructural power was given a substantial boost by O'Connor's work (1973) on *The Fiscal Crisis of the State*. His taxonomy of categories of state expenditure was extended by Saunders (1980, 147–8) to identify three (overlapping empirically but theoretically distinct) key state functions:

(1) The sustenance of private production and capital accumulation by, for example, providing a non-productive infrastructure (such as a road system), aiding the creation and restructuring of the built environment (through land-use planning), investing in the development of human capital (through education and training) and orchestrating

demand (through the letting of contracts, for example).

(2) Assisting the reproduction of labour power through providing items of collective consumption, such as subsidized housing, and cultural facilities such as parks and museums.

(3) Maintaining social order and cohesion by coercion (the police function), supporting the 'surplus population' through a range of social services, and promoting ideological support for the mode of production (using, for example, the education system).

Clark and Dear (1984, 42–3) suggested a reordering of the three, arguing that *securing social consensus* (maintaining social order and cohesion) was the state's primary responsibility, since 'only when those relations are established can production and exchange take place with any degree of continuity': the other two are contingent upon order and cohesion being present. They involve, respectively, *securing the conditions of production* by regulating public investment intended both to increase production and to ensure the reproduction of the labour force, and *securing social integration* by ensuring the welfare of all, especially those termed 'the subordinate classes' by Clark and Dear (43).

The state exists to promote and to legitimate the creation of wealth through the making of profits, therefore, which is the fundamental dynamic of the capitalist mode of production (Jessop, 1990, 341): 'The core of the state apparatus comprises a distinct ensemble of institutions and organizations whose socially accepted function is to define and enforce collectively binding decisions on the members of a society in the name of their common interest or general will.' The state's ideological function (part of its securing of social consensus) is undertaken to ensure that its decisions are accepted as in the 'common interest or general will'. The other two functions are aimed at ensuring that both major interest groups within the mode of production – under capitalism, the owners of capital and land on the one hand and of labour on the other – obtain a reasonable service from the state: that the former achieve adequate profit levels and the latter a satisfactory quality of life.

Whereas these three functions outline the necessity of the state within a capitalist society, they neither preclude other activities by those controlling the state apparatus nor circumscribe their roles to certain actions only. Many who achieve some power within the state apparatus use their position to advance other goals that are not directly relevant to either the promotion or the legitimation of capitalism – indeed, they may campaign for power within the apparatus on platforms which advance such other functions. To the extent that pursuing those goals is consistent with ensuring the health of the capitalist operations, then their operation is entirely feasible. When they have substantial negative impacts on either or both of wealth accumulation via profit-making and legitimation of such activities through adequate welfare support, however, then their potential is reduced.

Threats to the capitalist state are frequently categorized as originating in one of two types of crisis (as defined by Habermas, 1976). A *rationality crisis* results from the state's failure to ensure satisfactory wealth accumulation strategies on behalf of the owners of capital and land, and is expressed in a lack of investment within the state's territory. A *legitimation crisis*, on the other hand, reflects dissatisfaction with the state's efforts to protect and enhance their interests among the owners of labour, and is usually represented by struggles involving labour in conflict with both capital and the state. The state is called upon to resolve these crises to the relative satisfaction of the conflicting parties – which may involve it using the forces of repression at its command (which are usually employed to resolve legitimation rather than rationality crises, as with many instances of state treatment of workers' strikes, though see the section below on Third World states).

The two types of crisis are not independent – a substantial rationality crisis is likely to engender a legitimation crisis too, because a lack of investment will lead to a decline in the number of jobs available and an increasing welfare burden on the state, calling for higher levels of taxation on a declining productive base. Similarly a legitimation crisis, with substantial worker unrest, will have negative consequences for wealth accumulation because of the likely interruptions to production, and hence to profit-making. The state has to balance the competing demands of capital and labour to try and ensure the absence of such crises; when they do occur it must seek a resolution which does not tip the balance too far (e.g. resolution of a legitimation crisis by granting substantial concessions to labour can stimulate a rationality crisis).

Societies are complex organizations and their theoretical division into two conflicting groups – the owners of capital and land versus the owners of labour – must be enhanced by realizing that each is itself divided into several groups (such as the owners of productive and of finance capital). As Jessop (1990, 357) argues, even when it is seeking a resolution in the collective interests of one of the larger groups (capital or labour), the outcome is likely to favour one or more subgroups over others, thus creating a further potential problem for it to resolve: any state resolution of a conflict is likely to contain within itself the seeds of the next, because it alters the balance of advantage and power within society.

The economic context of state action

The state must intervene in order to resolve crises within capitalist societies – a role which Jessop (1990, 35) suggests it cannot avoid: 'the state's role has now become vital for accumulation, [and] it cannot solve economic crises simply by withdrawing or refusing to intervene. At best it can reorganize how it intervenes.' But how do those crises arise?

Economists have long been interested in long-run trends in the performance of capitalist economies, and in the apparent regularity of cycles of about fifty years' length within those trends. These have more recently been appropriated by other social scientists (including geographers, e.g., Hall, 1988), and used to inform their understanding of trends in other aspects of society (as in the study of geopolitical long waves, see Taylor, 1990). They are used here to advance our understanding of state actions within their own territories.

The nature of such cycles has recently been portrayed in substantial graphical detail by Berry, who concluded from his detailed analysis of US data that 'within the inherently high noise levels of history, prices and economic growth move in synchronized rhythms in which alternating stagflation crises and deflationary depressions separate cycles of deflationary and inflationary growth' (1991, 10).[3] Three of his graphs (Figures 3.4, 3.5 and 3.6) illustrate those rhythms in the United States: Figure 3.4 shows substantial fluctuations in growth rates in real GNP, with peaks approximately fifty years apart, and Figure 3.5 shows similar trends in three other indices of economic performance – though with the last few decades being somewhat deviant. Finally, Figure 3.6 shows the inverse correlation within those long waves between growth in real GNP on the one hand and wholesale prices on the other.

Berry's data refer to the USA only, but the cycles which these identify are global in their impact, not local, as suggested by a large number of other writers. Goldstein (1988) has recently published data compilations and analyses for a number of countries and variables, which clearly illustrate the global trends in capitalism's progress over the last five centuries. The problems that those cycles stimulate – both in the troughs and on the peaks – are global also, though the political responses are almost always local, within the territorial confines of individual states.

A variety of explanations has been advanced for the existence of long waves (and, according to some at least, for their necessity within a capitalist economy). These are reviewed in Berry's book, with no clear decision as to which is superior, although he concludes that whereas the provision of the physical infrastructure for successful accumulation follows the fifty-year cycle postulated by Kondratieff, the driving force for this is provided by the 25- to 30-year cycles of investment identified by Kuznets (1953): the implication is that the physical infrastructure can cater for two investment booms.

A powerful, extremely succinct, case for the existence of regular crises within the capitalist mode of production is provided in the essay by Harvey (1985) already extensively quoted (for a more detailed presentation, see Harvey, 1982). Having identified the origin of profits in the use of capital to employ 'living labour', the necessary 'class struggle' that this involves, and the 'perpetual revolutions' needed to achieve the techno-logical and organizational changes that will yield competitive advantage to individual capitalists, Harvey concludes (1985, 131–2) that the system is unstable, because it 'embodies powerful and disruptive contradictions that render it chronic-ally crisis-prone'. This is because:

The system has to expand through the application of living labour in production whereas the main path of technological change is to supplant living labour, the real agent of expansion, from production. Growth and technological progress, both necessary features of the circulation of capital, are antagonistic to each other. The underlying antagonism periodically erupts as full-fledged crises of accumulation, total disruptions of the circulation process of capital.

He terms the resulting surpluses of both capital and labour – reflected in both productive capac-ity and unemployment – a situation of *overaccu-mulation* (others term it underconsumption) in which 'surplus capital and surplus labour power exist side by side with apparently no way to bring the two together to accomplish socially useful tasks' (132).

Resolution of the overaccumulation problem is necessary for an upturn in the cycles depicted in Figures 3.4–3.6. The surpluses that cannot be absorbed (i.e. reused in new contexts) must be devalued, even destroyed: capital as money can be devalued through inflation; capital as pro-ductive plant and stocks of commodities can be sold at less than their assumed value. Workers' wages and real incomes can similarly be de-valued, creating conditions which favour new investment; many workers may be made redun-dant, or encouraged to accept a 'voluntary redundancy–early retirement' package. That, according to several commentators (Hall, 1988, following Schumpeter, 1934), almost certainly involves the adoption of new technologies, re-quiring new infrastructures and new labour skills. The new infrastructures may well be estab-lished in different areas from those which they are replacing, for a variety of reasons (see Massey and Meegan, 1979). This continues the cycles of uneven development noted by Harvey (1982), Smith (1984) and others, and has led many (see, for example, Hall and Preston 1988) to link

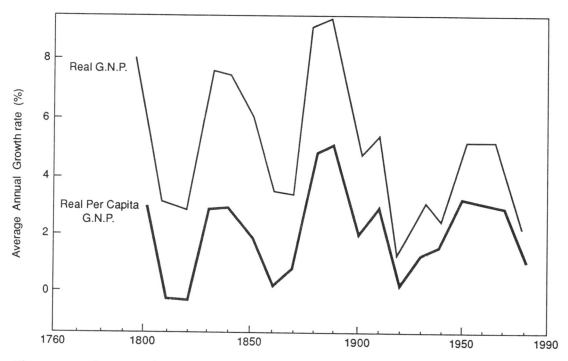

Figure 3.4 *Changes in the average annual rate of growth in real GNP in the USA, 1790–1980, as decadal averages. Source: Berry (1991, Figure 1).*

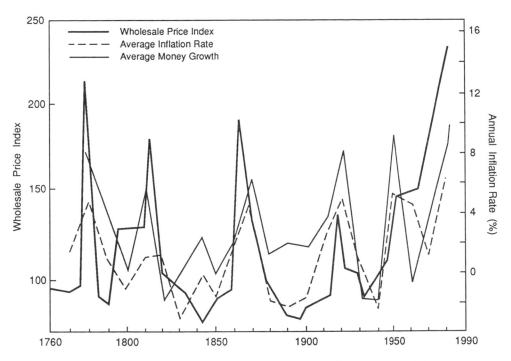

Figure 3.5 *Changes in the wholesale price index, the average inflation rate and the average money growth in the USA, 1760–1982. Source: Berry (1991, Figure 4).*

recovery from the Kondratieff downturns with both the introduction of new technologies and the creation of new geographies (see Massey and Meegan, 1981).

Most of the various aspects of the nature of Kondratieff long waves and their implications for societal restructuring are brought together by Marshall (1987), as demonstrated in Figure 3.7.[4] Marshall's diagram presents the current century as dominated by 'monopoly capitalism', in which the production process (termed by others the regime of accumulation) is dominated by 'Fordism' – characterized by mass production of goods in large factories by disciplined work-forces. There have been two such major Kondratieff cycles, peaking around fifty years apart. Each has been characterized by overaccu-mulation leading to recession, but the reasons for the decline in the rate of profitability are unclear, as Webber (1991) illustrates in his comparison of six competing theories.

To a number of observers, the recovery from the major downturn following the 1960s post-war boom has involved the replacement of monopoly capitalism and Fordism by what Harvey has termed 'flexible accumulation', 'dis-tinguished by a remarkable fluidity of pro-duction arrangements, labour markets, financial organization and consumption' (Harvey and Scott, 1989, 218: this represents a new 'mode of regulation' to others). Appreciating this new form of society is central to an understanding of the way in which the state has acted in recent decades to counter the downturn of the 1970s/ 1980s and to promote the expected upturn of the 1990s. For this, it is useful to take the fourteen points by which Lash and Urry (1987) character-ize Fordism (their term for it is 'organised capitalism') and its replacement regime of accumulation, which they prefer to call 'dis-organised capitalism'.[5]

Table 3.1 summarizes Lash and Urry's four-teen points, illustrating the major differences be-tween organized and disorganized capitalism in the nature of both industrial organization and class struggle, and the role of the state in medi-ating between the two. The main shifts involve the fragmentation of economic activity and its wider spread through the world. This is stressed in Table 3.2 (Lash and Urry, 1987, 16), which contrasts the two regimes of accumulation with that preceding them – which they term 'liberal capitalism' and Marshall (see Figure 3.7) terms 'competitive capitalism' – on three criteria. They emphasize the spatial changes accompanying each transition, in terms of both the pattern of uneven development (what they call changes within each territory) and the means of communication ('transmitting knowledge and executing surveillance'). The nature of those changes, and the economic and social dislocations involved in the shift from one regime to the next, are the context for our discussion of the nature and role of the state in promoting and legitimating both organized and disorganized capitalism.

Twentieth-century crises and the state in the core of the capitalist world-economy

The discussion so far has identified certain key features of capitalist dynamics – periodic crises of over-accumulation calling forth economic restructuring which may well involve a rewriting of the map of uneven development. Figure 3.7 suggests two such crisis periods during the twentieth century (the late 1920s/1930s and the 1970s/1980s). The state played a major role promoting recovery from the first, leading to the so-called 'post-war boom' of the late 1950s and 1960s. Recovery from the second is, presumably, nascent: certainly the state has been extremely active in creating a context for it. The present section looks at the very different roles of the state during those two periods, emphasizing both the commonality of approach among various states, and also the substantial differences between them, reflecting their particular histories and their governments' interpretations of the predicaments faced.

The major downturn in rates of economic growth in the 1920s and 1930s (Figures 3.4 and 3.5) was accompanied by mass unemployment in very many countries, and consequent problems of poverty for large numbers of people in the core of the world-economy – although for many their position relative to that of their contempor-

aries in the Third World remained good. The response of the state, more rapidly and comprehensively in some than in others, was to introduce policies aimed at: (1) providing relief for individuals and families suffering substantial hardship, especially those with no means on which to fall back; and (2) promoting economic policies which would ensure full employment in the future, so that the problem could never reappear to anything like the same extent. Thus were developed the welfare state and the corporate state – though in most cases the foundations were already laid and so were just very substantially extended in the 1930s.

In most countries, as illustrated below, the foundations of the welfare and corporate states were laid before the Kondratieff trough of the 1920s/1930s, but those foundations were neither deep nor complete. In general, those in control of the state apparatus had yet to be convinced of the desirability of a substantial and permanent welfare-corporate state structure. They saw the state playing a relatively minimal role in the internal economic and social affairs of a country, beyond the preservation of law and order and the protection of private property rights (which were held by relatively few), and emphasized statecraft in the international sphere as the main political activity: with regard to the burgeoning capitalist economy, the role of the state was to protect local interests overseas, through military, imperial and diplomatic policies and practices. Geopolitics was thus a political activity rather than an academic subdiscipline (as illustrated by the political career of the geographer Halford Mackinder in the United Kingdom: Blouet, 1987; Parker, 1982). Elections, then, were as likely to be fought on foreign policy issues (such as trade policy) as on domestic economic topics: only with the universal electoral franchise did the well-being of individuals, especially the less affluent, become the major determinants of voters' interests, and thus politicians' concerns. Only then was the 'nightwatchman state' transformed into the 'welfare-corporate state'.

The welfare state

The transition in the role of the state during this period is encapsulated by Esping-Andersen

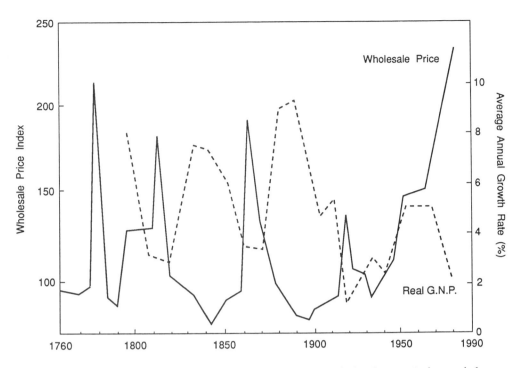

Figure 3.6 *The negative relationship between changes in the wholesale price index and the rate of growth in real GNP in the USA, 1760–1982. Source: Berry (1991, Figure 5).*

(1990, 1): 'What once were night-watchman states, law-and-order states, militarist states, or even repressive organs of totalitarian rule, are now institutions predominantly preoccupied with the production and distribution of social well-being. To study the welfare state is therefore a means to understand a novel phenomenon in the history of capitalism.' Quantitative indicators of that transition are given in the introductory section to this chapter: but what were the reasons for and the nature and extent of the growth of state welfare provision?

Esping-Andersen's (1990, 26–32) comparative analysis of various welfare states led to the identification of three clusters of 'regime types'.

(i) The liberal welfare state makes state benefits available mainly to low-income, working-class people who become, in effect, state dependants. These states developed in societies with a very strong work ethic; the entitlement rules for state benefits are therefore strict, the level of benefit provided is usually modest, and recipients may be stigmatized for occupying such a dependent position. The state is seen as a 'provider of last resort' for those who have failed to meet their subsistence needs in the market: it will probably both guarantee only a basic minimum of support, and encourage individuals (through subsidies) to provide their own welfare by membership of private schemes (as with health insurance). Australia, Canada and the United States are quoted by Esping-Andersen as typical liberal welfare states.

(ii) The conservative and strongly corporatist welfare state was never obsessed with market efficiency as a fundamental goal, and the granting of social rights was thus rarely contested – in clear contrast to the first category. Status differentials predominate in such societies, however, so that people's rights are closely linked to their class and status. Welfare policies have little redistributive effect (from rich to poor) as a consequence.

These societies – Austria, France, Germany and Italy are cited as paradigm exemplars – are also strongly imbued with religious beliefs,

127

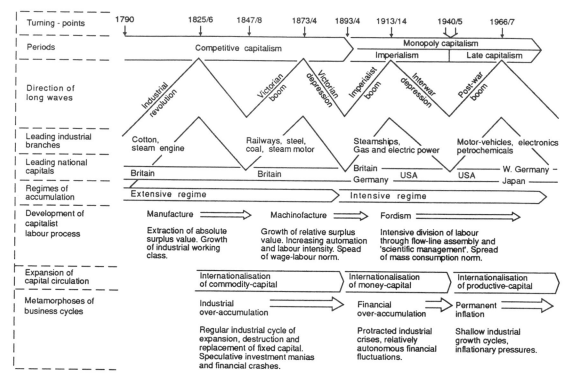

Figure 3.7 *Long waves and their major features since 1790. Source: Marshall (1987, Figure 3.1).*

which influence welfare policy through the commitment to traditional notions of 'family values'. They provide little in the way of welfare for non-working wives, for example; family benefits encourage motherhood; and child-care facilities are few: 'the principle of "subsidiarity" serves to emphasize that the state will only interfere when the family's capacity to service its members is exhausted' (Esping-Andersen, 1990, 27).

(iii) The social-democratic welfare state comprises countries committed to universal entitlements, in a system designed to promote 'an equality of the highest standards, not an equality of minimal needs as was pursued elsewhere'. This implies

first, that services and benefits be upgraded to levels commensurate with even the most discriminating tastes of the new middle classes; and, second, that equality be furnished by guaranteeing workers full participation in the quality of rights enjoyed by the better-off. . . . This model crowds out the market, and consequently constructs an essentially universal solidarity in favor of the welfare state. All benefit; all are

dependent; and all will presumably feel obliged to pay. (Esping-Andersen, 1990, 27–8)

This cluster is the smallest of the three, according to Esping-Andersen, and is dominated by the Scandinavian countries.

As with much social science theory, Esping-Andersen's (1990, 28–9) model is one of ideal-types against which real examples can be compared.[6] Many of the latter will occupy intermediate positions:

welfare states cluster, but there is no single pure case. The Scandinavian countries may be predominantly social democratic, but they are not free of crucial liberal elements. Neither are the liberal regimes pure types. The American social-security system is redistributive, compulsory and far from actuarial. At least in its early formulation, the New Deal was as social democratic as was contemporary Scandinavian social democracy. And European conservative regimes have incorporated both liberal and social democratic impulses. Over the decades they have become less corporativist and less authoritarian.

Nevertheless, states do differ substantially in the

Table 3.1   *The characteristics of organized and disorganized capitalism (after Lash and Urry, 1987)*

	Organized capitalism	Disorganized capitalism
1	Concentration and centralization of capital	Deconcentration of capital
2	Separation of ownership from control: bureaucratization	Growth of a service class
3	Growth of a managerial, scientific and technological intelligentsia	Decline in size of the industrial working class
4	Growth of collective organizations in the labour market (trades unions and professional organizations)	Decline of large-scale collective bargaining and replacement by local bargaining arrangements
5	Growth of corporatism involving the state, large capitals and worker organizations	Increasing independence of major monopolies from nation-states: decline of corporatism
6	Imperialism	Spread of capitalism into Third World and increased global competition
7	Growth of the state to promote capitalism: class nature of politics	Political class dealignment
8	Ideological glorification of science and technology	Cultural fragmentation
9	Concentration of industry within a few sectors and nation-states	Wider range of capitalist sectors in various nation-states
10	Dominance of manufacturing industry	Increased importance of service industries
11	Regional industrial concentrations	Lesser regional concentration
12	Economies of scale and large plant size	Declining plant size and growth of subcontracting
13	Dominance of large industrial cities	Decline of large industrial cities
14	Ideological importance of nationalism and aesthetic modernism	Growth of postmodernism

Source: Lash and Urry (1987, pp. 3–4 and 5–7)

quality of the social rights provided, the nature and depth of their social cleavages, and their inter-relationships among market, state and family.

Esping-Andersen (1990, 69) argues that the clustering of welfare states reflects differences in their 'historical legacies of conservative, liberal and socialist principles'. To illustrate those differences, and produce an empirical clustering, a range of variables was chosen to reflect the strength of each set of principles in eighteen advanced industrial countries. Two variables – the number of major pension schemes available (indicating the presence of status differences among occupations), and the amount of expenditure on pensions for government employees (a measure of what she terms 'etatist paternalism') – were selected to represent *conservatism*. For *liberalism*, the three indicators measured the proportion of state benefits subject to means testing, and the importance of the private sector in the

Table 3.2   *Temporal and spatial changes in liberal, organized and disorganized capitalism (after Lash and Urry)*

	Predominant temporal, spatial and organizational structures	Spatial changes within each territory	Means of transmitting knowledge and executing surveillance
Liberal	Collapse of large-scale world empires and growth of weak nation-states	Growth of tiny pockets of industry in new cities and expansion of commercial centres	Handwriting and word of mouth
Organized	Growing domination of about ten western economies through colonization	Development of regional economies around major cities: regional inequalities large	Printing
Disorganized	Development of global capitalism and world economy	Decline of industrial cities: growth in smaller places: development of service industries	Electronically transmitted information reduces time–space distances and increases power of surveillance

Source: Lash and Urry (1987, 16)

provision of both pensions and health care. Finally, two variables were associated with *socialism*: the percentage of the population entitled to certain benefits, and the degree of equality in benefit provision.

Countries were divided into three groups on the basis of aggregate scores on each group of variables (Table 3.3). On the conservative dimension, Austria, Belgium, France and Germany all scored highly on both variables: they comprise the regimes in which inequalities are greatest, scoring highly on neither liberalism nor socialism. Four non-European countries score best on liberalism, reflecting that their situations of 'social rights are unusually underdeveloped' (Esping-Andersen, 1990, 75). Finally, the four Scandinavian countries (along with The Netherlands) score highest on socialism, reflecting the characteristic universality of social rights there. Within this Table, the four countries considered in more detail in the later case studies indicate: (1) that the United States is among the

welfare states most committed to market provision, and with very little socialist or conservative influence; (2) that the United Kingdom similarly has few elements of the conservative regime, but is 'middle of the road' on both liberalism and socialism – it has some commitment to universal benefits but also substantial reliance on market provision; that (3) Germany is very high on conservatism but average on the other two scales; and (4) that New Zealand (pre-1991!) has few conservative and liberal components to its welfare state-system, but has more commitment (certainly than the United States) to socialist ideals.

Esping-Andersen's (1990, 29) attempt to account for these variations begins by asking:

What is the explanatory power of industrialization, economic growth, capitalism, or working-class political power in accounting for regime types? A first superficial answer would be: very little. The nations we study are all more or less similar with regard to all but the variable of working-class mobilization. And

Table 3.3   *The clusters of welfare states (after Esping-Andersen)*

Index score	Conservatism	Liberalism	Socialism
Strong	Austria	Australia	Denmark
	Belgium	Canada	Finland
	France	Japan	Netherlands
	Germany	Switzerland	Norway
	Italy	United States	Sweden
Medium	Finland	Denmark	Australia
	Ireland	France	Belgium
	Japan	Germany	Canada
	Netherlands	Italy	Germany
	Norway	Netherlands	New Zealand
		United Kingdom	Switzerland
			United Kingdom
Low	Australia	Austria	Austria
	Canada	Belgium	France
	Denmark	Finland	Ireland
	New Zealand	Ireland	Italy
	Sweden	New Zealand	Japan
	Switzerland	Norway	United States
	United Kingdom	Sweden	
	United States		

Source: after Esping-Andersen (1990, 74)

we find very powerful labor movements and parties in each of the three clusters.

The more sophisticated answer focuses on the nature of both working-class political mobilization and the political coalitions built during the transition from a rural economy to a middle-class-dominated society.[7] Thus (32):

In the corporatist [i.e. conservative] regimes, hierarchical status-distinctive social insurance cemented middle-class loyalty to a peculiar type of welfare state. In liberal regimes, the middle classes became institutionally wedded to the market. And in Scandinavia, the fortunes of social democracy over the past decades were closely tied to the establishment of a middle-class welfare state that benefits both its traditional working-class clientele and the new white-collar strata. The Scandinavian social democrats were able to achieve this in part because the private welfare market was relatively undeveloped and in part because they were capable of building a welfare state with features of sufficient luxury to satisfy the wants of a more discriminating public. This also explains the extraordinary high cost of Scandinavian welfare states.

The last point suggests that economic affluence is a necessary condition to the ability to afford a substantial welfare state, along with a political coalition in which the relatively wealthy accept its costs (see also Castles, 1978; Johnston, 1984a).

So what role does political influence play in determining the nature of the welfare state regime? With regard to the size of the social wage – the portion of the national resource distributed by social criteria (i.e. through the state) rather than by the market – Esping-Andersen found that the age of a country's population and its level of economic development were the main determinants: as populations age and become more affluent, the size of the social wage increases and more people become dependent on the state. In more detail, however, the importance of the state in the delivery of pensions reflects inter-country variations in political power: 'left-party power is decisive for de-commodification [i.e. state rather than market provision of benefits], full-employment efforts, and general social democratization. It is also clear that Catholic parties and a historical legacy of authoritarian statehood influence corporatist and etatist biases' (Esping-Andersen, 1990, 137–8). In general, the electoral strength of Catholic parties leads to strong conservative elements in the welfare state whereas 'strong labor movements appear to be a good guarantee against liberal welfare-state stratification' – strong social democracy leads to socialist principles in welfare state organization. Nevertheless, labour movements differ substantially among themselves. Esping-Andersen's earlier (1985) work on the three Scandinavian social democracies showed that Norway and Sweden had a more universal system of benefits than Denmark because, it was suggested, whereas the former two had 'vertical' unionism (i.e. one union would cover an entire industry), Denmark (like the United Kingdom) had 'horizontal' unionism (with separate unions for different occupations within an industry) and a labour movement dominated by the craft unions which were not favourably inclined towards universalism in the provision of benefits (see also Quadango, 1988).

The stress on strong labour movements and political parties which mobilize working-class support points to the importance of struggle by the relatively underprivileged in their search for the benefits that a welfare state brings. Whether that struggle is successful depends on the political context in which it occurs – as Castles (1973) has made clear for Scandinavia, where he links the success of social democratic parties to '[t]he relative failure of the Bourgeois image of society' (142). Occasionally, however, concessions may be yielded by the bourgeoisie without any struggle (even to forestall the development of such struggle), as when democracy was introduced in Spain in the late 1970s without much explicit lobbying for it.

Struggle to obtain political benefits through a welfare state need not be through a party political structure, however, depending on the nature of the local circumstances. In the United States, for example, there has never been a substantial

socialist party operating either at the federal level or in most of the states. Instead, 'poor people's movements' have been organized outside the partisan structure. Piven and Cloward (1971, 3) have argued that 'when mass unemployment leads to outbreaks of turmoil, relief programs are ordinarily initiated or expanded to absorb and control enough of the unemployed to restore order'. Afterwards, they argue, as order is regained, unemployment falls and the potential for turmoil diminishes so the relief programmes are contracted, even withdrawn. There may be some residual groups who require welfare state support; these are usually those 'who are of no use as workers' and 'their treatment is so degrading and punitive as to instil in the laboring masses a fear of the fate that awaits them should they relax into beggary and pauperism'. Thus 'periodic expansion and contraction' of relief programmes is 'made necessary by several strains towards instability in capitalist countries' (4). That conclusion may reflect the particular circumstances of the United States, however, and the lack of a major political party that will promote the interests of the relatively poor continually, mobilizing their political support for its programmes, whatever the current economic circumstances. The main difference between the United States and the Scandinavian countries in the size, strength and relative permanence of their welfare states, therefore, may reflect major differences between the two in the nature of political mobilization strategies and the degree of representation of 'poor people's interests' (see also Quadango, 1988).

King (1989) has generalized this conclusion further. He identifies several competing explanations for the growth of welfare states, and concludes that (239):

Firstly, welfare state policies and institutions follow economic transformation in Western societies – this is a necessary condition for their emergence. Secondly, the factors determining their original adoption differ from those facilitating their subsequent expansion. Thirdly, to a significant extent, the welfare state represents accommodations between capital and labour mediated through and by the state; such accommodations have frequently taken a corporatist form though some polities have resisted this arrangement. Fourthly, welfare policies and institutions have themselves influenced the processes of conflict and capital–labour accommodation underlying them.

Thus the growth in welfare provision began with the Kondratieff downturn in the late nineteenth century, so that by the 1930s most western democracies had accident, old-age, sickness and unemployment insurance policies. They lacked a coherently formulated overall view of the welfare state and its long-term role within capitalism, however. This came after the next Kondratieff downturn, when strong left-wing political parties which enjoyed prolonged incumbency of elected power were able to implement 'a range of policies collectively constitutive of the welfare state (and thus significantly constraining the power of capitalism) and allowing an expansionary role for the state' (242). Thus, struggle led first to short-term concessions through temporary relief programmes, and subsequent political strength through the ballot box enabled the translation of those temporary gains into a more permanent structure. The former was characteristic of the United States in the 1930s (as Piven and Cloward's case studies, 1971, 1977, illustrate), but the latter was absent, hence the lack of a substantial welfare state there in the 1950s. (For further discussion of the American case, see the vignette on the United states below, p. 154.)

In their seminal contribution to the appreciation of voting patterns in western European countries – what they term electoral cleavages – Lipset and Rokkan (1967) argue that in most countries the party structure has been 'frozen' since the 1920s. Thus in those countries which had not experienced substantial support for social democratic parties by, say, the Second World War, their likelihood of success (or of success for the voter interests that they mobilized) thereafter was remote. To this extent the variations in the nature and the extent of the welfare state apparatus established in different countries by that time should also have been frozen. As Harrop and Miller (1987) have pointed out, much has happened in many countries in the decades since Lipset and Rokkan wrote, with the growing dealignment of electorates (see Johnston, 1987, 1990). Nevertheless, in each of the four countries discussed in the vign-ettes below, the parties which dominate the political agenda now are those which did so in the 1950s.

The corporate state

The welfare state was a major political response to the legitimation crisis of the 1920s/1930s. Alongside its provision of welfare benefits for those needing support – either because of their dependency situation on others (such as children and the aged) or because of their inability to earn sufficient to meet their (socially-defined) basic needs – the state became actively involved, increasingly so after the 1920s/1930s downturn, in the creation of employment and in planning the economy. Until then it had largely played a permissive role only, benefiting most those with greatest political power, as in the struggles over protection and freetrade. The reaction to the rationality crisis of the 1920s/1930s was the growth of the corporate state.

Initial reactions to the recession in many countries were to blame labour for the problems of capital – and thus, ultimately, for its own problem of unemployment. Wages were too high, it was argued, and workers would have to accept lower rates if their employers were to compete successfully and to ensure continued job provision. The state's role, according to this view, was to assist employers in negotiating new contracts with their workers – for example, by helping them to resist strikes (as was common in the UK in the 1920s). This ideology rested on 'Say's Law' – many economists' belief that the free operation of market forces (including the price of labour) would ensure that the economy always returned to an equilibrium position characterized by full employment. In that context, all that the state had to do was ensure that individuals and collectives (such as trades unions) did not hinder the free operation of markets.

Two arguments which countered Say's Law emerged. The first, later known as *Keynesianism* after the leading British economist of the time, was that the state should stimulate short-term demand during a recession, and thus encourage investment, production and job-creation. This would generate multiplier processes, with wider employment and other consequences. Instead of the state seeking to reduce its spending (and also its borrowing to finance that spending), therefore, it should increase expenditure. Keynes rejected Say's Law arguing, as Galbraith (1989, 222) puts it, that: 'The modern economy . . . does not necessarily find its equilibrium at full employment: it can find it with unemployment – the underemployment equilibrium. Say's Law no longer holds; there *can be* a shortage of demand. The government *can and should* take steps to overcome it.' If unemployment brought down wages – and people accepted lower incomes in order to obtain jobs, even if this meant a lower quality of life than heretofore (though a better one than the situation if they were unemployed) – this could lead to an equilibrium at even lower levels of output and employment, and a reduction of effective demand, without which investment would not be forthcoming. Nor would low interest rates – to encourage borrowing – necessarily lead those with capital to invest in the mobilization of productive forces, if they saw no likely profitable outcome. They would just store their capital – invest it unproductively (in works of art, for example).

The solution according to Keynes, again as expressed by Galbraith (1989, 234–5), was that: 'There remained one – just one – course. That was government intervention to raise the level of investment spending – government borrowing and spending for public purposes. A deliberate deficit. This alone would break the underemployment equilibrium by, in effect, spending – willfully spending – the unspent savings of the private sector.' Government borrowing was to be used to stimulate demand through programmes of public works, for example, alongside those of welfare payments which would sustain consumption. (Its debts would be repaid later, with interest, from the income obtained from taxing the greater wealth that would be created once full employment equilibrium had been reached and the free market set free again.) The Keynesian solution was thus not a socialist one, for the state to replace the market, but rather a neo-liberal one in which the state periodically stimulated the market to ensure that it could meet the collective goals of full employment and widespread relative prosperity.

The second counter to Say's Law and the 'minimal state' conception went much further than Keynes, arguing for much greater and continuous government involvement in the free market. This argument drew on the case for socialism based on the growing – Marxist-based – appreciation of the permanently exploitative basis of market capitalism. The dynamics of capitalism do not operate in the interests of the working class, it was argued, but rather in those of the owners of capital whose ability to shift their investments (either to other locations in the same country or internationally) substantially threatens the interests of workers and their dependent communities. Further, such shifts are likely to have the greatest impact on those groups of workers who have won the greatest concessions from their employers: 'class struggle within a territory may force capitalists or labourers to look elsewhere for conditions more conducive to their respective survival. . . . Capital flight . . . is as typical a response to working-class victories within a territory as is individual worker mobility to avoid the more vicious forms of capitalist exploitation' (Harvey, 1985, 147–9).

Socialism was seen as the solution to this problem; public ownership of the means of production, distribution and exchange through the state would ensure that wealth was created in the interests of all and that the planning of wealth creation would avoid the cyclical crises that are endemic to capitalism. The argument never won sufficient support within an advanced capitalist country for it to be approved (in full) at the ballot box, though it was introduced through imposition in some less advanced countries (see below). But sufficient people in some places were convinced of parts of the case for moves towards socialist goals to be initiated, as parts of the programmes of electorally successful left-wing parties. In some countries, these involved nationalization (taking into public ownership, usually with compensation) of what were considered to be 'the commanding heights of the economy'. Elsewhere, a new ideology of state operations – later termed *corporatism* – was introduced, with much greater government involvement in planning for and regulating the private sector economy than acceptance of Say's Law called for.

Corporatism (frequently termed neo-, quasi-,

liberal or societal corporatism, according to Grant, 1985, 1) is concerned with the relationships between the state on the one hand and 'interest groups based on the division of labour in society' on the other. Those groups are based on neither their members' preferences for certain collective ends (as with political parties, for example), nor on common ethnic or other characteristics. And, as Cawson (1985, 4) points out, 'producer groups (including trade unions as well as employer and professional organizations) are more likely than consumer groups to develop significant corporatist relationships with state agencies' because the former are more easily mobilized than are consumer groups, in most situations.

Corporatism's fundamental characteristic is the inter-penetration of private and public power, with the participating interest groups not only influencing state policy but also assisting in its implementation. Selected interest groups are thus incorporated by the state into the decision-making arena, in order to obtain their collaboration: the quid pro quo for the privilege of being included in the corporatist arrangements is willingness to accept the outcomes and to promote them. This leads to Grant's (1985, 3–4) definition of corporatism as:

a process of interest intermediation which involves the negotiation of policy between state agencies and interest organisations arising from the division of labour in society, where the policy agreements are implemented through the collaboration of the interest organisations and their willingness and ability to secure the compliance of their members. The elements of negotiation and implementation are both essential to . . . understanding . . . corporatism. The arbitrary imposition of state policies through interest organisations, without any prior negotiations, does not constitute liberal corporatism . . . equally, the negotiation of understandings, with no obligation on the part of interest organisations to secure the compliance of their members, does not constitute a corporatist arrangement . . .

This implies that those in control of the state apparatus choose to share power, in order to enhance their control and to aid their attempts, through successful policies, to win control again. To some this implies that those who win power through the ballot box then allow its partial exercise by other groups who do not submit themselves for voter approval, which is contrary to the basic tenets of liberal democracy. Failure to achieve the goals of policies implemented through corporatist bargaining arrangements is thus likely to pose a substantial threat to the state's legitimacy – the price of sharing power with the unelected and not then promoting the welfare of large segments of the electorate is likely to be a loss of power, perhaps to parties (or other particular actors) which are not attracted to the corporatist model.

Corporatism originated in the sharing of power between the state and the church in many parts of Europe (Esping-Andersen, 1990). The general tradition was enhanced, according to Grant (1985, 6), in the circumstances of post-World War Two reconstruction, as in Austria where there was 'a new willingness to attempt to bridge differences between capital and labour in the task': it was also strengthened, he argues, in those countries where substantial nationalization had weakened the relative power of big business, producing a more even balance between capital and labour and thus the willingness of both to share power.

The particular focus of most corporatist arrangements is macroeconomic policy, involving what Atkinson and Coleman (1985, 26) describe as 'the disciplining of wage labour in the administration of incomes policy . . . and the enhancement of regime governability through the control of demand overload. . . . Thus the tripartite board responsible for developing macroeconomic policies, from tax expenditures to wage restraints, has come to be seen as *the* classic corporatist institution.' This is needed, according to Grant, because governments accepted the Keynesian policy objective of full employment. It tends to stimulate 'wage-push inflation', the control of which calls for some form of incomes policy that can be worked out and then delivered through the corporatist mechanism. The key interest groups working with the state in the corporatist structure are therefore the large employers (and their collective bodies) and the major trades unions.

Incomes policy can extend beyond decisions on the level of remuneration to a wide range of issues relating to the terms and conditions of employment, incorporating most of industrial relations. Grant (1985, 13–14) illustrates this with reference to training. Individual firms are unlikely to invest in the type and amount of training which is in the country's long-term interests unless they can get immediate and substantial returns for their money, but government imposition of a standard policy, or provision of a standard subsidy, is unlikely to be closely attuned to the needs of different sectors of the economy. Allowing employers and the unions to participate in the creation of a training policy, on the other hand, both ensures that it is tailored as far as possible to the needs of each of the three groups, and also ensures that it will be supported by those who are party to it. This is especially important at times of economic restructuring, during which many skills become redundant and large segments of the labour force have to be retrained before they can obtain employment in either new production processes or new service industries.

In an extension of the corporatist model, Saunders (1985) has promoted what has become known in the UK as the 'dual-state thesis', which separates political processes relating to production issues (i.e. to macroeconomics) from those relating to consumption issues (i.e. the provision of public services). The former, he argues, are more readily addressed through corporatist arrangements, whereas the latter involve conflict in the electoral arena. Furthermore, he claims that the corporatist processes tend to operate through the central state, whereas the conflicts over public services focus on local governments. His own data verify the claim that corporatism is the dominant mode of mediation in issues relating to production interests, but are less clear on the second: corporatism can operate to some extent at local levels (as illustrated by the 'boosterism' characteristic of state and city governments in the USA).

In the corporate state, therefore, the state's commitment to the pursuit of full employment and high levels of welfare – which in turn call for competitive advantage in the world-economy – involves the incorporation of major production interest groups in the decision-making processes. The state thus becomes part of the regional alliance for its territory. It shares its power with the other groups, in the belief that such power-

sharing is the best way of achieving its political aims (including approval for its own performance, which is crucial in liberal democracies where the hold on power is subject to regular public accountability and voters are increasingly influenced by government performance: Pattie and Johnston, 1990).

Just as the nature and extent of the welfare state varies substantially among the core countries of the world-economy, so too does the nature and depth of the corporate state. Lehmbruch (1982) has identified three groups of countries according to their level of corporatism. His paradigm exemplar of *strong corporatism* is Austria, characterized by a co-operative incomes policy and the constitutional right of the major interest groups involved (business, agriculture and labour) to give opinions on all parliamentary bills and to be represented on its advisory committees: indeed, he suggests that this representation is so strong, with the committees dominated by the interest groups, that Austria's government system is 'parliamentary corporatism' (17). Sweden and The Netherlands are other examples of strong corporatism.

Lehmbruch identifies Denmark and West Germany as exemplars of *medium corporatism*, with the UK as a 'borderline case' (19). Denmark differs from Austria and Sweden in the greater pluralism of its trade union structure, with antagonisms, for example, between blue- and white-collar unions and with the survival of traditional craft unions (Esping-Andersen, 1985). The participation of interest groups in parliamentary activities is less well developed in West Germany than in Austria, as they have no right of consultation in the former. Corporatist arrangements in the UK are hampered by 'intensely held beliefs about the sovereign power of Parliament and parliamentary government' (21): further, he considers that the attempts to develop a clear tripartite structure under Labour in the 1960s and 1970s 'largely failed'.

France is chosen to represent *weak corporatism*. A structure exists but has not been implemented because of 'ideological intransigence on the part of the labour unions', which, with the exception of one union's relationship with the Communist Party, have few links with the political parties. Further, Lehmbruch argues that

French unions are generally weak and divided, their membership is relatively small and their leaders have little power. Business, too, is not as co-ordinated as in the United Kingdom.

The free economy and the strong state

The welfare state and the corporate state were designed in the 1930s and 1940s to ensure full employment and widely-shared continuous growth in prosperity. The 'post-war boom' of the 1950s and 1960s suggested that this had been achieved (Figure 3.7) and gave rise to considerable optimism, not only for the future of those already-prospering countries, but also for those which had yet to achieve the 'take-off into sustained growth' (Rostow, 1971). But that optimism was rapidly diffusing by the 1970s, as serious economic difficulties were again experienced in countries where they were thought to be a thing of the past. Their initial appearance and cause were somewhat concealed by the problems engendered by the oil price rises of the early 1970s, but by late in the decade it was clear that most of the world's major economies were in severe difficulties. The welfare state and the corporate state had assisted countries to escape one major recession it seemed, but had failed to prevent them entering another five decades later.

Not surprisingly, the appearance of the new recession stimulated much debate about its causes and the relevance of the by-then traditional Keynesian policies of state-led demand management to the solution of the problems of underconsumption and large-scale unemployment. Whereas some retained a belief in the efficacy of corporate state policies and their commitment to the welfare state, others questioned their relevance to the new set of problems. Solutions were suggested which involved very substantial dismantling of both, with their replacement by a strong (yet significantly reduced) state whose major role was to facilitate the full operation of the free market.

Critiques of the welfare and corporate states are many and wide-ranging. They are examined here under four headings.

The culture of dependency and the welfare state

The crux of the argument against the welfare state is cogently presented by Bennett (1989), whose goal of promoting a reconsideration of the role of geography in what he terms a 'post-welfarist world' is considered in a separate essay (Johnston, 1992). He argues that there has a been a major recent shift in what he terms the 'culture of the times', away from a 'framework of welfarism: of improving the quality of life through collective and government intervention . . . [with] a focus on relative deprivation and social, as well as spatial inequality' (Bennett, 1989, 272).

The problem with 'welfarism' is that it promotes a 'general culture of expectations' (Bennett, 1989, 277) and a confusion of needs with desires. The welfare state's original mission was to care for the most disadvantaged, but the culture of 'entitlement' which it encouraged led to increased demands for state intervention to increase the lot of those suffering 'relative deprivation'. Rising prosperity was associated with rising expectations, which became the basis for the operation of the welfare state.

This in turn leads to the conclusion that the lack of particular material goods is not only a misfortune, but a *social injustice*. If there is a social injustice then individuals have an obligation to the state to enable it to relieve injustice; and this moral duty, with the legislation of the welfare state, is transformed into a *legal right*, or 'principle of entitlement', for those who are relatively disadvantaged. (278)

Many of the indicators of relative deprivation developed by scholars such as Townsend (1979), therefore, 'are not measures of needs but of inequality in the fulfilment of wants, and those perhaps primarily of inequality in "style of life". By this argument, deprivation can apply to any difference in habit, preference, capacity, even personal character' (Bennett, 1989, 278).

With specific reference to spatial variations in levels of relative deprivation, Bennett (1989, 279) continues that:

The social context of geography leads naturally to regional inequalities being identified as a major focus: inner city versus suburbs; rural versus urban; North versus South in Britain; 'north' versus 'south' in international development; and so on. . . . Differences exist because of geography. Geographical as well as other restrictions on choice, differences in lifestyles, differences in quality of life, access to certain goods and services can be identified. Because they can be identified, under a welfarist view, *morally* they should be overcome or at least ameliorated.

Under a welfare state the goal has become the removal of all differences, thereby, he asserts, bringing all people to the same condition in the aspect of life being considered; any variations between people and places are indicators of relative deprivation, which the welfare state has been created to eliminate.

A consequence of this culture of entitlement, according to Bennett (1989, 285), is that: 'If every difference in outcome is translated into an entitlement to state intervention, then clearly the only end point is total state intervention in everything. This is certainly incompatible with the market economy.' It is also incompatible with individual freedom according to other critics of socialism (e.g. Popper, 1945, and Hayek, 1944 – whom Bennett does not cite). Until the 1970s there was, in the UK at least, what Bennett terms a 'welfare compromise' (286) between the operation of capitalist markets and the state's desire to promote 'beneficent planning of life and work'. Since then the compromise has broken down. Thus:

It is no longer credible intellectually or politically to argue for many previous forms of government management or intervention. Mass public housing, demand-inelastic systems of health care, education objectives which seek to equalize outcomes rather than potentials, 'full employment' Keynesianism, regional policy, even the existence of local government, are each examples where policies have been developed over the past forty years, but where there have been significant and recurrent failures. (287)

and

Thatcher, in particular, has sensed that in Britain a 'sea change' has occurred in which old structures no longer have their previous relevance or power. . . . Thatcher has been able to translate the image of capitalism from 'dark satanic mill' to shopping mall: capitalism has become seen as the means of creating and distributing the good things in life. Where social theory [which Bennett links to the 'relative deprivation thesis' and with socialism] emphasizes the negative aspects of capitalism's capacity to create new wants and hence new 'relative deprivation', the emergent 'culture of the times' has been happier to see the market as both the

creator and the provider of new wants. Rather than markets being seen as an inhumane and exploitative system, socialism and even corporatist social democracy have come to be associated with the odious and paternalistic treatment of individuals. . . . Where the Thatcher era has heralded consumer choice and economic change, social theory and socialist politics has sought to defend the mode of production and to trap people in labour-intensive work practices and unattractive jobs vulnerable to technological change; the spirit of market freedom of individuals has heralded a consumer and service economy which has offered the release from the least attractive toils and labours, and has seemed to offer the potential to satisfy many of people's most avaricious dreams. (286)

Bennett's promotion of 'New Right' views is typical of many. Some virulently attack the assumed beneficence of governments which underpinned the welfare and corporate state: 'The "New Right" . . . must stress unremittingly the enduring moral bankruptcy of government. It must constantly compare the burden borne by the taxpayer, to fill the governmental trough from which the interest groups are feeding, with the benefits received by the swine at the trough' (Anderson, 1985, 123). Others are more thoughtful. Saunders (1985, 166–7), for example, after reflecting on his major research interest in housing provision, argues that:

From what I have read, seen and experienced, it seems clear to me that in Britain (though not necessarily elsewhere) ownership of a home is important for many people, not only economically (in terms of the capital gains which may accrue, the potential for transfers of wealth between generations which it creates, the opportunity to use personal labour to enhance the value of the property through DIY, and so on), but also culturally and socially. Ownership of a home brings with it some limited opportunity for autonomous control – over where you choose to live, how you choose to live, and so on. The popular and widespread desire to achieve owner-occupation can in my view mainly be explained in terms of a search for a realm of individual autonomy and freedom outside the sphere of production.

He then generalizes from that conclusion:

Gradually, and not without considerable discomfort, I came to realise that if it was right and desirable for people to participate in the market in order to accommodate themselves so too was it right and desirable for them to assume more control in other parts of their lives wherever possible in the sphere of consumption.

Thus he supports privatization policies for health care and schooling, for example, as long as there is no consequential stigmatizing and penalizing of the poor who, for good reason, are unable to participate in a market-based provision of major social services. His goal is an end to 'the continued domination of our lives by people of institutions we do not understand and cannot control' (172).

These modern New Right authors, who condemn the welfare state and its corporatist arrangements, are repeating arguments made decades ago. As Hayek (1985, vii–viii) reiterated, 'The socialist argument continues despite its many defeats – on the impossibility of economic calculation under collectivism, the fallacious claims for the use of markets under socialism, the incompatibility of liberty and state direction of the economy, and many others.' He claims that 'The argument is being won by the new liberals of the late 20th century.' Esping-Andersen (1990, 33) is less sure, arguing that

The risks of welfare-state backlash depend not on spending, but on the class character of welfare states. Middle-class welfare states, be they social democratic (as in Scandinavia) or corporatist (as in Germany), forge middle-class loyalties. In contrast, the liberal, residualist welfare states found in the United States, Canada and, increasingly, Britain, depend on the loyalties of a numerically weak, and often politically residual, social stratum. In this sense, the class coalitions in which the three welfare-state regime-types were founded, explain not only their past evolution but also their future prospects.

The variations among welfare states in their response to the recession of the 1930s are likely to lead to variations in the strength of the New Right in the 1990s, therefore.

Bureaucratic inefficiency and overload

The welfare and corporate states are characteristically operated by a large, bureaucratic state apparatus, which is increasingly being attacked as inefficient – because of its size, because of its immunity to competition, and because of the ever-increasing demands on it made through the 'drive to universalism' at ever higher levels of perceived relative deprivation.

Analysts of bureaucracies – and of state apparatuses in general – suggest that their quasi-autonomous status allows those controlling them to develop strategic plans for their bureaux' futures which are not necessarily consistent with the initial reason for their creation. Owners and managers of firms in the private sector aim to maximize profits (within certain constraints, such as the degree of risk they are prepared to take). Bureaucracies do not make profits, so what do their managers seek to maximize? Most public choice theorists argue that the goal is to maximize bureau size, and especially its budget, because (Dunleavy, 1991, 155):

larger budgets help bureaucrats push up salaries and fringe benefits (such as pensions), since increased responsibilities merit higher remuneration. They improve officials' promotion prospects, since budget scale increases normally trigger regrading directly or via increases in staffing and bureaucratization. Officials can more easily divert resources into creating perks . . . Bureaucrats in larger agencies have enhanced public reputations for influence, and higher status among their peer group.

Winning large budget increases may be seen as a mark of ability also, and can give the officials concerned greater powers of patronage: they also provide flexibility for restructuring the bureau and for achieving the officials' goals with regard to key personnel. (Downs (1967) argues that bureaucrats seek to maximize on five criteria: income; power; prestige; security; and convenience. Niskanen's (1973) list includes: salary; perks; public reputation; power; patronage; ease of making change; ease of managing the bureau; and bureau output.)

Because bureaucrats' behaviour has developed along these lines, they have created an environment in which bureaux are expected to grow: 'Sponsors expect to be presented with proposals for enlarged funding . . . If no increase is asked for, the sponsor will be thrown into confusion', in part because if the sponsor is not rewarding an agency with a larger budget, its political strength is in dispute too (hence, for example, the fights that British Cabinet Ministers have with the Treasury each year in the contest for more resources).

The strength of bureaucracies is almost invariably enhanced by their monopoly situation. Because of the lack of competition, and thus of yardsticks against which to compare their efficiency, many bureaucracies are able to obscure the real costs and benefits of their operations (assuming that they know them). They may enter into contractual arrangements with private-sector suppliers of goods and services which further obscure that situation, and can lead to substantial inefficiencies also: this is often argued to be the case with defence ministries, for example, and their 'membership' of the 'military-industrial complex' made notorious by Eisenhower.

Critiques of the operation of bureaucracies from the New Right perspective focus on their inefficiency. Dunleavy (1991, 161) concludes his (not favourably inclined) exegesis of their arguments that 'Essentially, all bureaucrats everywhere seek to maximize their budgets by radically oversupplying outputs' and quotes Niskanen (1973, 33) to the effect that 'All bureaus are too large. For given demand and cost conditions both the budget and the output of the bureau may be up to twice that of a competitive industry facing the same conditions.'

More sympathetic critiques of bureaucracies argue differently. Dunleavy (1991, 209) himself, for example, contends that:

The characteristics of public service employment systems make it likely that the welfare of higher-ranking bureaucrats is closely bound up with the intrinsic characteristics of their work, rather than near-pecuniary utilities. Rational bureaucrats therefore concentrate on developing bureau-shaping strategies designed to bring their agency into line with an ideal configuration conferring high status and agreeable work tasks . . .

From this he concludes that bureaucrats will by-and-large accept the arguments that monolithic welfare state agencies should be either privatized or split up (or both) because these policies allow them to reshape their bureaux in line with their personal preferences.[8] This is consistent for different reasons, with the New Right case that putting as much as possible of the welfare state into the market place will make it more efficient and effective, as well as accountable.

A further argument against the welfare state and the large bureaucracies which populate it is closely linked to Bennett's claim that, because the state is called upon to cater for ever-increasing aspirations, it must inevitably fail. Held (1987) has identified two interpretations of this case. He terms the first interpretation the *overload thesis* (drawing, among others, on Brittan, 1977), illustrated by the flow diagram in Figure 3.8. His analysis begins with two basic components of the liberal-democratic, welfare-corporate state: the commitment to Keynesian economic policies to deliver full employment and prosperity; and the role of the state as an arbiter between competing groups operating within its democratic structures. Its success in these two, especially the former, leads to higher aspirations, which are then transmitted to the state through the electoral and corporate processes. At the same time, the 'decline in deference' associated with prosperity and democracy means that the relatively deprived are less willing to accept their positions than in the past, and are more ready to use their political strength to promote demands. Politicians are prepared to respond to those demands, thereby courting (electoral) popularity: they may well offer more than they can deliver, which only serves to increase expectations. Inability to deliver is rarely accepted by governments, however, who instead seek to retain popularity by 'appeasing' the various interest groups with further promises. The size of the state bureaucracy thus increases, which not only stimulates inefficiencies, but also precludes firm economic management: 'Public spending becomes excessive and inflation just one symptom of the problem' (232). As the state expands, however, so it erodes the wealth-creating private sector and increasingly impinges on individual initiative. A vicious cycle is initiated (cells 4–11 in the figure, inclusively) 'which can be broken only by, among other things, "firm", "decisive" political leadership less responsive to democratic pressures and demands' (233).

Adherents to the overload thesis argue 'that the form and operation of democratic institutions are currently *dysfunctional* for the efficient regulation of economic and social affairs, a position broadly shared with the New Right' (233). A Marxist interpretation of the same phenomena (after Habermas, 1976) stresses the *legitimation crisis for the state* as a necessary consequence of the inherent contradictions

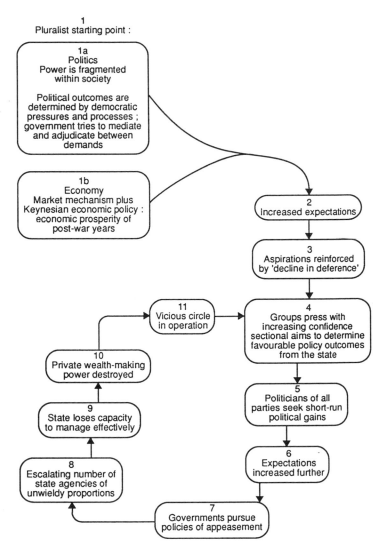

Figure 3.8 *The crisis of the overloaded government. Source: Held (1987, Figure 7.2).*

within capitalism: Held also illustrates this dia-grammatically (Figure 3.9).

This second argument builds on the inherently cyclical nature of growth and recession typical of a capitalist liberal democracy, in which the state both manages the economy to promote private profit-making and mediates competing claims for resources. The state is continuously involved in regulatory activities designed to avoid rationality crises, while at the same time it negotiates with interest groups through corporate structures to ensure their support. A major way of avoiding

rationality crises in the private sector is for an ever-growing proportion of responsibility for the economy and civil society to be assumed by the state, which must be financed through increased taxation and borrowing. Those policies operate counter to the requirement to promote accumulation, however, because high rates of taxation depress initiative and substantial state borrowing forces up the price of money: as a consequence, state action stimulates a 'situation of almost permanent inflation and crisis in public finance' (235).

Faced by such fiscal problems, the state has a rationality crisis: it cannot develop a coherent strategy which will meet all the demands placed upon it. The crisis will emerge whatever the current government's ideology. A right-wing government cannot drastically reduce its spending on public services (and hence its taxation demands, thereby encouraging initiative) because of the potential disruption which trades unions and other groups might press;[9] a left-wing government cannot promote high levels of spending and taxation because of the impact on investment, especially in an increasingly global economy in which mobile investment may rapidly move away from perceived socialist liberal-democratic states, with drastic impacts on

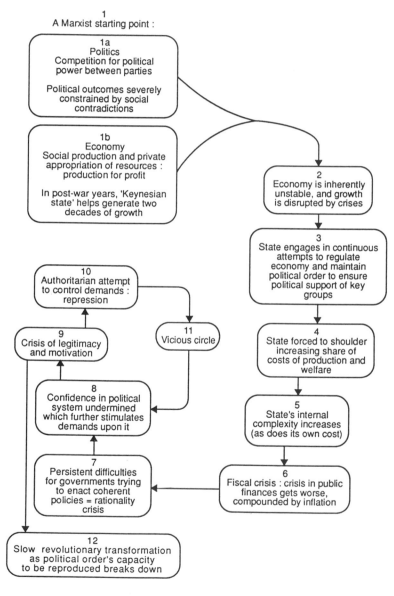

Figure 3.9 *The legitimation crisis of the liberal democratic capitalist state. Source: Held (1987, Figure 7.3).*

the local economy. Thus, Held concludes: 'the result is a pattern of continuous change and breakdown in government policy and planning (e.g. a 'stop – start' approach to the economy, the fluctuating use of an incomes policy) . . . governments of different persuasions come and go, and policy chops and changes, (235).

According to Held's version of the Marxist model, a consequence of the rationality crisis (box 8 in Figure 3.9) is that as the state becomes more involved; so more aspects of life become potentially (if not actually) politicized, and the demands on it increase further. If it is perceived to fail, then it may face a legitimation crisis also, as increasing proportions of the population lose faith in the state's ability to meet their needs. The likely outcome, Held suggests, is either the emergence of a 'strong state' which takes authoritarian measures to repress dissent and defuse the crisis forcefully, or a revolutionary transition to a new social formation as the political order breaks down. The former is not a viable strategy, at least in the long term, according to the model, because the authoritarian repression only furthers the legitimation crisis: eventually the political order will collapse.[10]

Held identifies several common threads to these two theoretical positions, pointing out that: 'both overload and legitimation crisis theorists claim that state power is being eroded in the face of growing demands: in one case these demands are regarded as 'excessive'; in the other they are regarded as the virtually inevitable result of the contradictions within which the state is enmeshed, (237).

He identifies no clear evidence of either a worsening crisis in the state's legitimacy, however, nor that the state's power is eroding. Particular governments may be vulnerable, but not the state itself. This, he suggests, is because dissensus has not been translated into extensive criticism of the existing economic and political order, and 'the widespread scepticism and detachment of many men and women in their attitude to traditional forms of politics has not given way to any clear demands for alternative kinds of institutions: there is a clear absence of images of alternatives, except among margarial groups' (239). Furthermore, governments themselves have been able to prevent substantial ero-

sion of their power by continuing to negotiate either compliance or consent from a large segment of their subject populations. They do this using what Offe (1984) calls 'strategies of displacement' which:

disperse the worst effects of economic and political problems onto vulnerable groups while appeasing those able to mobilize claims most effectively. This is *not* to argue that politicians or administrators necessarily desire or intend to displace the worst effects of economic problems onto some of the least powerful and most vulnerable of society. But if politics is the 'art of the possible', or if . . . elected governments will generally try to ensure the smoothest possible continuity of the existing order (to secure support, expansion of economic opportunities, and enhanced scope for their policies), then they will see little option but to appease those who are most powerful and able to mobilize their resources effectively. (Held, 1987, 240)

The political capacities of governments are substantial, because of their ability both to reward supporters and to marginalize opponents (yet not antagonize them too far).

The perils of monopoly and the corporate state

The other three critiques of the welfare and corporate state are in part components of the previous two, hence the briefer presentations. They are introduced separately because as components of the general critique they have been important in stimulating reformist ideas.

The creation of monopolies is a 'natural consequence' of the operation of a capitalist economic system: Engels saw monopolies as 'necessary transformations' of capitalism, and their replacement by the state as subsequently necessary also (Harvey, 1982, 137). The centralization and concentration of capital into fewer and fewer hands is an outcome of competitive success. Within any sector of the economy, the successful firms progressively eliminate their competitors until they control production therein (individually or oligopolistically). The surplus reaped from that monopoly (or near-monopoly) may not be reinvested in the sector, however, because the potential returns are low: instead it is used to diversify the firm's interests in other sectors. Thus, success in one sector leads to the search for it in others, so that economic power becomes concentrated in a few successful corporations whose interests cover many industrial and service sectors, and many countries too.

Monopolies emerge from competitive success, as the reward for efficiency. But monopolies are usually inefficient, because they lack the spur of competition! The owners and managers of monopolies can exact surplus value through their power over prices without necessarily continuing to invest in efficiency increases, and so economic growth may slow down.

Countering the economic power of monopolies calls for similar strategies by their employees. Most trades unions began as organizations to represent the employees (or just certain groups of employees defined by occupation) at one workplace, and slowly they either federated or amalgamated so that groups of workers in particular occupations in the same industry were represented by the same union – which could then work for equality in their treatment across the industry. With the concentration and centralization of capital, however, has come an increasing concentration and centralization of unions. It becomes ever-more difficult to separate the interests of one group of workers in an industry from those of others, and with employers no longer concentrating on one industry the interests of several groups of workers also become more interdependent: all want the same treatment from the same employer. Thus unions have merged, presenting the major employers in the corporate state with the united front of major workers' organizations.

Large corporations faced by large trades unions are dealing with monopoly power, for if the unions used their strike weapon they could substantially damage an economy in ways less likely to occur if there were more employers and unions. Corporations may react to this potential threat by a willingness to meet union demands – which they can then counter by using their monopoly power to raise prices – rather than risking industrial action and a loss of income. This, according to the critics, is a further element of inefficiency in monopoly capitalism – with the losers (as Brittan, 1977, 194) argues, being 'other workers, who are either non-unionised or are in weaker unions'.

Under corporatism, therefore, the state is involved in advancing the inefficiencies of a capitalist system in which monopoly employers deal with monopoly unions. The state becomes compromised in the situation, and indeed may encourage employers to accept union demands: strike action may then be seen (and even presented)[11] as a threat to the state's legitimacy (Brittan, 1977, 196):

> Governments live in the knowledge that particular groups may hold up essential services unless their demands are met. The fear that people may starve, or freeze, or be faced with a mass breakdown in sanitation and health in the face of a showdown with one or more key unions has become the dominant force behind too many political decisions.

Thus the corporate state sustains monopolies, and so contributes to the growing inefficiency of the economic system.

Profit is the lifeblood of capitalism: the potential to make profits is its stimulus, and investment will flag, the economy will stagnate, unemployment will grow, and the state will face rationality and legitimation crises when that stimulus is weak. To avoid those crises, the state must ensure the presence of the stimuli so that people do invest in the search for profits. Investment involves taking risks. Profit is the return on a risk to the individual who invests capital in the expectation of gaining more: a high salary, perhaps performance-related, is the reward for the people who manage that capital – both those who invest it for individual owners (as with banks and pension funds), and those who use it to create saleable goods and services. If profits and salaries are low, then risk-taking is perceived as not worth it. The same is true if the profits and salaries are taxed highly by the state.

Under welfarism, the state needs a large income in order to pay for the services that it provides: as the demands on it increase, so its income must grow, with clear implications for the levels of taxation on incomes and profits. (Indeed, because the provision of state services is labour-intensive, then their costs will probably grow more rapidly than those of goods and services provided through the private sector, with further taxation consequences: strong unions within the public sector could drive those costs up even further.) Thus the welfare state – especially the liberal welfare state – can create its own rationality crisis by depreciating the value of risk-taking.

According to this argument, the welfare-corporate state contains within itself the seeds of its own destruction, because it cannot stimulate the creation of sufficient profit. It must create more incentives, which means better returns on investment for those with available capital and better salaries for those who manage it. This means lower taxes, so that people who take risks see more of the benefits. And lower taxes means reducing the size of the welfare state.

Associated with this case is a linked argument that the welfare state's monopoly is a major restriction on individual freedom. Within a country people are socialized into a wide range of cultures which influence their relative demand for goods and services over and above those of basic subsistence: their requirements and priorities vary. As the welfare and corporate state becomes involved in more and more aspects of economy and society, however, so people's ability to exercise choice is reduced. The state rarely offers much choice, and certainly less than the private sector does. Indeed, it is virtually forced to 'regress to the mean' by focusing its spending on that which is acceptable to the greatest mass of the population (which will return the incumbent government to power at the next election). The state is thus a force for conformity and 'sameness' rather than for liberating human variability; as such, as it grows, so it is a further constraint to economic and social development.

A possible solution to this situation is, as far as possible, to have not one state within a country but many, providing a market in 'tax and service packages' (as, for example Tiebout, 1956, suggested for local governments in the United States). Such a market is difficult to construct and even more difficult to operate, however, because of the constraints on information and mobility for many of its consumers: as US experience suggests, the market can readily be manipulated to the benefit of the affluent and at the cost of the poor (see Johnston, 1984b). Its operation at the international scale is precluded by constraints on the mobility of labour and, to a lesser extent, of capital which are necessary to the operation of national 'regional alliances'. Thus, the only alternative is to reduce the state, to put as much as possible into the private sector and let the consumers choose. The taxation burden will then be reduced and people will have more to spend on their choices. Reducing the size of the welfare state therefore increases personal freedom.

The scourge of inflation

'Inflation is a disease, a dangerous and sometimes fatal disease, a disease that if not checked in time can destroy a society' (Friedman and Friedman, 1980, 298). Inflation is a consequence of the rate of growth of the money supply in an economy outpacing the rate of growth of production of goods and services for sale; it is not a disease of capitalism alone, because other modes of production (socialism and communism, as practised in the twentieth century) also use money as the indicator of value. When inflation is high, some (those on fixed incomes, for example) suffer more than others. There is a general impact too, because high levels of inflation are usually accompanied by high rates of interest for borrowing, which inhibit investment and thus expansion of the productive base. Prices must rise, too, and there is pressure for increased wages and salaries to match the price rises and to maintain the relative standard of living of those able, through their collective bargaining power, to win increases. All of this militates against risk-taking by investors, who instead keep their capital liquid by investing in items that are more likely to retain their value and can be easily disposed of.

Many arguments have been advanced to account for inflation rates at different times and places; among the identified culprits are trades unions which force up wages, and business people who force up prices – both groups are presented as rapacious and acting against society's interests (and thus, in the long term, against their own too). Low labour productivity is also blamed, combining both poor performance by workers with low investment by capitalists. All these may be proximate causes of the rate of inflation, according to Friedman and Friedman (1980), but the origin must be the growth of the money supply, for which governments are responsible. Governments are rarely, if ever, willing to accept the blame, however (Friedman and Friedman, 1980, 298):

Government officials always find some excuse – greedy businessmen, grasping trade unions, spend-thrift consumers, Arab sheiks, bad weather, or anything else that seems even remotely plausible. No doubt, businessmen are greedy, trade unions are grasping, consumers are spendthrifts, Arab sheikhs have raised the price of oil, and weather is often bad. All these can produce high prices for individual items; they cannot produce rising prices for goods in general. They can cause temporary ups and downs in the rate of inflation. But they cannot produce continuing inflation for one very simple reason: none of the alleged culprits possesses a printing press on which it can turn out those pieces of paper we carry in our pockets: none can legally authorize a bookkeeper to make entries on ledgers that are equivalent of those pieces of paper.

The cause of inflation is the relatively rapid growth in the money supply, and this is controlled by governments, hence the Friedmans' (1980, 309) conclusion that 'excessive monetary growth, and hence inflation, is produced by governments'.

High levels of government spending *per se* do not lead to inflation, however, because if they are financed by either or both of taxation and borrowing, then prices will remain relatively stable: there would simply be a switch of spending power from the private to the public sectors. But neither is it very popular politically: high levels of taxation are not electoral vote-winners, and high levels of public borrowing tend to force up interest rates and so make money more expensive for private individuals (for mortgages, for example). Inflation is created, according to Friedman and Friedman, by governments creating more money, not necessarily by printing more but by increasing the volume in circulation – by, for example, the sale of government bonds (including from one state apparatus – the Treasury – to another) and by deficit financing.

The only cure for inflation, according to this analysis (one which became increasingly accepted during the 1970s and 1980s), is for governments to slow the rate of growth in the money supply to that of production within the economy. The initial side-effects of such a policy will be an apparent reduction in the demand for many goods and services, because the money is not available with which to purchase them, and as a consequence output and employment may fall too. Neither unemployment nor an apparent reduction in personal purchasing power are popular with the electorate, but once stability has been ensured, prices should be stable, investment will return, jobs will be created and growth without inflation made possible. The initial side-effects are unavoidable, it is argued, and must be accepted if a long-term cure is to be sustained.[12]

These four arguments have been used, either separately or together, to counter the size of the welfare-corporate state in many countries, as a means of promoting a way out of the recession that marks the Kondratieff downturn of the 1970s and 1980s. In different places, and at different times in the same place, different arguments have been given greater weight.

All four are interrelated. The argument may begin, for example, with the case for tackling inflation as central to the cure, which implies reducing the rate of growth of the money supply. A major reason for this growth, as the other critiques suggest, is expansion of the welfare state, which does not have to be curtailed if the volume of government spending can be covered by taxation and borrowing – though high levels of both are likely to be electorally unpopular. Furthermore, high taxation levels and high interest rates are negative incentives to entrepreneurial investment and risk-taking, and so contrary to the perceived need to stimulate the free market.

Curbing government spending requires increased control on the money supply, and thus on expenditure on the welfare state. This stimulates calls to tackle the culture of dependency and encourages acceptance of the overloaded, inefficient, and increasingly ineffective state theses advanced by other critics. A healthy economy requires a free economy, with much less state intervention. Production and consumption decisions should as far as possible be made in a freely-operating market-place, which means removing constraints imposed by monopolies, including the monopoly trades unions.

A free economy is the best source of economic prosperity for all, it is thus argued, which means a much reduced state apparatus from that erected to counter the economic and social problems of the major downturn in the 1930s. This does not mean a weak state, however. It must be strong as a regulator and protector of the free market. But strength and size are not to be equated.

Achieving such a goal, by introducing policies and then seeing them through, calls for considerable political will.[13] It requires an ability not only to analyse the problem and identify a viable cure, but also to 'sell' the diagnosis and prognosis through an ideological programme which legitimizes a major shift in perceptions of what the state is for and why.

Kondratieffs and a welfare-corporate state cycle

The arguments just outlined could be interpreted as no more than a reaction to the concepts of a welfare and a corporate state which, if successful, will leave the full flowering of those structures as a historical peculiarity of the mid to late twentieth century. Alternatively, we might suggest that just as there are cycles in economic life, so there are cycles in political structures.

This latter interpretation has been promoted by Paci (1987), drawing on ideas outlined by Hirschman (1982) in an extension of his classic work on 'exit, voice and loyalty' (Hirschman, 1970). For several decades, writers have seen the growth of the welfare state as 'a unilinear historical process' (Paci, 1987, 184). The recent attacks on the welfare state have challenged that interpretation, but rather than represent what has occurred as a one-off 'parabola of the welfare state' (192), Paci suggests a 'hypothesis of a long-waved cyclical development of the complex system of social protection' (195). He suggests that each switch, presumably correlated with a Kondratieff downturn, involves a reallocation of social responsibility not just between two (market and state), but among three sectors, with the third being 'reciprocity, whether within such traditional frameworks as the extended family, local communities, charities, and corporate fraternities, or within modern frameworks based on voluntary association' (185). He reinterprets the history of welfare provision in Great Britain within this context, concluding that 'the medical and hygiene sector, like the whole sphere of public assistance, more generally passed through major phases, each characterized by a diverse mix of three fundamental allocative mechanisms' (191): the current situation is represented not just as a switch from welfare state to market, but as a recovery of the market alongside the revival of the voluntary sector.

Paci's fundamental hypothesis that 'the differential development of social policies over time and across countries can be broadly described as an experimental process of defining boundaries between three different resource-allocation mechanisms: the market, voluntary or traditional non-profit institutions, and the welfare state' (182–3) calls for a broader study than that undertaken here: the voluntary sector has been relatively unimportant for most of the twentieth century in the countries on which this chapter is concentrating, and it is too soon to know whether it will become important again during the next Kondratieff upturn. Paci suggests that the 'current expansion of voluntary and market sectors' is 'a venerable impulse that cannot simply be denied' (195): an attempt at the falsification of such a hypothesis is not consistent with the general goal being pursued here.

Four vignettes

As stressed throughout this chapter, the rise, the detailed nature of its implementation, and the critique of the welfare-corporate state during the twentieth century have varied substantially from place to place. The full extent of those variations cannot be explored here. Instead, four brief vignettes illustrate their nature.

New Zealand

New Zealand rates low on Esping-Andersen's conservatism and liberalism scales, and in the middle range on socialism (the last because of an absence of universalism in its provision of welfare benefits). Nevertheless, the country erected a very substantial welfare-corporate state in the 1930s, and then rapidly dismantled much of it during the 1980s – in both cases, under governments of the (relatively left-wing) Labour Party.

The depth of the 1930s recession was marked in New Zealand by extremely high rates of unemployment. Sutch (1969, 218) suggests that at the maximum about 100,000 men were unemployed (no estimate of female unemployment was possible): the country's total population then was only 1.5m. This stimulated much social unrest, especially in the big cities. To counter it, the government established an Employment Promotion Fund, which at the peak in late 1933 was providing sustenance through work for nearly 80,000 men – 45,000 of them part-time and over 30,000 full-time (Lloyd Prichard, 1970, 377). Much of this was in public works, including the planting of extensive forests in the central North Island, which thirty years later were the foundation for a thriving forest products industry.

The Employment Promotion Fund was one part of the extensive welfare state established by Labour during the 1930s. As in many other countries, the developments were built on earlier foundations – such as the National Provident Fund, established in 1910 to provide 'insurance against the hazards of poverty' (Sutch, 1966, 146). The first Labour government, elected in 1935, had an extended pension system, a free health system and a free education system as central planks to its election manifesto. A Social Security Bill was the last to be introduced during its first three-year term – Chapman (1981, 344) describes it as 'the most significant measure to come before what had already proved to be a transforming Parliament'. Its proposals were central to the 1938 election campaign, in which victory 'marked the peak of achievement for the New Zealand Labour Party in the twentieth century' (Chapman, 1981, 347), and its implementation on 1 April 1939 marked 'the defeat of poverty' for the people (Sutch, 1966, 241).

Subsequent amendments to the Act extended the scope of social security provision. A Royal Commission of Inquiry reported in 1972 that six types of assistance had been included: cash benefits at flat standard rates (plus allowances for dependants) as of right for all those eligible; emergency benefits for those in need not entitled to the standard benefits; supplementary assistance for those whose 'particular needs or reasonable commitments' cannot be met by the standard provision; medical and pharmaceutical benefits for all, with free public hospitals; universally applied benefits (no means tests) for those over 65 and for dependent children; and social work and counselling help available to all (including potential) beneficiaries. This produced a system consistent with principles of meeting need, guided by the belief that (Royal Commission on Social Security in New Zealand, 1972, 165): '*The community is responsible* for giving dependent people a standard of living consistent with human dignity and approaching that enjoyed by the majority, irrespective of the cause of dependency.' (Interestingly, the statement then asserts that this community responsibility should not be discharged in such a way as to 'stifle personal initiative' nor hinder individuals' attempts to promote their own personal standards.)

A corporate state was erected alongside the welfare state. The Labour Party attributed the 1930s recession to the fall in export income because of the drop in world demand for New Zealand's primary products, which then made the purchase of needed imported items impossible. Protection of the export income was needed, along with support for new industries that would reduce import demand. A Marketing Department was established in 1936 for certain primary industries (not the major ones). This was followed by, for example, a Wool Commission, established in 1952. It set a 'floor price': when the market price fell below this, the Commission purchased and stored the wool; it marketed it later when prices were higher and put the returns in an account used to buy in the wool when the market price fell below the floor. The result was a virtual guaranteed price scheme (Hawke, 1981).

The role of workers in the corporate state was cemented by policies relating to trades unions. The Industrial Conciliation and Arbitration Act of 1936 (updating the original 1893 Act) made membership of a union compulsory. The Court of Arbitration, to which the reference of disputes involving registered unions was also compulsory, was empowered to impose certain conditions (such as a forty-hour, five-day week) as elements of settlements, and to fix national minimum wages. (For fuller details, see Roth, 1973.) In the 1960s, when a series of Development Conferences

(Industrial, Export, Agricultural Production, etc.) culminated in the National Development Conference of 1969, the unions were among the major interest groups represented.

This very extensive welfare-corporate state structure was substantially dismantled during the 1980s, largely by the Labour government which was in power from 1984 to 1990, although some of the erosion of the corporate state was begun by the National Party governments which preceded it (see, for example, James (1986), on moves to end detailed regulation – of the size of potato which could be sold – in 1979). The structure erected in the 1930s, and for long the envy of many in other countries, was coming under increasing attack. Franklin (1985, 2), for example, argued that the New Zealand economy must be restructured and that:

This involves destroying the privileges New Zealand has progressively created as it has created the welfare state. Restructuring means facing up to inequalities, and we will not face up to the fact. The consequences of denying what is happening are: minimal growth in living standards, unacceptable limits of indebtedness, and a rancorous, divisive, society. The welfare state, of which New Zealanders are so unreasonably proud, is now a cause of their difficulties and no longer a cure.

He suggested that New Zealand's egalitarian policies had created a 'Hirschian paradox' (after Hirsch, 1976) because they had not removed the demand for 'positional goods', consumption items which allow the affluent to maintain their 'relative advantage' within society; 'At the present level of affluence, most incomes cover the basics of life, and relativity of income becomes a crucial objective' (64).

Franklin's analysis of the problems faced by the New Zealand state is summarised in Figure 3.10. The system of 'managed welfare capitalism' comprised two economic sectors – the exposed, subject to international market forces, and the insulated, which was sheltered from them. The latter can meet the demands for more and more positional goods only for so long as the exposed sector remains internationally competitive. If the latter fails, then the managed welfare state will come under increasing pressure.

If the exposed sector fails to cope with the changing international market and economic growth slows down; if government spending is maintained by deficits; if wages push occurs in the sheltered sector – the result is inflation, increasing the competition for positional goods as a hedge against inflation. If the sheltered sector is exposed to imports, income relativities between those rendered unemployed by the competition and those whose incomes are secure are widened. Hirsch seeks a solution in the spread of altruism, but alone that is insufficient. A restructuring of the economy to place people in more internationally competitive jobs is also required, entailing a replacement of relativities derived from protection with those derived from competition. Capitalism must be managed, but not as in New Zealand without reference to the global economy. (66)

New Zealand was in a cul-de-sac, he claimed. Escape required: economic restructuring focusing on 'sunrise industries' that utilized the country's technical and managerial talents; removal of the wage-push to inflation by the corporate wage-bargaining processes in the sheltered sector especially; and encouragement of competition. He concludes by setting New Zealand's cul-de-sac in context (178):

the cul-de-sac is not a geographical one. It is one we ourselves have created in response to a major phase in the development of capitalism and its associated geopolitics. A phase that is now at an end. During the phase we have borrowed ideas, practices, institutions and forged trading links that gave us a view of ourselves and the world in which we live. That world has changed. To hold any position at all in it we must change: change those ideas, those practices, those institutions we still regard as our achievements but which in fact have become our bonds.

Franklin was not alone in his ideas, and the country's fourth Labour government, elected in 1984, did much over the next three years to take the country out of the cul-de-sac by dismantling major elements of the corporate state. The welfare state was relatively secure, Labour politicians argued, and would be strengthened once the economic reforms had delivered renewed prosperity. The National Party government elected in 1990 moved immediately to dismantle parts of the welfare state also, however. Within two months the universal family allowance had been abolished, pensions were frozen, and the levels of unemployment, sickness and single-parent benefits cut by twenty-five per cent. Prescription charges were tripled, and those for visits to general practitioners' surgeries

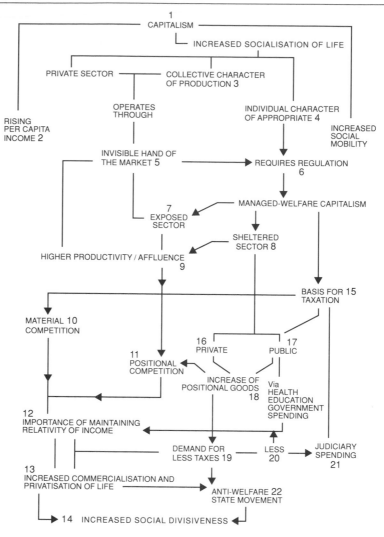

Figure 3.10 *The 'Hirschian paradox' producing the crisis of the New Zealand state.*
Source: Franklin (1985, Figure 3).

EXPLANATION

Capitalism **1**, mobilises the factors of production, by removing restraints upon the free movement of labour and capital to produce rising living standards **2**. The collective character of production **3** (a requirement of the industrial matrix) contrasts sharply with the individual character of appropiation **4**. The unacceptable social consequences of early capitalism **5** leads to managed capitalism **6**.

The overall effect of government intervention is the spread of general affluence **9**, which allows all classes to satisfy their demand for material goods, provided some are prepared to wait longer than others. The demand for positional goods **11**, not so readily satisfied, mounts, and underlines the importance of maintaining relative income **12**, despite high absolute living standards. A growing commercialisation and privatisation of life **13** produces increased social divisiveness **14**. This constitutes in outline the Hirsch thesis. Hirsch did not draw sufficient attention to the fact that welfare expenditure leads indirectly to the same result.

In New Zealand managed capitalism created an exposed sector **7**, subject to market forces, and an insulated sheltered sector **8**. Taxation and redistribution stimulated the economic activity of the sheltered sector **8**, in both its private and public divisions **16** and **17**. An increased supply of positional goods resulted **18**, and this initially made for social consensus rather than division. Through the creation of positions in the public bureaucracies and in the expanding firms of the sheltered sectors, and through advancement via the education, health and social services, a new segment of the population moved into higher-paid, well-protected positions, where it entered into positional competition. The increased demand for material goods can be satisfied by the sheltered private sector **8** and **16** so long as the exposed sector **7** remains competitive. But the increased demand for positional goods **11** emphasises the importance of maintaining relativity of incomes **12**, and leads to a demand for reduced taxation **19** from those who are paid out of taxes, or whose income is derived from the protection society affords those in the sheltered sector. A demand for less expenditure upon the welfare matters grows **20** and **21**. Anti-welfare state attitudes harden **22**.

If the exposed sector fails to cope with the changing international market and economic growth slows down; if government spending is maintained by deficits; if wages push occurs in the sheltered sector - the result is inflation, increasing the competition for positional goods as a hedge against inflation. If the sheltered sector is exposed to imports, income relativities between those rendered unemployed by the competition and those whose incomes are secure are widened. Hirsch seeks a solution in the spread of altruism, but alone that is insufficient. A restructuring of the economy to place people in more internationally competitive jobs is also required, entailing a replacement of relativities derived from protection with those derived from competition. Capitalism must be managed, but not as in New Zealand without reference to the global economy.

increased. In its first full budget seven months later, the universal pension payable from age sixty was abolished (with means-testing introduced for all under seventy), and 'user pays' charges were imposed for health and education. Almost 40 per cent of New Zealanders were receiving some social benefit payments prior to that budget, in which the government's aim was to make the top half of all income-earners responsible entirely for their own health and retirement, plus the education of their dependants. (Apart from the chronically ill, for example, all New Zealanders above the stated income level would have to pay for their first five outpatient visits and first ten days in hospital, each year. Political debate centred on the pension scheme however. It was only introduced – by a National government – in the mid-1970s, but was paid for out of current taxation and thus not based on an actuarily-sound fund. The number of pensioners had doubled in the fifteen years since the scheme's introduction, and its annual cost by 1990 was twice that of unemployment benefits and family allowances combined. The new government claimed that the country's demography made the scheme unviable.)

The architect of the mid-1980s economic changes was the Labour Party's Finance Minister, Roger Douglas. He became convinced of the need for restructuring (as described by James, 1986, and Collins, 1987) and in 1980 published his personal manifesto entitled *There's Got to be a Better Way*. That 'better way' became known as Rogernomics. It involved:

(1) Major restructuring of the public sector, in the search for greater efficiency, with the State-Owned Enterprises Act of 1986 indicating that each of the agencies covered was to 'operate as a successful business' and be 'as profitable and efficient as comparable businesses that are not owned by the Crown' (Collins, 1987, 69); some agencies were required to raise revenue through marketplace operations (Gregory, 1987).
(2) Abandonment of the government's role in wage-fixing and its pursuit of incomes policies, together with major changes in the industrial relations structure (Boston, 1987).
(3) Reform of the taxation structure, with re-

ductions of income tax, especially at the higher levels, and a switch to indirect taxation with introduction of a sales tax (the Goods and Services Tax) (Scott, 1987).
(4) Removal of the protection afforded to the sheltered sector of the economy (what Collins, 1987, terms the 'fortress economy'), including abolition of exchange controls. (This is illustrated for the crucial agricultural sector in Roche, Johnston and le Heron, 1993; see also Britton, le Heron and Pawson, 1992).

His thinking was closely mirrored by that of the civil servants in the country's Treasury Department, who produced a report for the incoming government (New Zealand Treasury, 1984) which criticized: the long-standing practice of attending to short-term difficulties rather than addressing long-term problems; the size and inappropriateness of much of government intervention in the country's economy; the instability in macro-economic policy that had characterized previous decades, with the emphasis on fine-tuning, stop–go measures; and the resistance to fundamental restructuring typical of New Zealand politicians. They thus welcomed Douglas's commitment to, and intention to implement, fundamental change.

From a welfare-corporate state, therefore, New Zealand was being rapidly moved to a segment of a global free-market economy. (In more journalistic language, Collins identified the components of 'Rogernomics' as: opening up to the world; an orgy of deregulation; the neutral and profit-making state; incentives to get rich; 'targeting' the welfare state; flexible wages and flexible workers; and the fight against rising prices.) The impacts were immediate, with the sheltered industries (including much of agriculture) suffering considerably, unemployment increasing rapidly, and major shifts of wealth towards the financial services and related industries, as well as to urban property-owners. The October 1987 stock-market crash affected the country substantially, however, and although Collins could write in 1987 regarding Douglas that 'his policies have been in tune with what many people wanted' (170), he also recognized that despite their cynicism about big government prior to the 1984 reforms it may be that 'New Zealanders do not approve of governments which step back and allow unemployment to rise' (171). Douglas himself resigned from government in 1989, and Labour lost the 1990 general election by a landslide: within nine months, the government that replaced it was the most unpopular ever recorded by opinion polls in the country.

West Germany

West Germany is one of the most corporatist of states, according to Lehmbruch's (1982) comparative analysis, though less so than Austria because the major German interest groups do not have access to parliamentary power (Marin, 1985). It has also developed the reputation since 1950 as one of the most stable of the democracies at the core of the world-economy (Paterson and Smith, 1981). (In 1990, the formerly separate states of West and East Germany were amalgamated. This vignette is concerned only with West Germany – the Federal Republic of Germany – although the roots of its welfare and corporate state long predate the division of 1945.)

The German welfare-corporate state originated during Bismarck's rule in the late nineteenth century (Mangen, 1989; Smith, 1989). The basis of the welfare state is a belief that the welfare of individuals contributes to the welfare of society as a whole, and it lacks the moral imperatives typical of some other countries. The state is not the provider of welfare, in general terms, but rather the overseer of the provision of social rights by other agencies – hence the preferred term 'social state': the voluntary and private sectors are much involved, with the federal state as the 'provider of last resort'. There are over 1,500 separate funds which provide insurance against loss of earnings through sickness, for example, although there is a federal agency concerned with unemployment payments. This German 'social state' has two major disadvantages, according to Mangen (1989). First, because it is closely tied to the labour market, 'those with previous poor employment status do rather less well out of it, as is evidenced

by the inferior financial position of women pensioners when compared with men' (173). Secondly, 'the welfare system relies on a heavy bureaucracy and rigid legalism to govern the many contractual agreements between a myriad of autonomous funders and providing agencies', leading to a 'growing disillusionment with its large-scale, insensitive and bureaucratic nature'.

The corporate state has been based on a belief in 'steering rather than planning, of indirect rather than direct control, of working through the market forces rather than attempting to set them aside' (Smith, 1979, 194). This has been achieved by the pursuit of economic consensus involving full co-operation among government, trades unions, and commercial and industrial organisations.[14] Furthermore, those interest groups are themselves integrated: Dyson (1989, 149), for example, points to the 'close, collusive, interlocking network of relations between banks and industry' which acts as its own self-regulating and crisis management structure. The unions are not as fragmented as in other countries. The German Trade Union Confederation (DGB) was established after the Second World War with only sixteen member organizations; one more was added in 1978, and together they represented nearly 8 million workers in the mid-1980s. Markovits and Allen (1989, 294) point out that the 'one industry – one union' and 'one plant – one union' principles mean that unions do not compete among themselves and could represent all workers, whereas the legal mechanisms within which unions operate ensure that most of the issues in which they have become involved are legal disputes rather than political conflicts. The unions, which represent workers to employers and state in the national corporate structure, are complemented by works councils that handle local issues.

This combination of characteristics meant that West Germany approached 'the organized capitalist ideal type', according to Lash and Urry (1987, 28: see above, p. 129). The country had strong industrial cartels, strong trades unions, and a well-developed welfare state. But:

Where Germany has deviated from the ideal-type is in the *mode* of organization and especially the organization of the state; Germany in particular is deviant here in terms of *who* was the central agent of organization.

In Germany it was *capital* which, in an unparalleled one-sided manner, was doing the organizing. The dominant classes were not only key in bringing about state organization through high levels of protectionism from the 1870s, but it was the 'top' itself which organized the 'bottom' in the Bismarckian social insurance legislation.

This, they argue, was a dominant class response to working-class organizational strength: the corporate and welfare states were a means of avoiding the potential threat of conflict with the proletariat.

The economic success and social stability of West Germany over the last four decades has created fewer strains and conflicts than those visible in many neighbouring, less successful, countries. Nevertheless, a 'change of course' (*Wende*) since 1982 has involved some challenge to the welfare and corporate structure. Economic planning has been affected by, for example, the rapidity of technological change, the development of new industrial sectors, the growth of the global market, the inflationary periods of the 1970s (pushed by the oil-price rises), and relatively high interest rates. The consequence, according to Dyson (1989, 150), has been 'a new critique of policies for being too late, limited and ineffective; and a recognition that the explanation for this condition of inertia and *immobilisme* might lie in the characteristics of the West German state.' Nevertheless, a policy style which 'values stability, predictability and order' (161) was retained, focused on the independent role of the Bundesbank and its (largely successful) policies for countering inflation. By producing price stability, the Bundesbank has been able to influence the development of European Community economic and monetary union very strongly.

West German trades unions may not have been 'bashed' in the way that their contemporaries were in the UK and USA during the 1980s, but they have not avoided the general mood of the decade which favoured reduction in union power and influence (Markovits and Allen, 1989). Perceived labour market rigidities which resulted from union actions were increasingly resisted by employers, who in 1980 challenged the 1976 Co-Determination Law which substantially extended workers' rights in the workplace. This promoted a DGB walkout from the 'Concerted Action' structure with the employers and the government, which thus failed to become institutionalized as anticipated (for fuller details, see Lash and Urry, 1987). The power of the unions has also been somewhat reduced because they have not become as strong in the new industries of the 1980s as they were in those previously established, because their position in the corporate state structure has been renegotiated by the government (Jessop, 1989), and because they have been challenged by new political forces, such as the women's, ecology, peace and other movements. The unions are no longer widely perceived as the major 'progressive force' in West German politics and the union movement is increasingly divided – between, for example, the 'activists' and the 'accommodationists' (Markovits and Allen, 1989):[15] nevertheless, the unions are closely involved in the negotiations for a new corporate structure, unlike the situation in the UK in the 1980s, even if the apparent goal is to divide the labour force into a core–periphery structure with an increasing proportion marginalized in the latter (Jessop, 1989).

Bismarck's social insurance policies were introduced in the face of opposition from workers as well as groups in parliament. They were part of his nation-state-building process: part of an attempt to build political loyalty, not a socialist-inspired measure to protect the exploited working classes. His success played a major role in the country's economic expansion. But in this policy area too, the 1980s' West German governments promoted retrenchment through proposed reductions in the generosity of the welfare system. Mangen (1989, 175) has identified twelve strategies in pensions policies and thirteen in health policies between 1977 and 1985, for example, each designed to reduce the levels of spending and to increase the recipients' personal contributions: prescription and dental fees were increased, for example; 'comfort drugs' were no longer covered; and the scope of optical fees was restricted. This was done to reduce the level of public spending and was promoted as a prudent line, given the increase in other demands on the state because, for example, of the growth in unemployment. The policies are politicized because of the growing 'welfare constituency comprising not just the twenty-five per cent of the electorate now believed to get its main source of income from the welfare state but also the large bureaucracy involved in the delivery of welfare services: younger people are increasingly concerned over whether the state can afford to be so generous to them in their later life' (Mangen, 1989, 176). In part, therefore, the demographic trend of an ageing population (as in New Zealand) is being used as the reason for revising the welfare state-system: without it, the Germans are being told, increased social security burdens could harm economic prospects and political and social stability. (The increasing proportion of older people in the population can itself be claimed as an indicator of the success of welfare state policies: thus success is bringing difficulties in its wake!)

Although West Germany has not experienced the anti-welfare and corporate state backlash typical of some other countries in the response to the latest Kondratieff downturn, its corporate and welfare structures have come under close scrutiny and have been subject to detailed proposals for reformulation. As Smith (1981, 174–5) put it, the West German consensus was predicated on post-war economic recovery and continued economic growth thereafter: a changed economic situation could well bring that consensus into question. (May 1992 heralded what may be perceived as a clear breakdown of the consensus. German – largely West German – public-sector workers went on strike for a substantial pay award to compensate for the high levels of inflation being produced, according to some analysts, as a consequence of the economic costs of integrating the former East Germany into the Federal Republic's economy. The country's public transport system was paralysed and garbage was uncollected.)

The United States

The United States of America is widely recognized as relatively weak in both its welfare state and its corporate state structure. Indeed, several writers have suggested that corporatism is notable by its absence from the USA. Wilson (1982,

219–20) defines corporatism as 'a form of government which rejects and supplants representation based on geographic units of approximately equal numbers of voters', and argues that it is unacceptable to the American conception of a republican democracy. Corporatism, he continues, is based on: (1) an integrated coalition of interest groups; (2) the ability of those groups not only to speak for but also to make binding commitments on behalf of the individuals they represent; (3) all government decisions are made only after close consultation with the major interest groups; (4) all groups involved are committed to the corporatist structure, and are prepared to modify themselves and their actions accordingly; (5) politicians and other decision-makers pay more attention to pressures coming through the corporatist system than to those (many emanating directly from the people) mobilized through the party system of representative democracy; and (6) policies are against *laissez-faire* and promote government intervention at the micro- as well as the macro-scale of the economy. The USA, however, has divided interest groups, politicians who are more concerned to represent their constituents, and a lack of national commitment to corporatist economic planning.

Wilson (1982, 225) contends that corporatism is absent from the United States because 'fragmented interest groups face divided governments'. The latter division reflects both the federal structure of government, linked to the constitutionally protected States' rights, and the separation of powers (among the executive, the legislature and the judiciary) within both State and federal governments. Corporatist strategies have been attempted, for example by the 1930s National Recovery Administration – established under the 'core idea . . . that businesses should band together to increase efficiency and prevent harmful competition' (Lewis-Beck and Squire, 1991, 108) – and by Nixon in the 1970s, but they failed to become a fixed component of the American political scene. According to Wilson (1982, 235), corporatism is unnecessary because 'the conventional approach to economic management, relying on monetary and fiscal methods rather than on neo-corporatist negotiations, is still possible in the United States because that country has the good fortune not to have a strong labour movement. In the last analysis, the United States has yet to be driven to corporatist solutions because . . . there is no socialism in the United States.' Salisbury (1979) suggests that, as a consequence, the federal government can plan without corporatist mechanisms simply by playing off one sector against another; 'there has generally been enough slack within each sector that a relatively free market process could operate at least some of the time, to bring downward pressure on prices.' Nevertheless, his final conclusion is that 'there is not much planning in America' (229). In addition, trades unions in the United States have largely confined themselves to negotiations at workplaces (and are increasingly constrained to individual workplaces, as Clark, 1989, illustrates: see below) and the market-place for labour (Quadango, 1988) and they have not become closely involved in the political mobilization strategies of either major political party.

Despite such conclusions, the United States federal government has sought increasing involvement in macro-management of the economy since the 1930s – but without incorporating the main interest groups to its policy initiatives. Owens (1986) illustrates this with the regulation, and later deregulation, of financial services and institutions. Perhaps surprisingly to the outsider, he shows that: 'Largely as a result of policies adopted in the 1930s, financial institutions and services in the United States are amongst the most highly regulated in the world, and amongst the most punctiliously regulated industries in the country' (172).

Financial regulation began in the 1860s with the National Currency Act. There was no central bank until the Federal Reserve Act of 1913, which created a federal regulatory mechanism that national banks were required to join. The call for more regulation gathered support in the 1920s and 1930s when many thousands of institutions failed – 2,298 banks in 1931, with deposits of $1.6bn, for example. The Reconstruction Finance Corporation was established in 1933 to make short-term finance available to banks, while at the same time congressional hearings uncovered many examples of either or both of incompetence and corruption in banking operations. Investment banking was separated from commercial banking, interest rate ceilings and limits on the type of deposit which could earn interest were imposed, and minimum capitalization levels were set to halt the establishment of small, hardly-viable banks. In addition, a Federal Deposit Insurance Corporation was established. The federal government guaranteed deposits up to a certain limit: membership was optional for State banks, but the attractions were too great to pass-up. Other financial institutions (such as those providing mortgages) were similarly regulated to protect depositers' interests. The outcome was a series of quasi-cartels, each with a protected segment of the financial market (such as mortgages, insurance, banking, etc.), but with the institutions in each largely prevented from competing with those in the others.

From the 1950s on, attempts were made to by-pass these government-regulated cartels, especially as variations in the interest rates on offer increased. In 1980, Congress responded with the Depository Institutions Deregulation and Monetary Control Act, which removed the banks' monopoly on interest-paying deposit accounts. Other actions increased the amount of competition between the sectors, and removed restrictions on their activities. This led, for example, to savings and loans institutions (traditionally the source of mortgage loans) diversifying into other forms of lending, with disastrous effects at the end of the 1980s when many experienced very substantial liquidity problems (calling for very large government insurance payments) as a consequence of the failure of their lending policies.

The regulatory mechanisms introduced in the 1930s brought stability and confidence to the American financial services industry: deregulation in the 1980s reintroduced the sort of instability characteristic of the pre-regulation period (as illustrated by bank failures – 42 in 1982, 48 in 1983 and 79 in 1984 after an average of only 6 per annum between 1943 and 1981: Owens, 1986, 199). The removal of ceilings on interest rates has increased the cost of money, and 'In order to meet these new costs, institutions are encouraged to make risky loans and investments which promise high return, in the knowledge that should their institutions fail most of their deposits (currently [1986] up to $100,000 per deposit) will be fully insured' (201). This global instability has been accompanied by geographical variability. When banking was closely regulated, inter-State banking was largely prohibited, and local deposits were used to fund loans to local borrowers. With deregulation, however, money was increasingly shifted into the markets offering the highest returns, which may be distant from where they were deposited: this may have been more efficient in terms of maximizing returns, but it clearly contributed to geographical variation in the availability of finance – as Owens (1986, 206) put it, 'the increasing national orientation of capital and credit markets seems certain to further marginalise many areas of the country and thus further encourage the development of a dual economy'. Texas, Arizona and California boomed as money flowed in, but the frostbelt areas declined even further – stimulating calls for greater local efforts, by local governments and other 'regional alliances' for example, to boost local fortunes.

With regard to the welfare state, Roosevelt's New Deal in the 1930s introduced social security with what de Swaan (1988, 204) calls 'the American big bang'. There were welfare precedents, however, notably in the generous treatment of war veterans and their dependants which originated with the Civil War and has continued ever since: the Veterans' Administration is currently responsible for spending about $30bn. annually (about 2.6 per cent of the federal budget). Lewis-Beck and Squire (1991) hypothesized that the Roosevelt administration policies built on those of his predecessor (Hoover) rather than produced any major innovations, and they tested this for five policy areas: economic stabilization; economic regulation; income redistribution; use of the police power; and growth and centralization of government. They found very little convincing evidence of changes in policy outcomes between the two administrations: Roosevelt certainly tried to achieve change through his New Deal policies, but was largely unsuccessful, despite the legislation passed. Nevertheless, the 1935 Social Security Act did lead to unemployment insurance provisions in all States: it was followed in 1939 by old-age insurance (see Quadango, 1988), in 1956 by disability benefits, and in 1965 by health care for the aged and for welfare clients

(Medicare and Medicaid). But the equivalent of a national health service (or insurance scheme) has never been established: interestingly, though lagging behind many European countries in welfare provision, the United States has led in the development of a mass, largely free, education system.

As indicated earlier, although American governments introduced relief programmes in the 1920s and 1930s, responding to public pressures and mounting disorder (as Piven and Cloward, 1977, demonstrate), these were not brought together in an integrated welfare-state structure. Instead they withered, because the political need had gone and politicians no longer felt disposed to support large spending programmes, with their implications for taxation, in order to protect their electoral positions. Thus, as Piven and Cloward (1971, 117) conclude their discussion of the New Deal relief programmes:

> The cycle was complete. Turbulence had produced a massive federal direct relief program; direct relief had been converted into work relief; then work relief was cut back and the unemployed were thrown upon state and local agencies . . . What remained were the categorical-assistance programs for the impotent poor – the old, the blind, and the orphaned. For the able-bodied poor who would not be able to find employment or secure local relief in the days, months and years to come, the federal government had made no provision.

Turbulence returned in the 1950s and 1960s, associated with the civil rights movement. As a consequence of changes in agricultural practices in the South, large numbers of blacks had migrated to the big cities, where they experienced high unemployment levels and family break-up was substantial. To counter the unrest that followed, the 1960 Democrat administrations of Kennedy and especially Johnson introduced the welfare programmes that became known as the 'Great Society'. This was done, Piven and Cloward argue, simply to moderate that political unrest and protect the party's control of both Congress and the Presidency. The outcome was not a structured welfare state, but rather a series of individual programmes whose size was substantially cut by the Reagan administrations in the 1980s. Some slight liberalization of welfare practices was the only long-term outcome, Piven and Cloward conclude, leading them to draw the moral (1971, 338) that 'a placid poor get nothing, but a turbulent poor sometimes get something'. April–May 1992 may have witnessed the beginning of another period of turbulence by the poor – specifically the black urban poor – with the large-scale riots, looting and arson in Los Angeles (copied slightly in several other cities) which, though triggered by a court decision finding four policemen not guilty of assaulting a black (despite the beating being recorded on video), was widely interpreted as a black protest against underclass status, especially relative to the successful Korean community in Los Angeles against which some of the violence was directed.

The federal nature of the American system, the separation of powers, and the States' rights enshrined in the Constitution, together make it difficult for the national government to act in many spheres, should it want to, and thus add a further constraint to the achievement of a coherent welfare-state structure rather than an *ad hoc* series of relief remedies. This is illustrated by challenges to the New Deal policies in the 1930s and by the difficulties that have been faced in implementing the Great Society programmes equally throughout the country. Six New Deal Acts passed by Congress and approved by the President between 1933 and 1936 were found unconstitutional, for example. One of them established a comprehensive pension system for railroad workers, which the Supreme Court ruled was irrelevant to the conduct of inter-State commerce and so the issue lay outside Congress's jurisdiction: later Acts were not challenged, however, in part because of the resolution of a conflict between the President and the Supreme Court in the former's favour, and so the scheme was eventually implemented. Also ruled unconstitutional was the first (1933) Agricultural Adjustment Act, which was to use revenues from taxing foodstuffs producers to pay other farmers to reduce their production: agricultural production was ruled to be outside the federal government's jurisdiction, although it was later brought within it under a 1937 Act. Regarding welfare, the federal government is prevented by the States' Rights clause of the Constitution from making direct payments to

citizens. All it has been able to do is encourage State governments, through substantial subsidies, to operate welfare systems – such as Aid to Families with Dependent Children and Medicare. Because of major cultural differences among States, the implementation of such policies has been spatially very variable, with southern States both the strictest in determining eligibility and making the lowest payments (Johnston, 1991, gives fuller details).

Trade unions are generally weaker in the United States than in most other countries in the core of the world-economy. The 1935 National Labor Relations Act (the 'Wagner Act') guaranteed workers the right to organize and bargain collectively, and established a National Labor Relations Board to administer that law: it was modified by the Labor Management Relations Act (the 'Taft-Hartley Act') of 1947. The chosen milieu for implementing those rights was the individual plant, and the outcome made it (Clark, 1989, 28):

relatively more difficult to organize labor in the United States than in other advanced countries. Unions have had to organize firms on a plant-by-plant basis, seeking representation rights for workers in different communities across the country. While they may be successful in some plants, they are required to organize all plants of a firm if they are to have a coordinated national agreement with a particular firm. . . . the union movement has had to cope with a federal legal structure deliberately designed to fragment its national power.

Under the 1947 legislation, States were encouraged to develop their own labour laws, and some 40 per cent have introduced what are known as 'right to work' acts which severely limit unions' ability to create closed shops (see also Peet, 1983; Johnston, 1991).

Clark (1989) has analysed both the election processes whereby unions have sought the rights to represent workers at individual workplaces, and the congressional (notably what he terms the 'southern veto') and court-based mechanisms used in the Reagan years in particular to reduce union power. Regarding the latter, for example, he has traced the impact of the 'employment-at-will' principle from its first enunciation by the Tennessee Supreme Court in 1884: the principle treats employment relations like all other trading relationships, and gives employers the right to hire and fire at will if the worker has no contract of employment. Over the succeeding century, many States have evolved exemption provisions whereby the employment-at-will principle can be overridden. Nevertheless, the rights being introduced in the exemptions are individually rather than collectively phrased, and Clark concludes that as a consequence unions will decline in importance, even in the northern States, as they are replaced by governments acting as the guarantors of workers' working conditions and remuneration. The outcome represents (238) 'a retreat from concerted political action involving coalitions of classes and elites to a rights theory of society where progress is measured in terms of individuals' freedom as opposed to their social rights derived from their membership of unions.' As such, the potential for corporatist development is even further reduced.

The United States provides an excellent example of Esping-Andersen's (1983) argument that the nature of a country's welfare and corporate structure reflects its historical and political context very strongly. The nature of the state apparatus which evolved in the USA from 1776 on, and the absence of labour as a strong political force, both constrained the development of some aspects of a welfare-corporate structure and meant that the claims for such were often muted and ineffective. Piecemeal elements of a structure have been created in response to events – both economic (the 1930s recession) and political (the civil rights movement of the 1960s) – but there has been no co-ordinated approach which would justify a claim that the country is either corporatist or welfarist.

The United Kingdom

The origins of Britain's welfare state lie with the Liberal governments of the first two decades of the twentieth century, when pensions and insurance schemes were introduced.[16] Relief from unemployment and other sources of poverty had been provided (often minimally) by local governments since early in the previous century and this was supplemented by mutual societies and the trades unions. Through the first four decades of the twentieth century a variety of contributory schemes was introduced to extend this earlier provision, and these were amalgamated into a comprehensive system by the first majority Labour government (1945–51) acting on the basis of the seminal Beveridge Report of 1942. Beveridge proposed a comprehensive national insurance system covering retirement pensions, widows' benefits, sickness, industrial injury and unemployment benefits, maternity benefits and grants, death grants, and invalidity allowances and pensions. Entitlement to all of these was bought by compulsory contributions to the national insurance scheme when in work: alongside it was a supplementary benefits system, to which access was controlled by means-testing. Family allowances were available to all, and a variety of tax concessions (for contributions to private pension schemes, for example) was also offered to encourage people to contribute to their own welfare. Though modified in a variety of ways, this system remains the basis of the British welfare state.

Alongside the system of welfare payments, the UK government went further than many others with its creation in 1949 of a National Health Service, whose fundamental tenet was that treatment was to be available free at the point of need. Although private health care systems remained, the great majority of the population from then on received free medical treatment from general practitioners and free outpatient and inpatient treatment at public hospitals. Alongside this, from 1944 all children aged 5–14 (later fifteen, and then sixteen) received free education at state schools, and higher education places were funded for all those able to obtain entry through the competitive system based on academic performance alone.

With regard to the corporate state, co-ordination has been relatively weak within both the employers' organizations and the unions, and political attempts to negotiate with both together (in the introduction of wages and prices policies, for example) have rarely succeeded for long.[17] Nor has there been much national macro-economic planning, and certainly very little that has been effective. Instead, the post-1945 Labour government engaged in a substantial nationalization programme, bringing major sectors (both the public utilities – electricity, gas, water – and infrastructure – notably the railways) into public ownership, along with a small number of productive industries, notably coal-mining and iron and steel manufacture. With the exception of the last, these actions were generally accepted by subsequent Conservative governments (prior to 1979). Post-1945 governments also engaged in large-scale public housing provision, with over 30 per cent of all households being state tenants (most via local governments) at the peak of the programme.

During the late 1950s and early 1960s there was general optimism within the UK that the problems of organizing production to ensure full employment with low inflation had been largely overcome and that the 'fair' distribution of the fruits of growth were being tackled by the welfare state: Britons had 'never had it so good', according the Prime Minister of the time. But that optimism rapidly faded through the 1960s, and the Conservative government elected in 1970 intended a major restructuring of economic and social policies. It made a major U-turn in the face of growing economic problems in 1972, however, becoming more rather than less interventionist – it introduced a national wages policy, for example. The successor Labour government (1974–9) also perceived a need for major changes in, for example, the level of public spending,[18] but major industrial unrest in 1978 and 1979 led to its electoral downfall and replacement by a Conservative government led by Margaret Thatcher: this was determined (as was she, in particular) to achieve the major restructuring which its gurus of the New Right argued was necessary for a return to economic health.

Thatcherism, as it was soon called, was a complex phenomenon whose major articles of faith are summarized in the couplet that forms the title to Gamble's (1988) book – 'The free economy and the strong state'. The first element in this reflects the belief in the market as by far the best mechanism for determining what is produced and in what quantities, and at what price it is sold. The role of the state is to ensure that markets operate properly, but to intervene in them very little. The strong state is thus a regulatory state only – both at home and abroad.

The Conservative governments of the 1980s set themselves five major tasks, according to Gamble (1988, 121): (1) to restore the health of economic and social life; (2) to restore incentives; (3) to uphold Parliament and the rule of law; (4) to support family life; and (5) to strengthen Britain's defences. In the first of these, the initial premise on which all policies were based was the need for 'sound money', which required defeating inflation: for Thatcherites this involved controlling the money supply and reducing public spending (tasks for the strong state) and withdrawing from corporate bargaining structures over wages, which were seen as inflationary. The free market had to work in the labour market too, which meant eroding the monopoly power of the trades unions – a task highlighted by government intransigence during the year-long miners' strike of 1984–5 (Griffiths and Johnston, 1991; Johnston, 1991). The initial consequence of these policies was a rapid increase in unemployment as many manufacturing companies failed. There were occasional government rescue operations to assist ailing large firms (see Bonnet, 1985), but in general little was done to stem the tide of redundancies.

Growth in the number of unemployed increased demands on the welfare state, which also came under close scrutiny, accompanied by an ideological campaign designed to promote individual self-reliance and the use of incentives (through reduced taxation and the switch from direct to indirect taxation) to encourage individuals to provide for their own future welfare (as with the pressure on council tenants to purchase their rented homes at substantial discounts). Public provision of welfare was reduced in many cases, notably in the squeeze on local government spending; in others, greater efficiency in service delivery was sought by the introduction of a quasi-market within state organizations, as with the National Health Service, and by the requirement that local governments put certain services (such as street cleaning) out to competitive tender. A number of public utilities was sold by the state, through stock-market flotations which advanced 'popular capitalism': regulatory mechanisms were retained (see Veljanovski, 1990). The (to many surprising, given that the country was in deep recession) re-election of the Conservatives in

April 1992 saw the programme carried forwards: British Coal was to be privatized, according to the Queen's Speech, and private companies would be allowed to tender to run services on British Rail's tracks: meanwhile, the new Secretary of State for Education indicated that he expected a large number of schools to 'opt out' of local authority control and thus remove themselves from their traditional local democratic accountability to other than elected parent-governors.

The Thatcherite strategy for advancing capitalist accumulation involved 'rolling back the frontiers of the state' and putting as much of society's operations as possible in the market place. A strong state was required to protect that market, however (Gamble, 1988, 128). A vigorous defence and foreign policy was needed to protect the capitalist world, hence the stridency of Thatcher's (and Reagan's) opposition to the Soviet Union in the early 1980s and the welcome given to the developments there and in Eastern Europe in the early 1990s.[19] At home, the 'rule of law' was stressed, arguing that a successful economy required a disciplined society, in which certain virtues (termed 'Victorian family values' by some) were applauded: alongside this were changes in the generosity of aspects of the social security system, which particularly affected the young, especially the young unemployed (Weale, 1990). Throughout, there was a strong ideological campaign aimed at undermining opposition to the various policies, using, for example, a restructuring of the education system to make it more responsive to the needs of the 'enterprise society' that was being built.

In his evaluation of Thatcherism, Gamble (1990) concludes both that the claims that the Conservative governments introduced major shifts in British policy ignore many of the continuities from the mid-1970s on, and that 'By its own criteria for success the achievements of the Thatcher revolution can seem meagre' (337). The size of the public sector has not been substantially reduced, for example, even if its scope has. But the attacks on the corporate and welfare states, their inefficiencies and their barriers to enterprise and economic success, have been telling. According to some analysts, a major element of the ideological project within Thatcherism has

been to 'rid the country of socialism for ever' thereby to achieve an end to class politics. The latter was achieved in the late 1980s in some parts of the country at least (Johnston, Pattie and Allsopp, 1988; Johnston and Pattie, 1990).[20] More importantly, perhaps, the Labour Party has very substantially restructured itself and its politics to embrace many aspects of 'disorganized capitalism': 'The Labour Party was substantially remodelled during the 1980s and has begun to look more like a European socialist party, with looser ties to the trade unions, and firmer commitment to market economics and the European Community' (Gamble, 1990, 360).

These four vignettes have illustrated the great diversity in state actions with regard to the provision of welfare and the management of the economy during the twentieth century. They emphasize the difficulties involved in making any but the most general of conclusions. The need for corporate and welfare state structures was widely recognized long before the Kondratieff downturn of the 1920s and 1930s, though the extent to which such structures were in place by then differed considerably among countries. The role of the state in stimulating economic growth again and in protecting those suffering from the consequences of the crisis of underconsumption increased in most countries during the 1930s – though it is unclear whether it was economic policies or the preparations for war in the 1940s which really produced the revival (armaments production proved to be an excellent Keynesian stimulus to the economy). And from the mid-1970s on there has been a widespread recognition that recovery from the next crisis of capitalism required not more, but less (or, at least, very different forms of) state intervention: again, the interpretation of that general case has varied substantially from country to country.

Accounting for both the general and the specific interpretations is extraordinarily difficult. With regard to the former, one could claim that every solution to a crisis contains within it the seeds of the next crisis, so that what helps in one situation is an impediment in the next. Capital benefited from the corporate and welfare states in the 1930s and 1940s, but by the 1970s and 1980s found them no longer conducive to its goals. But why the corporate and welfare states then? Why not earlier, or later? And why were different corporate and welfare structures erected in different countries? Esping-Andersen (1989) has attempted an answer, but much of it suggests that the antecedent social and political structures are crucial influences on contemporary situations. We are thus only a little further forward: we make our own futures, but in the context of inherited cultural constraints rather than in conditions of our own choosing.

All that has been shown in this and the previous sections is that the twentieth century has seen both substantial shifts in the perceived and actual roles of the state in the countries at the core of the capitalist world-economy and substantial variations in the implementation of those roles. It may be that such reorientations were necessary to the continuation of the capitalist system at the particular periods of the century when those shifts occurred. What is just as clear, however, as many philosophers of science and social science stress (see Sayer, 1984), is that local circumstances are crucial in determining how those necessary mechanisms are implemented.

Crisis and the state beyond the capitalist core

This chapter has paid the majority of its attention to the countries at the core of the world-economy. It has argued that an understanding of the changing nature of the welfare and corporate states can be obtained through an appreciation of the major cycles that are characteristic of the capitalist world-economy. At each of the major downturns in that system during the twentieth century there has been a perceived need to re-structure the state in order to counter the rationality and legitimation crises that accompanied the economic and social difficulties. In the 1930s that restructuring involved greater direct state involvement in both promoting capitalist accumulation and supporting individuals' welfare – though with substantial variations between states in the nature of the corporate and welfare structures erected. Fifty years later, the argument accepted in most countries was that the role of the state should be reduced and that capitalism

would flourish, with positive consequences for the welfare of all, if state involvement were reduced: again, this argument was widely accepted, though with substantial variations in the exact details of what was done, where.

The core countries occupy only a minority of the world, however, so what of the rest, of the periphery which forms what is popularly termed the 'Third World' and of the countries which, for part of the period from 1917 on, experimented with a form of socialist organization – the 'Second World' – which was relatively independent of the capitalist system? In the following sections, it is argued that that fifty-year cycle model which so helps the appreciation of events in the core is largely irrelevant to understanding the periphery.

The Third World

In this segment of the capitalist world-economy there has been much short-term variability in the interpretation of the state's role in the promotion of capitalism's health; in general, the cyclical processes that many countries have experienced have been of much shorter duration than fifty years.

Countries on the periphery of the capitalist world-economy occupy particular niches in the pattern of uneven development that has been created and recreated to promote the interests of capital in general and capitals in the core countries in particular (Harvey, 1982; Smith, 1984). The processes of imperialism and colonialism have incorporated those countries within the structure, as areas where cheap resources (land and its products, plus labour) can be exploited. The states and elements of the middle classes there have been involved in this exploitation as facilitators of the flows of cheap resources and profits to the capitalist core.

Most of those countries are now politically independent. Some have democratically-elected governments; others have dictatorships, usually backed, if not staffed, by a strong military presence. In some analyses, the nature of the state (whether it is democratic or not) is a function of the level of a country's economic development, but others have shown that most Third World

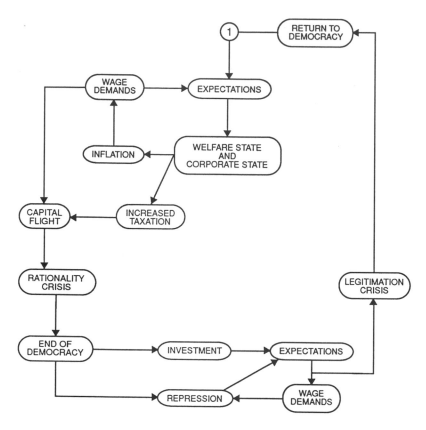

Figure 3.11 *The democracy–dictatorship sequence in Third World states.*

countries have experienced substantial political instability (Johnston, 1984, 1989; Taylor, 1989): permanent democracy is rare, and instead a series of switches in and out of democracy is usual.

The reason for this switching pattern is summarized in Figure 3.11. In Third World countries, as elsewhere, the state has to balance its 'promotion of accumulation' and 'legitimation of capitalism' roles. At point 1 in the diagram, the country concerned is a democracy. Its government has to react to the expectations of the electorate which put it there and will determine its future; it does this by creating welfare and corporate state structures. These actions may stimulate inflation if expectations are met in part by the government creating money to stimulate demand in the economy, and this generates further wage demands to sustain current expectations, let alone greater ones. The government must increase its taxation levels and borrowing

requirements to meet some of those demands, which make its country less attractive than others for investors, both external and local. The consequence is capital flight and a rationality crisis: there are fewer jobs to create the wealth from which to meet the rising expectations. The government cannot be too harsh in its treatment of the population because of its fears of electoral consequences. Political and civil disorder is likely, however, until eventually democracy is replaced – almost invariably by the military, probably covertly linked to one or more of the capitalist core states – in order to bring order and stability.

This new government is able to encourage investment, with its repressive policies towards the local population providing the stability needed to attract capital. Jobs are created and prosperity slowly grows, especially for the middle classes and to a lesser extent for those who obtain work: the welfare and corporate structures are weak, however, and so the expectations of many may not be met. The government may promise a 'return to democracy' once prosperity is restored and stability is ensured, but the growing resentment against the repressive policies may reflect an impatience with the pace of delivery, especially if the economic benefits of those policies accrue to a small number only. The government may well then face civil disorder, which is increasingly costly to police: eventually it may be forced to yield, and there is a reintroduction of democracy, returning the country to point 1 on the cycle.

There is no theoretical reason to set the timing of this cycle of alternating democracy and dictatorship within the fifty-year periods of the Kondratieff long waves. The empirical evidence is certainly that in many countries the transition has occurred much more frequently, though the frequency with which the cycle is completed will reflect local circumstances: acceptance of a dictatorship may be sustained for longer in some places than others, for a variety of reasons. Ghana is one of many countries that illustrate this alternation between democracy and dictatorship: during the last three decades, 'The brief periods when Ghanaian governments have been economically successful have coincided with good cocoa prices . . . Development plans are drawn up in expectation of a certain income from cocoa, but have to be periodically revised as prices fluctuate. Low prices mean harsh austerity measures and social unrest' (Osei-Kwame and Taylor, 1984, 579 and 587). Prior to independence in 1957, two general elections (both won by the Convention People's Party led by Kwame Nkrumah) were held, and these were followed by another in 1960. In 1964, however, Nkrumah declared Ghana a one-party state, initiating the first shift away from democracy in the face of a legitimacy crisis: the one-party state introduced socialist policies. Nkrumah was overthrown by a military coup in 1966, which reinstated a market economy. The military yielded to an elected civilian government in 1969 (led by Nkrumah's main rival, Busia, of the United Gold Coast Convention): this was removed in 1972 in a further crisis linked to the falling price of cocoa. In 1978 the military tried to introduce a 'no-

party union government' through a referendum, but this failed and there was another military coup in 1979. This held elections during that year, which were won by a pro-Nkrumah grouping (Nkrumah was by then dead). The government had great difficulties coping with the economic situation, however, and a further coup in 1981 (led by the same officer as in 1979) installed a further military dictatorship which, despite some challenges and plots, remained in power throughout the following decade.

Not all countries are paradigm examples of this 'alternating form of government' model for the Third World. In some, such as India and Jamaica, democracy has been retained throughout more than forty years of post-colonial independence – though in each there have been 'non-democratic interludes', such as the state of emergency introduced by Indira Ghandi to India in 1976, which were akin to dictatorship. Other countries have no experience of democracy: in some such as the oil-rich states of the Arabian Gulf, national wealth has been used to 'buy' the population's acceptance of a form of 'benevolent totalitarianism', whereas elsewhere a strong repressive apparatus and ideological policies designed to inhibit the growth of expectations have limited the demands for popular participation.

The Second World

The Second World was inaugurated by the Russian Revolution of 1917, consolidated over the next three decades, and substantially extended in the late 1940s by Communist Party takeovers in both Eastern Europe and Eastern Asia. For the next thirty years, the repressive state apparatus in those countries sustained the non-capitalist experiment, and substantial efforts were made to extend its territorial cover, notably in Asia and Africa. There were occasional popular uprisings against Communist Party rule and Soviet Union hegemony (Hungary in 1956, for example, Czechoslovakia in 1968, and Poland in 1980), but the repressive apparatus was very successful until the late 1980s when it crumbled in all Eastern European countries. Contemporaneously, some of those in control of the state in the Soviet Union realized the need to restructure prior to any popular revolt;[21] in China, the popular uprising characterized by the events in Tienanmen Square in Beijing in 1989 was quelled by force.

Why the uprisings, why their failure prior to the late 1980s, and why the substantial restructuring since – including a full embrace of capitalism throughout the Soviet Union and Eastern Europe since late 1989 and the adoption of western democratic state apparatus forms in several countries? Because the countries involved in the socialist experiment were not incorporated in the capitalist world-economy, Kondratieff cycles are not the basis for an answer. Figure 3.12 outlines an alternative.

The socialist experiment was built on a belief in central planning to ensure that all citizens had their needs met equally and that all contributed equally, according to their abilities, to meeting those needs. Politics, as defined in the capitalist democracies, were irrelevant because all citizens were committed to the same goal – a prosperous society lacking class divisions. The state, comprising the representatives of all the people selected through the party dedicated to the socialist ideal, determined how the country's resources would be used, to produce what, and where: it also set prices and determined wages. All resources were to be collectively owned and

allocated, which removed any need for market mechanisms.

The ideal was not fully implemented, in that some residual private ownership of resources (notably land) was allowed in some countries, and some market trading was tolerated, if discouraged. State planning dominated, however, and so there could be no rationality crisis, only a legitimation crisis: Figure 3.12 suggests how that emerged. The socialist state faced rising expectations from its population, in part because it could not prevent people from becoming aware of material standards in capitalist countries and in part because it promoted them: it encouraged hard work with the incentive of high living standards (elements of which were achieved in some of the countries at least, as illustrated by their high life expectancies: Johnston, 1989), but not with the incentive of high wages with which to exercise choice of material benefits. It failed to meet those expectations and to motivate people to create more wealth, however: agricultural and manufacturing targets were often not attained, and governments had to import basic requirements, which involved some participation in the capitalist system. In addition, a large proportion of the productive capacity, especially in the Soviet Union, was used to sustain the very large military arm of the state apparatus, used in the repression of its citizenry, in promoting the socialist experiment elsewhere, and in preparation for the feared confrontation with the superpowers of the capitalist world. The socialist states, and the Soviet Union in particular, were as much victims of 'imperial overstretch' as the leading capitalist states (Agnew and Corbridge, 1989: see also Chapter 5 and Epilogue below).

In part to counter its failure to meet the rising expectations and in part to sustain its military operations, the socialist state created inflation through the 'printing of money'. This generated wage demands, which stimulated a perceived need for more repression. As this cycle, depicted in Figure 3.12, continued, so the probability of a legitimation crisis, and a possible popular revolt which the military either could not or would not repress, increased. It came in the late 1980s. In most of Eastern Europe the states were unable to react and the removal of Soviet Union backing for their military apparatus and strict interpret-

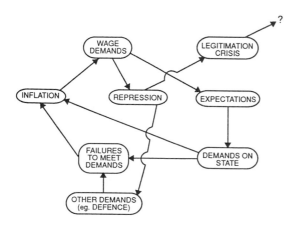

Figure 3.12 *The legitimacy crisis of the Second World state.*

ation of the 'party line' heralded a very rapid yielding to the demands for both popular democracy and a market economy. In the Soviet Union, where the economy was in disarray and the federal state was faced by popular demands for independence from a number of republics that are the homes of strong nationalist movements, steps were taken (*glasnost* and *perestroika* as they are popularly known) to introduce a market economy and to allow some democratic forms, while strongly resisting the secession movements (see Epilogue): the failed hard-line communist coup of August 1991 substantially accelerated the pace of economic and democratic reform, while at the same time further encouraging the break-up of the Soviet Union into separate independent republics, but with many latent national problems as a consequence. The result is a series of countries experiencing the dismantling of their equivalents of the welfare and corporatist-regulated states.[22]

Conclusions

This chapter has covered a great deal of ground and yet, as stressed at the outset (p. 117), the temptation to overgeneralize from its material must be resisted. Despite its length, there are many aspects of the state, its apparatus and their operations, which have been ignored here. All that can be offered as a conclusion, therefore, is a hypothesis.

It has been shown that the state is big in capitalist societies and has become bigger through the twentieth century, but is now under substantial attack from some quarters. Whether it is too big and too powerful or not is inevitably a value judgement. As Rose (1984, 3) argued:

If government actions are viewed as intrinsically good, then the more that government does, the better society will be. If government is thought of as bad, then the bigger that government becomes, the worse will be its effects upon society. There is no consensus, within nations or internationally, about whether a given amount of government is too much, not enough or just about right.

Nevertheless, throughout much of the core of the capitalist world-economy there has been a strong anti-government move in recent decades, as there has been in the (now disappeared) Second World: in the Third World, too, agencies of core-state governments (such as the International Monetary Fund) have encouraged the adoption of 'less government policies'. The argument presented here has suggested a link between changing views of the role of the state and the Kondratieff cycles that characterize the capitalist system. This is not to suggest that those changing views are in any sense 'natural' or 'predetermined'. Rather, the argument is that, given (1) the regularity of major crises in capitalist 'progress', and (2) the salience of the state to the perpetuation of the capitalist system, then it is likely that (3) changes in the nature of state activity will be proposed as elements of attempts to resolve the crises. Each crisis, then, throws up a proposed solution and, given the tight interdependence of the capitalist world-economy, the same solution (in outline, though almost certainly not in detail) is likely to win supporters everywhere. But perhaps we need to wait another fifty years before that hypothesis can be tested.

Notes

1 Thus the presentation here is firmly set within the realist philosophy of science as set out by Sayer (1992).

2 Mann argues differently in the quotation above. My contention here is that in modern democratic states at the core of the world-economy although the state does, for example, tax our incomes at source without our explicit consent (nobody has asked me whether it can deduct income tax from my monthly pay cheque, and my employer is required to do it), nevertheless that action is based on implicit consent because the state could be precluded, by democratic decision, from taxing me in that way. The difference is between individual and collective consent: the former is not given, but the latter is.

3 It is worth noting that Berry reached this conclusion after starting from a position of extreme scepticism.

4 Note that Berry (1991, 62–3) criticizes Marshall's portrayal for 'lacking a firm empirical footing' and incorporating representations of long waves that bear 'little resemblance to the evidence on price movements presented by non-Marxist investigators' and so are 'figments of the Marxists' wishful

thinking: for them, the "facts" must fit-their pre-formed notions of crises in capitalism.'

5 Note that to them 'disorganized' does not imply 'chaotic', which is a frequent vernacular synonym (what they refer to (Lash and Urry, 1987, 8) as 'high-entropy random disorder' for disorganization) but rather 'a fairly systematic process of disaggregation'. Disorganized capitalism is characterized, for them, by a proliferation of small units rather than the dominance of a few very large ones.

6 The use of ideal-types is typical of much social science research influenced by the work of Max Weber: see Schutz (1972).

7 In many ways, this is similar to Lipset and Rokkan's (1967) essay exploring differences in electoral cleavages among many of the same countries, which Esping-Andersen does not cite.

8 This was illustrated in the UK in the early 1990s when many local managers in the state health and education services welcomed the opportunities to 'opt out' from central control and manage their own units within a cash budget independent of any higher authority.

9 The situations in both New Zealand and the UK in the 1980s in part counter this argument. The right-wing governments were able to attack the welfare state because they had first 'neutered' the power of the trades unions, whose strength was also rapidly eroded by the high levels of unemployment.

10 In the UK in the 1980s this potential 'crisis of rational administration' was tackled by aggressive anti-union policies, by major attacks on opposition political parties (who contributed to their own electoral weaknesses), and by the ideological promotion of an 'enterprise economy'.

11 This is illustrated by the 'winter of discontent' in the UK in 1979, which contributed to the downfall of the Labour government then, and the National Union of Mineworkers' strike in 1984–5 (Griffiths and Johnston, 1991).

12 This argument was used substantially by the British government in the early 1980s, and again in the early 1990s, contending that until inflation was defeated prosperity could not be guaranteed: the means were to be endured to achieve the necessary end.

13 Friedman and Friedman (1980), for example, argue that the effects of reducing the rate of growth of the money supply will not be enjoyed immediately, and that the 'pain' of the immediate side-effects, such as high unemployment, must be endured in order to reach the long-term benefits of stable prices.

14 Dyson (1986, 139) writes of 'An elite of bankers [who] continue to enjoy a salient position on the borderlines of foreign and economic policies. They enjoy close informal relations with the governmental elite and mediate in the relations between government and industry.'

15 There is a strong parallel here with the divisions within Germany's Green Party, between those who accept the need to compromise with other political movements in order to achieve some gains (the 'realos', or realists) and those unprepared to yield on their principles (the 'fundos', or fundamentalists).

16 The Liberal Party was seeking support from the nascent Labour Party through the promotion of 'Lib–Lab Pacts' in individual constituencies: the Liberal Party would not contest a seat that Labour hoped to win, in return for Labour MPs' voting support at Westminster.

17 This is illustrated by the wages policies introduced by the Heath government after 1972 and the 'social contract' between government and unions which underpinned the early years of the 1974–9 Wilson and Callaghan Labour governments. Trade union leaders were never invited to discussions at 10 Downing St during the twelve years of the Thatcher governments – indicative of the intention to reduce union power very substantially and to restrict that which remained to negotiations on wages and conditions at individual workplaces.

18 Major public spending cuts as an anti-inflationary policy are usually dated to 1976 when the Labour government negotiated a large loan from the International Monetary Fund, which insisted on such measures.

19 Indeed, Margaret Thatcher claimed responsibility for promoting the onset of democracy in several East European countries, and was thanked by their representatives at a staged session of the 1990 Conservative Party Conference.

20 The growing 'North–South divide' within Great Britain was a major geographical consequence (probably unintended) of these policies (Hudson and Williams, 1980; Johnston, 1991).

21 The August 1991 attempted coup by members of the Communist Party leadership who resisted *glasnost* and *perestroika* failed largely because of the mobilization of popular resistance, especially by the Russian President in Moscow, and the unwillingness of the army to be involved in the forced repression of Soviet citizens.

22 One plan for the transition to a market economy in the Soviet Union envisaged it taking 500 days only; this was resisted by some politicians because of the large-scale, at least temporary, hardship that it

would probably cause (there were estimates that 30 per cent would be unemployed, for example). Given the inability of the state to provide major support to the sufferers, and the already substantial unrest because of failed economic policies, it was believed that such a rapid transition would bring about a further legitimation crisis.

References

Agnew, J. A. and Corbridge, S. 1989: The new geo-politics: the dynamics of geopolitical disorder. In R. J. Johnston and P. J. Taylor, editors, *A World in Crisis? Geographical Perspectives*. Basil Blackwell, Oxford, 266–88.

Anderson, M. J. 1985: Of the state. In A. Seldon, editor, *The 'New Right' Enlightenment*. Economic and Literary Books, Sevenoaks, 115–24.

Atkinson, M. M. and Coleman, W. D. 1985: Corporatism and industrial policy. In A. Cawson, editor, *Organized Interests and the State: Studies in Meso-Corporatism*. Sage Publications, London, 22–44.

Bennett, R. J. 1989: Whither models and geography in a post-welfarist world? In B. Macmillan, editor, *Remodelling Geography*. Basil Blackwell, Oxford, 273–90.

Berry, B. J. L. 1991: *Long-Wave Rhythms in Economic Development and Political Behaviour*. The Johns Hopkins University Press, Baltimore, MD.

Blouet, B. W. 1987: *Halford Mackinder: A Biography*. Texas A&M University Press, College Station, TX.

Bonnet, K. 1985: Corporatism and Thatcherism: is there life after death? In A. Cawson, editor, *Organized Interests and the State: Studies in Meso-Corporatism*. Sage Publications, London, 85–105.

Boston, J. 1987: Wages policy and industrial relations reform. In J. Boston and M. Holland, editors, *The Fourth Labour Government: Radical Politics in New Zealand*. Oxford University Press, Auckland, 151–83.

Brittan, S. 1977: *The Economic Consequences of Democracy*. Temple Smith, London.

Britton, S., le Heron, R. B. and Pawson, E. J., editors, 1992, *Changing Places: A Geography of Restructuring in New Zealand*. New Zealand Geographical Society, Christchurch.

Castles, F. G. 1978: *The Social Democratic Image of Society*. Routledge and Kegan Paul, London.

Cawson, A. 1985: Varieties of corporatism: the importance of the meso-level of interest mediation. In A. Cawson, editor, *Organized Interests and the State: Studies in Meso-Corporatism*. Sage Publications, London, 1–21.

Chapman, R. M. 1981: From Labour to National. In W. H. Oliver and B. R. Williams, editors, *The Oxford History of New Zealand*. Oxford University Press, Wellington, 333–68.

Clark, G. L. 1989: *Unions and Communities under Siege: American Communities and the Crisis of Organized Labor*. Cambridge University Press, Cambridge.

Clark, G. L. and Dear, M. J. 1984: *State Apparatus*. George Allen and Unwin, Boston, MA.

Collins, R. 1987: *Rogernomics: Is There a Better Way?* Pitman, Wellington.

Cox, K. R. 1989: The politics of turf and the question of class. In J. Wolch and M. Dear, editors, *The Power of Geography*. Unwin Hyman, Boston, MA, 61–90.

Douglas, R. 1980: *There's Got To Be a Better Way*. Fourth Estate, Wellington.

Downs, A. 1967: *Inside Bureaucracy*. Little Brown, Boston, MA.

Driver, F. 1991: Political geography and state formation: disputed territory. *Progress in Human Geography* 15, 268–80.

Dunleavy, P. 1991: *Democracy, Bureaucracy and Public Choice*. Polity Press, Cambridge.

Dunleavy, P. and O'Leary, B. 1987: *Theories of the State: The Politics of Liberal Democracy*. Macmillan, London.

Dyson, K. 1986: The state, banks and industry: the West German case. In A. W. Cox, editor, *State, Finance and Industry*. Wheatsheaf Books, Brighton, 118–41.

Dyson, K. 1989: Economic policy. In G. Smith, W. E. Paterson and P. H. Merkl, editors, *Developments in West German Politics*. Macmillan, London, 148–67.

Esping-Andersen, G. 1985: *Politics against Markets: The Social Democratic Road to Power*. Princeton University Press, Princeton, NJ.

Esping-Andersen, G. 1990: *The Three Worlds of Welfare Capitalism*. Polity Press, Cambridge.

Franklin, H. 1985: *Cul-de-Sac: The Question of New Zealand's Future*. Unwin, Wellington.

Friedman, M. and Friedman, R. 1980: *Free to Choose*. Penguin Books, Harmondsworth.

Galbraith, J. K. 1989. *A History of Economics: The Past as the Present*. Penguin Books, Harmondsworth.

Gamble, A. 1988: *The Free Economy and the Strong State: The Politics of Thatcherism*. Macmillan, London.

Gamble, A. 1990: The Thatcher decade in perspective. In P. Dunleavy, A. Gamble and G. Peele, editors, *Developments in British Politics 3*. Macmillan,

London, 333–58.

Goldstein, J. S. 1988: *Long Cycles: Prosperity and War in the Modern Age*. Yale University Press, New Haven, CT.

Grant, W. 1985: Introduction. In W. Grant, editor, *The Political Economy of Corporatism*. Macmillan, London, 1–31.

Gregory, B. 1987: The reorganization of the public sector: the quest for efficiency. In J. Boston and M. Holland, editors, *The Fourth Labour Government: Radical Politics in New Zealand*. Oxford University Press, Auckland, 111–33.

Griffiths, M. J. and Johnston, R. J. 1991: What's in a place? An approach to the concept of place as illustrated by the British National Union of Mineworkers' strike, 1984–85. *Antipode* 23, 185–213.

Habermas, J. 1976. *Legitimation Crisis*. Heinemann, London.

Hall, P. 1988: The intellectual history of long waves. In M. Young and T. Schuller, editors, *The Rhythms of Society*. Routledge, London, 37–52.

Hall, P. and Preston, P. 1988: *The Carrier Wave*. Unwin Hyman, London.

Harrop, M. and Miller, W. L. 1987: *Elections and Voters*. Macmillan, London.

Harvey, D. 1982: *The Limits to Capital*. Basil Blackwell, Oxford.

Harvey, D. 1985: The geopolitics of capitalism. In D. Gregory and J. Urry, editors, *Social Relations and Spatial Structures*. Macmillan, London, 128–63.

Harvey, D. and Scott, A. 1989: The practice of human geography: theory and empirical specificity in the transition from Fordism to flexible accumulation. In B. Macmillan, editor, *Remodelling Geography*. Basil Blackwell, Oxford, 217–29.

Hawke, G. R. 1981: The growth of the economy. In W. H. Oliver and B. R. Williams, editors, *The Oxford History of New Zealand*. Oxford University Press, Wellington, 369–95.

Hayek, F. A. 1944: *The Road to Serfdom*. Routledge, London.

Hayek, F. A. 1985: Socialism, liberalism and the young. In A. Seldon, editor, *The 'New Right' Enlightenment*. Economic and Literary Books, Sevenoaks, vii–viii.

Held, D. 1987: *Models of Democracy*. Polity Press, Cambridge.

Hirsch, F. 1976: *Social Limits to Growth*. Harvard University Press, Cambridge, MA.

Hirschman, A. O. 1970: *Exit, voice and loyalty*. Harvard University Press, Cambridge, MA.

Hirschman, A. O. 1982: *Shifting involvements: private interest and public action*. Princeton University Press, Princeton, NJ.

Hudson, R. and Williams, A. M. 1980: *Divided Britain*. Belhaven Press, London.

James, C. 1986: *The Quiet Revolution: Turbulence and Transition in Contemporary New Zealand*. Allen and Unwin, Wellington.

Jessop, B. 1989: Conservative regimes and the transition to Post-Fordism: the cases of Great Britain and West Germany. In M. Gottdiener and N. Kominos, editors, *Capitalist Development and Crisis Theory: Accumulation, Regulation and Spatial Restructuring*. St Martin's Press, New York, 261–99.

Jessop, B. 1990: *State Theory: Putting Capitalist States in their Place*. Polity Press, Cambridge.

Johnston, R. J. 1984a: The political geography of electoral geography. In P. J. Taylor and J. W. House, editors, *Political Geography: Recent Advances and Future Directions*. Croom Helm, London, 133–148.

Johnston, R. J. 1984b: *Residential Segregation, the State and Constitutional Conflict in American Urban Areas*. Academic Press, London.

Johnston, R. J. 1987: Dealignment, volatility and electoral geography. *Studies in Comparative International Development* 22, 3–25.

Johnston, R. J. 1989: The individual in the world economy. In R. J. Johnston and P. J. Taylor, editors, *A World in Crisis? Geographical Perspectives*. Basil Blackwell, Oxford, 200–28.

Johnston, R. J. 1990: Lipset and Rokkan revisited: electoral cleavages, electoral geography and electoral strategy in Great Britain. In R. J. Johnston, F. M. Shelley and P. J. Taylor, editors, *Developments in Electoral Geography*. Routledge, London, 121–42.

Johnston, R. J. 1991. *A Question of Place*. Basil Blackwell, Oxford.

Johnston, R. J. 1992: Meet the challenge: make the change. In R. J. Johnston, editor, *The Challenge for Geography*. Basil Blackwell, Oxford,

Johnston, R. J. and Pattie, C. J. 1990: The regional impact of Thatcherism. *Regional Studies* 24, 479–93.

Johnston, R. J., Pattie, C. J. and Allsopp, J. G. 1988: *A Nation Dividing?* Longman, London.

King, D. 1989: Econimic crisis and welfare state recommodification: a comparative analysis of the United States and Britain. In M. Gottdiener and N. Kominos, editors, *Capitalist Development and Crisis Theory: Accumulation, Regulation and Spatial Restructuring*. St Martin's Press, New York, 237–60.

Kuznets, S. 1953: *Economic Change*. W. W. Norton, New York.

Lane, J.-E. and Ersson, S. O. 1991: *Politics and Society in Western Europe*, Sage Publications, London.

Lash, S. and Urry, J. 1987: *The End of Organized Capitalism*. Polity Press, Cambridge.

Lehmbruch, G. 1982: Introduction: neo-corporatism in comparative perspective. In G. Lehmbruch and P. C. Schmitter, editors, *Patterns of Corporatist Policy-Making*. Sage Publications, London, 1–28.

Lewis-Beck, M. S. and Squire, P. 1991: The transformation of the American state: the New Era–New Deal test. *Journal of Politics* 53, 106–21.

Lipset, S. M. and Rokkan, S. E. 1967: Cleavage structures, party systems and voter alignments: an introduction. In S. M. Lipset and S. E. Rokkan, editors, *Party Systems and Voter Alignments*. The Free Press, New York, 3–64.

Lloyd Prichard, M. F. 1970: *An Economic History of New Zealand*. Collins, Auckland.

Mangen, S. 1989: The politics of welfare. In G. Smith, W. E. Paterson and P. H. Merkl, editors, *Developments in West German Politics*. Macmillan, London, 168–91.

Mann, M. 1984: The autonomous power of the state: its origins, mechanisms and results. *European Journal of Sociology* 25, 185–213.

Mann, M. 1986: *The Sources of Social Power, Volume 1: A History of Power from the Beginning to AD 1760*. Cambridge University Press, Cambridge.

Marin, B. 1985: Austria – the paradigm case of liberal corporatism. In W. Grant, editor, *The Political Economy of Corporatism*. Macmillan, London, 89–125.

Markovits, A. S. and Allen, C. S. 1989: The trade unions. In G. Smith, W. E. Paterson and P. H. Merkl, editors, *Developments in West German Politics*. Macmillan, London, 289–307.

Marshall, M. 1987: *Long Waves of Regional Development*. Macmillan, London.

Massey, D. and Meegan, R. 1979: The geography of industrial reorganisation. *Progress in Planning*, 10, 155–237.

Massey, D. and Meegan, R. 1982: *The Anatomy of Job Loss*. Methuen, London.

New Zealand Treasury 1984: *Economic Management*. Government Printer, Wellington.

Niskanen, W. A. 1973: *Bureaucracy: Servant or Master*. Institute of Economic Affairs, London.

O'Connor, J. 1973: *The Fiscal Crisis of the State*. St Martin's Press, New York.

Offe, C. 1984: *Contradictions of the Welfare State*. Hutchinson, London.

Osei-Kwame, P. and Taylor, P. J. 1982: A politics of failure: the political geography of Ghanaian elections, 1954–1979. *Annals of the Association of American Geographers* 72, 574–89.

Owens, J. E. 1986: The regulation and deregulation of financial institutions in the United States. In A. W. Cox, editor, *State, Finance and Industry*. Wheatsheaf Books, Brighton, 172–230.

Paci, M. 1987: Long waves in the development of welfare systems. In C. S. Maier, editor, *Changing boundaries of the political*. Cambridge University Press, Cambridge, 179–200.

Parker, W. H. 1982: *Mackinder: Geography as an Aid to Statecraft*. Clarendon Press, Oxford.

Paterson, W. E. and Smith, G., editors 1981: *The West German Model: Perspectives on a Stable State*. Frank Cass, London.

Pattie, C. J. and Johnston, R. J. 1990: Embellishment and detail? *Transactions, Institute of British Geographers* NS 15, 205–26.

Peet, J. R. 1983: Relations of production and the relocation of United States manufacturing industry since 1960. *Economic Geography* 59, 112–43.

Piven, F. F. and Cloward, R. A. 1971: *Regulating the Poor*. Vintage Books, New York.

Piven, F. F. and Cloward, R. A. 1977: *Poor People's Movements*. Pantheon Books, New York.

Popper, K. R. 1945: *The Open Society and its Enemies*. Routledge and Kegan Paul, London.

Quadango, J. 1988: *The Transformation of Old Age Security*. University of Chicago Press, Chicago.

Roche, M. M., Johnston, T. and le Heron, R. B. 1992: Farmers' interest groups and agricultural policy in New Zealand during the 1980s. *Environment and Planning A*, 25.

Rose, R. 1984: *Understanding Big Government*. Sage Publications, London.

Rose, R. 1985: The significance of public employment. In R. Rose, editor, *Public Employment in Western Nations*. Cambridge University Press, Cambridge, 1–53.

Rostow, W. W. 1987: *The Stages of Economic Growth*. Cambridge University Press, Cambridge.

Roth, H. 1973. *Trade Unions in New Zealand*. Reed Education, Wellington.

Royal Commission on Social Security in New Zealand, 1972: *Social Security in New Zealand*. Government Printer, Wellington.

Salisbury, R. H. 1979: Why no corporatism in America? In P. C. Schmitter and G. Lehmbruch, editors, *Trends toward Corporatist Intermediation*. Sage Publications, London, 213–30.

Saunders, P. 1979: *Urban Politics: A Sociological Interpretation*. Penguin Books, Harmondsworth.

Saunders, P. 1985a: Corporatism and urban service provision. In W. Grant, editor, *The Political*

169

Economy of Corporatism. Macmillan, London, 148–73.

Saunders, P. 1985b. Labour. In A. Seldon, editor, *The 'New Right' Enlightenment.* Economic and Literary Books, Sevenoaks, 163–74.

Sayer, A. 1984: *Method in Social Science.* Hutchinson, London.

Schumpeter, J. A. 1934: *The Theory of Economic Development.* Cambridge University Press, Cambridge.

Schutz, A. 1972: *The Phenomenology of the Social World.* Heinemann, London.

Scott, C. 1987: The 1985 tax reform package. In J. Boston and M. Holland, editors, *The Fourth Labour Government: Radical Politics in New Zealand.* Oxford University Press, Auckland, 184–97.

Smith, G. 1979: *Democracy in Western Germany.* Heinemann, London.

Smith, G. 1981: Does West German democracy have an 'efficient secret'?. In W. E. Paterson and G. Smith, editors, *The West German Model: Perspectives on a Stable State.* Frank Cass, London, 166–76.

Smith, G. 1989: Structures of government. In G. Smith, W. E. Paterson and P. H. Merkl, editors, *Developments in West German Politics.* Macmillan, London, 24–39.

Smith, N. 1984: *Uneven Development.* Basil Blackwell, Oxford.

Streeck, W. 1982: Organizational consequences of neo-corporatist co-operation in West German labour unions. In G. Lehmbruch and P. C. Schmitter, editors, *Patterns of Corporatist Policy-Making.* Sage Publications, London, 29–82.

Sutch, W. B. 1966: *The Quest for Security in New Zealand.* Oxford University Press, Wellington.

Sutch, W. B. 1969: *Poverty and Progress in New Zealand.* A. H. and A. W. Reed, Wellington.

de Swaan, A. 1988: *In Care of the State.* Polity Press, Cambridge.

Taylor, P. J. 1989: *Political Geography: World-Economy, Nation-State and Community.* Longman, London.

Taylor, P. J. 1990: *Britain and the Cold War: 1945 as Geopolitical Transition.* Belhaven Press, London.

Tiebout, C. 1956: A pure theory of public expenditure. *Journal of Political Economy* 64, 416–24

Townsend, P. 1979: *Poverty in the United Kingdom.* Penguin Books, Harmondsworth.

Veljanovski, C. 1990: The political economy of regulation. In P. Dunleavy, A. Gamble and G. Peele, editors, *Developments in British Politics 3.* Macmillan, London, 291–304.

Weale, A. 1990: Social policy. In P. Dunleavy, A. Gamble and G. Peele, editors, *Developments in British Politics 3,* Macmillan, London, 197–220.

Webber, M. J. 1991: The contemporary transition. *Environment and Planning D: Society and Space* 9, 165–82.

Wilson, G. K. 1982: Why is there no corporatism in the United States? In G. Lehmbruch and P. C. Schmitter, editors, *Patterns of Corporatist Policy-Making.* Sage Publications, London, 219–36.

4

Colonialism, post-colonialism and the political geography of the Third World

Stuart Corbridge

at the precise instant of India's arrival at independence, I tumbled forth into the world. There were gasps. And, outside the window, fireworks and crowds. . . . Now, however, time (having no further use for me) is running out. I will soon be thirty one years old. Perhaps. If my crumbling, over-used body permits. But I have no hope of saving my life, nor can I count on having even a thousand nights and a night. I must work fast, faster than Scheherazade, if I am to end up meaning – yes, meaning – something. I admit it: above all things, I fear absurdity.

Salman Rushdie (1981), *Midnight's Children*, p. 9.

Introduction

At first glance, these are saddening times to be writing about politics in the so-called Third World. Just thirty or forty years ago there was a feeling of optimism that 'development' and 'reason' (for which, read parliamentary democracy) would come quickly to the newly independent countries of Asia and Africa, and to the resurgent countries of Latin America and the Caribbean. In 1961 the talk was of a Development Decade which would see capital and knowledge transferred from the advanced countries to the backward countries, with democracy and party-building following rapidly in their wake. (The one exception to this was the socialist Third World, which had followed the deviant route of communism: Rostow, 1960.) In the 1990s the emphasis is on maldevelopment and political catharsis, at least for most Third World countries. The watchwords for Africa are now misery, famine and war, with desertification, dictatorship and AIDS also haunting the Dark Continent (Watts, 1989). In Latin America it is debt which exercises the popular and political imaginations, with drugs, environmental threats and death squads adding further visions of despair and dislocation. In Asia, meanwhile, the economic successes of the four little tigers of East Asia are tarnished by a failure to achieve representative democracy in this region. Further west, the rise of Islamic 'fundamentalism' is seen by many in the West as an obstacle to progress in South and West Asia, while regionalist tensions and communalism are also clearly at work in the Indian sub-continent. The assassination in 1991

of Rajiv Gandhi by Tamil separatists was not the sort of millennial celebration which President Kennedy had in mind when he launched the Development Decade; nor was the massacre in Tienanmen Square in 1989.

This chapter considers the rise and apparent fall of progressive Third World politics in two distinct ways. The first part of the chapter provides a narrative account of the incorporation of the Third World into the European world-economy. It also considers the rise of nationalist struggles in the periphery of the world-system, and the emergence of independent countries in the period (largely) since 1945. The regional coverage is necessarily broad and the approach taken is largely chronological. The second part is more analytical in style and raises several questions about the manner in which Third World politics is often written about and understood. It comprises five sections, all of which take their lead from Geoff Hawthorn's suggestion that the paradoxes of the Third World (at least to western eyes) have their roots in 'the combination, as one might describe it, of an incipiently modern economy and (behind the curtain of modern constitutions) an archaic politics; the coexistence of the economics of separate interests and a politics which looks more like that of ancient liberty' (Hawthorn, 1991, 28). A common point of reference in the chapter is the politics of British and postcolonial India.

In more detail, the first part of the chapter begins by considering the record of Third World countries since independence in terms of democracy and development (crudely defined). The remaining sections then discuss the sources of the 'Third World malaise' as they are affected by economic concerns and international economic relations, by geopolitical concerns and agendas, by local concerns relating to the building of state institutions and a vibrant civil society, and by those concerns raised by local oppositional movements (and in particular those movements which call into question the project of modernity in Third World social formations). The conclusion to the chapter reflects more freely on the problems posed to conventional political geography and political science by attempts to re-conceptualize the various discourses of politics and development in a diverse Third World. It also affirms the importance of understanding Third World politics in relation to the politics of the First and Second Worlds, and with regard to the changing international economic situation.

The commands of empire (writing history)

A narrative account of Third World politics in the twentieth century is bound to start some 400 years previously with the European voyages of discovery and conquest in the late fifteenth and sixteenth centuries. It is from this time that the modern paradoxes of the Third World begin to take shape – which is not to suggest that Third World countries are merely prisoners of their colonial histories. The integration of Latin America, Africa and Asia into the European world-economy had pressing implications for the economies of these regions, – for the structure of their merchandise trade and relations of production; for their political geographies at the level of 'national' boundaries, the state and civil society; and for the integrity of their populations and cultures. Within this overarching history, a 'very broad distinction can be made between three main forms of colonialism, each chiefly associated with a particular era, a particular continent and a particular set of colonial powers: the Americas, both rich and easy to control; Asia, rich but difficult to control; and Africa, for the most part poor and so scarcely worth controlling' (Clapham, 1985, 13).

Colonialism

The history of colonialism in Latin America and the Caribbean was especially brutal and involved the destruction of many indigenous societies. This was especially true in the Caribbean islands, where plantation crops were worked from early on with slave labour imported from Africa (and later with indentured labour shipped from the Indian sub-continent). The Caribbean countries quickly came to resemble the dual society/dual economy archetypes so often to be recalled elsewhere (Furnivall, 1948): at the top of local societies were the European settlers and landowners;

beneath them were transported and transplanted labouring families and some remants of the indigenous population. A similar fate befell parts of the Andean region (though less so in Peru and Bolivia), and the so-called 'white-settler colonies' of Chile, Argentina and Southern Brazil.

The economies of Latin America and the Caribbean were soon dependent upon structures of supply and demand which emanated mainly from Europe. Latin America was mined extensively for its mineral wealth, and its tropical regions were planted with crops including sugar, tobacco, coffee and bananas (Stein and Stein, 1970). Local labour forces were tightly controlled under systems of corvée labour (haciendas and latifundias), and local systems of transportation and communication came increasingly to be centred upon a few coastal cities. Although most of the Latin American colonies achieved independence in the nineteenth century, their means of integration into the world-economy continued to ensure their *de facto* dependence well into the twentieth century. Industrialization and urbanization were sponsored most vigorously in the southern temperate regions, where British capital helped to develop the railroads of Argentina and promote a local meat-packing industry designed to meet the food needs of Western Europe. In the late nineteenth century, US capital became more influential in Latin America, particularly so when the USA began to police the region as its own backyard: the Monroe Doctrine was promulgated as early as 1823. The USA also acted to secure those local landowners and urban elites who could be relied upon to support US interests in the region (Dunkerley, 1988). In so doing, the USA helped to make sure that the state, in many Latin American countries, would continue to be seen as the puppet of various feudal and non-national interests. In the 1960s it was widely believed that development would not come to Latin America until the feudal-imperialist alliance had been dismantled.

The situation in Asia was rather different to that which prevailed in Latin America and the Caribbean. The English East India Company was incorporated in 1600, just two years before its Dutch counterpart was restructured, and European political interest in Asia dates mainly from this time. (This does not mean that long-distance trade was not already important in the Indian Ocean region: Chaudhuri, 1990.) The Dutch interest in the East Indian islands was strenghtened in the seventeenth century when the Dutch fleet defeated the English fleet in several naval engagements in the region. The Dutch consolidated their control over the East Indies spice trade when the Dutch East India Company achieved supremacy in Java in 1752 (Bayly, 1989). Meanwhile, the English fell back to the east and west coasts of India, where they also engaged in competition with the French and the Portuguese. Both countries were aided by the collapse of the Mughal Empire at this time. The main interest of the Europeans in India was to service their trade with China. The Europeans were hopeful that they would take from India commodities which could be exchanged with the Chinese in return for Chinese silks, spices and teas. As things stood, the Chinese showed little interest in European commodities, preferring to receive European bullion for Chinese traded goods.

It was not until the eighteenth and nineteenth centuries that the British achieved dominion over large parts of the Indian sub-continent, and with it the defeat of France and Portugal (save for remnant territories such as Goa and Pondicherry). It was at this time, too, that the British began to move inland from their coastal forts at Bombay, Madras and Calcutta to secure those rural hinterlands which supplied the East India Company with its stocks of opium, tea, cotton, jute and indigo. Even then the British were unable to subdue the entire sub-continent of India and they resisted a temptation to rule India by means of a centralized colonial government and/or through massive European settlement. The British instead worked through local notables, confirming the power of village headmen, landlords and rent collectors (Yang, 1989), while seeking the support of India's Princes in those parts of the sub-continent which were to be ruled indirectly (and with the 'advice' of British Residents). This option was preferred still more strongly after the First War of Independence of 1857: a 'Mutiny' which prompted the British to rule India as a Crown Colony under the Queen-Empress Victoria (the title was

assumed in 1876), but to do so in a way which did minimal violence to the customs and econmies of most of the populations which fell under Imperial rule. Macaulay's Education Minute of 1836 was quietly retired, his desire to bring English-style education to the natives being dismissed as impracticable and undesirable; far better to prepare the natives for self-rule in the distant future while respecting their customs, and their ignorance, in the here and now (Rothermund, 1986).

Not all of Asia fell to the British, of course, even though India did come to be seen as the jewel in the crown of an Empire upon which the sun never set. Japan and China avoided direct colonial rule in the nineteenth century, as did Afghanistan, Thailand and Iran. (Of these countries, only Japan achieved a genuine independence from the European powers.) Further west, the Arab regions of the Middle East remained mainly under the control of the Ottoman Empire until the end of World War I. Thereafter, the British and the French achieved control over most of the Arab territories, and both sought to rule their new populations in conjunction with local dynasties (Hourani, 1991). As Clapham points out:

The impact of western colonialism was [here] mitigated not only by a degree of internal self-government, but also by a pronounced sense of regional identity expressed in political form through Pan-Arabism and in religious form through Islam. Unlike the state-centred nationalisms of Iran or Thailand, Arab nationalism cut across the boundaries established by the administering powers, in a way which – more than in any other area of the Third World – has challenged the pre-eminence of the postcolonial state as the framework for political activity. (Clapham, 1985, 16)

The conquest of Africa took a different form again to that evident in the Americas and Asia. Africa remained the Dark Continent until the very end of the nineteenth century; indeed, Foureau-Lamy's expedition to the geographical centre of Africa was not completed until 1900, just fifty years after the explorations of southern and south-central Africa undertaken by Livingstone. The coastal regions of west and east Africa had provided important trading posts in the maritime world-economy for several centuries, but beyond the coastal regions little was known about Africa by Europeans, save that it was a fecund source of slave labour. About 10 million African slaves were shipped to the Americas between 1500 and 1870 (Wolf, 1982). The prevailing assumption was that Africa was poor and lacking in resources, both physical and mental. This view was strengthened in the late nineteenth century, when geographers and other social scientists fell upon the discourses of social Darwinism and environmental determinism to help justify their discharging of the White Man's Burden in 'pre-historic' Africa: a continent without history or culture or purpose.

In the wake of the exploration of Africa, conquest and direct colonial rule came very swiftly and with little apparent rhyme or reason save that of geopolitics. In 1884–5 the major European powers agreed at the Congress of Berlin to stake out their own spheres of influence in Africa, and not to encroach on the zones of influence of one another. There followed a twenty-year 'Scramble for Africa' which left the French in control of most of north-west Africa, the British in charge of an easterly corridor running (more or less) from Egypt to South Africa, and the Portuguese in charge of south-west and south-east Africa. The Germans, Belgians and Italians also had their shares of Africa, along with the Spanish in the far north-west. Only Ethiopia (Abyssinia) remained free of an occupying power (until the Italian invasion of 1936), together with the small kingdoms of Swaziland, Lesotho, Rwanda and Burundi.

The division of Africa into separate colonies is a disturbing story which is well known to most students of geography (see also Pakenham, 1991). Lines on a map were often drawn with no regard for cultural or linguistic ties, or even with regard to relationships of kinship or lineage. It is a story of arbitrary division which continues to haunt parts of Africa.

Elsewhere, the incorporation of colonized populations into European culture and politics was less hasty, if no less deleterious in some of its consequences. In broad terms, the formation of colonial rule entailed four key points or activities. First, 'colonialism established territories and territorial boundaries where [sometimes] none had existed before' (Clapham, 1985, 18). Second 'colonialism established within each territory a

political order and the administrative hierarchy to run it. The ultimate basis of this political order was, invariably, force: there is no other way that a small group of alien rulers can establish control over a people not their own' (Clapham, 1985, 18). This point was made with particular clarity in the Jharkhand, India. After a rebellion by local Kol tribals in the early nineteenth century, 'the Government thought that strong measures, such as subjecting the rebels to capital punishment by blowing them from the mouth of a gun, combined with conciliation after an impression had been made, were absolutely necessary' (Basu, 1957, 18). Third, colonialism brought with it a series of ties which bound a colony into the wider networks of trade and production which defined the colonial world-economy (Washbrook, 1990). Finally, colonialism brought with it a culture of rule bound up with the imposition of an alien language. Claphan rightly remarks that: 'There is still no more striking, even shocking, reminder of the impact of colonialism in Africa than to cross an entirely artificial frontier and witness the instant change of language – and with it a myriad number of associated manifestations of culture and government – that results' (Clapham, 1985, 20).

Beyond these four points of reference, the imposition of colonial systems of rule varied sharply from place to place, from time to time, and with reference to the philosophies and machineries of power which defined a given colonial power. The point is well made if we briefly compare the *raisons d'être* of the two main colonial powers, Britain and France.

Nationalism and Imperial raisons d'être

The main objective of British colonial rule was to maintain a cheap and effective system of colonial administration which would police the colonized populations, and which would make possible the remittance to Britain of the profits of Empire. The British were not interested in the politics of assimilation – at least not in the non-white, non-settler colonies. The British idea of what it was to be British made this unthinkable. A 'black Briton' was an oxymoron. The British thus favoured systems of direct and indirect colonial rule which would aim not to disturb local populations unduly (or to 'develop' them as we would now say), while paradoxically aiming in the long run to prepare these subject nations for the possibility of self-rule.

In the nineteenth century, the British were able to negotiate this paradox by means of their superior force of arms, by an efficient machinery of colonial rule (epitomized in the Indian Civil Service), and by systems of governance which made a virtue of both negotiation with local elites at one scale and divide and rule at another (as with Hindus and Muslims in British India). In the twentieth century, this process of negotiation came to favour the British less and less. In 1885 the formation of the Indian National Congress (INC) hinted at a later and more direct challenge to British rule in India. At first, the leaders of the INC wanted a share in the government of India; they wanted to play the part of the 'black British', as some nationalists later described the aspirations of their forebears. Later on, under Gandhi and Nehru, the aim was to oust the British from India. The British had helped to manufacture a progeny which they could no longer control. Precisely because the British could not accept their colonized populations as their equals, they were condemned to promote a politics which must have recourse to the symbols of nationalism and anti-colonialism. This politics in turn ensured that British rule in India would become more and more expensive (as it would be elsewhere later on: Holland, 1985). Movements which sought to resist the importation of British goods (the *swadeshi* campaign), in combination with movements which increased the costs of policing in India (such as the Quit India campaigns of the 1940s), paved the way for a hasty retreat from India in August 1947 (Tomlinson, 1979). This retreat was also enabled by the attitudes of some in the British Labour Government of 1945–51, by the costs imposed on Britain by World War II, and by pressures placed on the UK by the USA (its banker) to move away from a colonization of spaces and towards a new 'democracy of things' (or commodity colonialism).

The *raison d'être* of French colonialism was subtly different to that of the British, which is not to say that the two systems of government did not share certain fundamentals in common. The main commonalities were a commitment to a system of economic imperialism, and a desire to create and produce colonial subjects.

Colonialism is often defined as a system of governance which seeks to defend an unequal system of global commodity exchanges. But colonialism was (and is) also about the production of ideologies and what Foucault would call 'subject positions'. This insight has been developed in recent years by Edward Said (Said, 1979), whose main focus has been upon the discourse of Orientalism. Said maintains that colonialism was only made possible by a discourse which sought to oppose a superior Europe to its inferior, Oriental, Other: a distinction which allowed, and even mandated, the exercise of colonial force to bring reason to peoples untouched by science or the Enlightenment. According to Said: 'without understanding Orientalism as a discourse one cannot possibly understand the enormously systematic discipline by which European culture was able to manage – and even produce – the Orient politically, sociologically, militarily, scientifically and even imaginatively during the post-Enlightenment period' (Said, 1979, 3). Orientalism thus 'produced' both the abject subject populations of the colonies and the West's own image of itself.

Timothy Mitchell has explored a related theme in his book, *Colonising Egypt*. Mitchell suggests that the British colonial project in Egypt involved detailed attempts to 'regulate the daily life of rural Egyptians' (Mitchell, 1991, ix). This attempt at regulation depended crucially upon surveillance; upon seeing, but on not always being seen to observe. For example, the streets of Cairo were forced to carry names and numbers to make them legible to the European mind, and strict controls were placed on the movement of Egyptian field labourers. The bodies of the colonized populations were also captured by new methods of military control, architectural order and schooling. Mitchell concludes that: 'Colonialism was distinguished by its power of representation, whose paradigm was the architecture of the colonial city (see also Metcalf, 1989; King, 1990), but whose effects extended themselves at every level' (Mitchell, 1991, 171; see also Arnold, 1988; Hefferman and Sutton, 1991; Pratt, 1992).

This analysis of colonialism is one that we will have cause to return to. It is a colonization of minds and bodies, of space itself (both macro and micro), as much as it is a colonialism of trade and war. It follows, too, that colonialism thus defined is much harder to transcend or to throw off; colonial attitudes and practices can easily survive the act of political decolonization. They might even inform the project of 'development'. But this is to jump ahead. If we return to the French project of colonialism, it is apparent that a shared commitment to a European conception of Empire did not preclude a French particularism. Indeed, it has often been remarked that French colonialism was more deep-rooted and violent than the colonialism even of the British, and that the French were correspondingly more reluctant than the British to let go of Empire after World War II.

The reasons for this have something to do with French attitudes to colonialism and/(as) assimilation. According to Clapham:

For the French, the key ideas were those of centralisation and assimilation, derived from the experience of nation-building in metropolitan France, coupled with an appeal to the egalitarian principles of the revolution of 1789. Indigenous social and cultural systems were, in principle, dismissed as worthless, but indigenous peoples were, in accordance with the Rights of Man, to be offered the chance of assimilation to the ideals of France herself, by acquiring French language, culture and nationality, and hence in time becoming indistinguishable from other Frenchmen. (Claphan, 1985, 21)

This project was no less paradoxical than the colonial project pursued by the British, but it was so in different respects. In theory, the French impulse to colonization eschewed racism, preferring to ground the subordination of the colonized populations in terms of the latter's failure to achieve a cultural system on a par with that of France. Colonialism was then meant to make good this failing by the messianic promotion of an ideology and a politics of assimilation – to the point that black Africans could achieve high positions in Parisian society in the 1920s and 1930s. In practice, however, the politics of assimilation required monetary support, and herein lay a contradiction for an Empire which was meant to run at a profit (or at least not at a

loss). A majority of French citizens were not keen for their governments to spend freely on the subject populations of the French Empire. Nor were most French men and women keen to welcome into French society individuals who they perceived not to be French. A philosophy of non-racialism is not the same thing as a local politics of non-racialism – a point which most French governments understood well enough, and which Fanon raised in his classic account of *The Wretched of the Earth* (Fanon, 1967).

The French were thus compelled to negotiate the politics of colonialism and nationalism in a manner which was not entirely unknown to the British and some other European powers. As compared to the British, the French did invest heavily in Empire, especially in the policing of Empire. The costs to France of a standardized and centralized imperial rule might be guessed at from one statistic: namely, that while in British Nigeria in the 1930s there was one European administrator for every 15,000 inhabitants, in neighbouring French West Africa there was one officer of Empire for every 4,000 inhabitants (Clapham, 1985, 23). In terms of geopolitics, the reluctance of the French to give up on Empire was expressed in the conflicts which took shape in Vietnam and Algeria in the 1950s, and which continued even after de Gaulle took power in 1958. In this respect, the French prefigured the ambitions – or the follies – of the Portuguese, whose own lack of democratic government at home encouraged a willingness to fight for an Empire abroad until the 1970s. As with the French and the British, the Portuguese were finally removed from Africa in the wake of well-organized local wars for independence (in this case for Angola and Mozambique). It then remained only for the British to supervise a transfer of power in Rhodesia/Zimbabwe (1980), for a century of European colonialism in Africa to be drawn to a close.

Some dependencies remain, of course, as they do in Asia (Hong Kong) and the Americas (the Falklands/Malvinas), but they are perhaps of lesser importance than the struggle which continues in South Africa between the black populations and a white settler population. They might also be eclipsed by the failure of many postcolonial countries to achieve those measures of democratization and development which they looked forward to at independence. Even in the revolutionary Third World, the socialist project as it has taken shape in China, Cuba, Tanzania and elsewhere has hardly met with the success that its leaders promised when power was first seized. The question we must now ask is, 'Why?' We must also begin to render problematic the terms in which our narrative has been stated, and the terms in which this demand for explanation – this simple and singular 'Why?' – is put.

The search for a text (writing politics)

Development and democracy

The seeming failure of many Third World countries to live up to the promises made for them at independence has a good deal to do with the nature of the expectations which surrounded them when power was transferred. These promises were of three main sorts.

First, there was the promise associated with victory itself. In so far as colonialism could reasonably be presented as a cause of the under-development of an ex-colonial country, the absence of colonialism could itself be accounted a mainspring of development. An absence could become a presence if the state in a postcolonial country pushed that country towards an era of modernity and modernization.

Second, there was a general expectation of progress in the West and the Third World in the 1950s and 1960s, not least in the wake of the rebuilding of Europe and Japan post-1945. The creation of the Golden Age of Capitalism (Marglin and Schor, 1990) between 1945 and 1970 (at least in the OECD countries) seemed to be based upon a Keynesian social democratic consensus (or Fordism) which could be rebuilt in the developing world. Even the socialist Third World could look with some favour upon the achievements of the Soviet Union since 1917 (and since 1945). Industrialization could be achieved rapidly by dint of the application of certain rules and resources propagated by a secular state.

Third, there was the expectation of high theory. Most accounts of Third World development and politics in the 1950s, 1960s and 1970s

were phrased in terms either of modernization theory or dependency theory. According to the theorists of modernization, development would come quickly to those latecomer countries which were prepared to accept the advice, technologies and capital on offer from the pioneer industrial countries (Taylor, 1979). A take-off into self-sustaining growth could be achieved in a developing country if and when the government of that country committed itself to saving between 10 and 15 per cent of a country's GDP, and to investing this surplus in a planned programme of public- and private-sector industrialization. Income inequalities would have to expand in the short-term, as would the gap between core areas and the periphery of a given space-economy; in the medium-to-long term, however, the benefits of growth and development would trickle down the social and central-place hierarchies of that country (Hirschman, 1958). Meanwhile, politics in the developing world would be modernized to the extent that Third World countries embraced the rules and regimes of a representative democracy, and in so far as Third World governments fostered a culture of political pluralism built upon functioning political parties. This was the diagnosis offered by Samuel Huntington and others in the 1960s (Huntington, 1968), and one can see in retrospect how neatly the work of Huntington dovetailed with the advice offered by Rostow in his 'non-communist manifesto'. The Third World was to be developed – to be modernized – by virtue of its attachments to a wider process of westernization. In the wake of this process of development, earlier and more primordial attachments of kin and tribe and place would wither away. In Durkheimian terms, countries which were once integrated mechanically and through face-to-face contacts, would now be integrated organically, through the impersonal mechanisms of the market, secular politics, urbanization, education and modern communications.

Set against this body of theory was dependency theory. At first glance more pessimistic than modernization theory, dependency theory (or some versions of it: Palma, 1989) also sought a unitary theory of Third World politics and a unitary model of and for Third World political tranformation. The main difference was that, for dependency theorists, the West, and western capitalism, continued to create the conditions of existence of dependent mal-development or underdevelopment in the periphery of the world-system. Systems of unequal exchange ensured that Third World resources and commodities were lost to the metropolitan regions, and that Third World economies remained under-industrialized and bereft of modernization (Emmanuel, 1972). To become 'western', the Third World would have to break with the West, either through a socialist rupture and/or by the imposition of economic and other controls which would minimize the contacts between the developed and the developing worlds. Modernity and modernization would then come to the Third World in western forms, but not by means of the West directly (Frank, 1969; Furtado, 1963).

The failure of most Third World countries to fulfil this three-fold set of expectations is hardly surprising. Most western countries have not fulfilled the expectations commonly made of them in the 1950s and 1960s; economic uncertainty and political dislocation are not unknown in the West (or in the socialist Second World). Nevertheless, the nature of the 'failure' of some Third World countries is not quite the same as that exhibited in many western countries, where representative democracies and high standards of living continue to be assured for the majority. Moreover, the nature of the success and/or failure of the Third World has hardly been uniform. Notwithstanding the grand narratives of modernization theory and dependency theory, the Third World has refused to fit into a simple conceptual grid of modernization and/or under-development. Most Third World countries have met with political successes and failures at different times, with regard both to development and democratization. At the same time, some countries in the periphery have performed better in some respects than could reasonably have been expected, while still others have frustrated their most sympathetic observers.

Since independence, the economies of the Third World have grown at rates of growth sufficiently different from one another to lead the World Bank and some other international organizations to move away from the appellation

'Third World'. The World Bank now prefers to speak of 41 Low-Income Economies (13 of which come from Asia, including India and China; 27 of which come from Africa; with just Haiti coming from the Americas); of 41 Middle-Income Economies (including Zimbabwe, Colombia and Malaysia); of 17 Upper-Middle-Income Economies (including Gabon, Brazil and South Korea); and of 25 High-Income Economies (including Singapore, Saudi Arabia and the United Arab Emirates) (World Bank, 1991; some ex-socialist countries are not included in this classification). Data from the World Bank also reveal that average per capita incomes are now higher in South Korea than in Portugal; while an average Kuwaiti family *appears* to be better off than its counterpart in the United Kingdom. Meanwhile, in nine LIEs in 1989, per capita incomes were less than $200 per annum. In some LIEs, too, life expectancy at birth in 1989 was as low as 48–50 years for women and 46–48 years for men; infant mortality rates also continued at rates in excess of 100 per 1,000 live births in twenty-four LIEs. In South Korea in 1989 the infant mortality rate was 23 per 1,000.

The position with regard to politics is no more unitary. If welfare and economic growth are not yet the same thing, development and democracy are not always linked either. South Korea and Taiwan are each ex-Third World countries according to some definitions, but in neither country have conditions yet been laid for a representative democracy. In still other Third World countries the goal of western-style democracy has been challenged or rejected as a desirable outcome. Some socialist developing countries have preferred to build people's democracies, disdaining the sham bourgeois democracies which they see as protective of the West's culture of possessive individualism. People's democracies are intended to involve people as decision-makers at all levels of society, including at the neighbourhood scale and within the factory, and not just by means of the ballot box. Still other forms of people's democracies have been tried in some Islamic republics, just as an attention to democracy has been used for rhetorical purposes in such one-party states as Hastings Banda's Malawi and Mobutu's Zaïre. The Third World, as Geoff Hawthorn reminds us, is 'still waiting for a text' (Hawthorn, 1991). Its very different pasts and peoples, together with its very different presents and problems, admits of no simple map to its political systems, successes and failures.

This does not mean that Third World politics can only be dealt with on a case-by-case basis, although a local expertise is always desirable. Hawthorn further points out that the Third World does have a past in common, if not a common past, and that its leaders and populations are striving typically to confront two problems: the problem of development, and the problem of state-building. Hawthorn suggests that it is in the many different attempts made by Third World countries to negotiate these two tasks or problems (problems which might often rub against one another) that we can find a key to some aspects of 'Third World politics'. I agree with him. In the remainder of this chapter, I examine why so many Third World countries find it so difficult to pursue a politics of democratic state-building and politics of development *at the same time*. I do this by examining four areas in which resources are available both to support and to frustrate these twin projects. The four areas I refer to are: the international economy; geopolitics; domestic politics with reference to state-building, and domestic politics with regard to oppositional movements.

Post-colonial countries in the international economy

The one commitment made in common by all Third World countries at independence was to 'development', or the 'development of their peoples'. To promote development it was in turn assumed that the state in postcolonial countries should be a strong state, or a developmental state, and herein there lies a potential contradiction.

This assumption of strength with regard to the polity and the state was not only a matter of ideology, as some in the counter-revolution in development theory and policy now like to assume (see Toye, 1987); it was not just a matter of a prevailing Keynesianism, or a related regard in the Third World for the achievements of the Soviet Union. It was also the case that many

postcolonial countries were starting from scratch, or close to it. The underdevelopment of their economies under colonial rule had been such that the main icon of development – manufacturing industry – was all but absent from their territories, along with the infrastructures which would be necessary to support a capability in manufacturing industry. Even India, which on one count was the world's tenth largest industrial power in 1947, was largely bereft of manufacturing industry at independence. Eighty per cent of its population lived in rural areas and earned a living mainly from farming activities (Corbridge, 1991). The major centres of industry were in the mining belt of central-east India, in the cotton belt of Bombay-Ahmedabad, and in the jute – exporting port of Calcutta. In Pakistan the absence of manufacturing industry was felt more keenly still; indeed, the new state of Pakistan placed its first capital at Karachi precisely in order to anchor its future as a modern economy and polity to the one urban centre with obvious connections to industry and modernity.

A presumption that the postcolonial state must take a lead in promoting development was not an unreasonable one for most Third World countries at independence. Private capital markets were often thin on the ground, together with private capital itself, and examples of market failure were legion. Postcolonial states were almost bound to commit themselves to a political economy of development which would make a virtue of state-led programmes for industrialization – usually in the form of import-substitution industrialization (ISI) – and which, in some cases, would embrace socialist programmes for the industrial development of a given space-economy. In this section of the chapter I want first to consider what resources were available to some Third World countries to support such a strategy (or strategies). I then consider the problems that are posed to such an ideology of development within a changing international political economy.

Resources for development

For those countries which came to independence in the 1940s, 1950s and 1960s, there were three broad sets of economic resources which could be drawn upon to support a politics of development based around state-centred industrialization.

A first set of resources was intellectual. The depression in the world-economy in the 1930s had led several eminent economists and politicians to call into question a philosophy of economics which sought refuge always in the market mechanism, and which derided state intervention in the economy as *dirigiste*. In 1936 Maynard Keynes published his great work, *The General Theory of Employment, Interest and Money* (Keynes, 1936). Keynes argued that workers in most advanced industrial countries were not minded to accept swingeing cuts in their nominal wages because of certain cultural expectations into which they had been socialized (and which were supported by the trade unions). Wages were thus sticky in a downwards direction and labour markets were not always bound to clear at a price where workers could price themselves into a job. Involuntary unemployment would then sap business confidence as aggregate demand in a given national economy started to fall. The 'animal spirits' of entrepreneurs might then fall into a state of depression, with the result that further investment in the economy would be withheld. The assets of entrepreneurs might instead be held in more liquid forms. This would be a recipe for recession, if not depression. If the situation was then added to by competitive devaluations amongst countries, the recession/depression would quickly be generalized to the world stage.

This is what had happened in the 1930s, at least according to Keynes (for a different view, see Haberler, 1987). To put it right, Keynes argued, and to make sure that the system was not attacked from within by socialist agitations, governments would have to pump-prime (and fine-tune) an economy in disequilibrium. Keynes thus pressed for public actions in support of a private, capitalist economy – an economy which, in his judgement, guaranteed the freedom of the individual (Moggridge, 1992). These public actions might be along the lines of Roosevelt's New Deal, or they might take the form of a more social-democratic commitment to welfarism and a possible welfare state. The degree and nature of state intervention would clearly vary from country to country. Finally, at the international

level, Keynes argued for a *de facto* system of world government which would ensure that mercantilist trading policies could not again break out between nation-states. As Britain's chief economist at the 1944 Bretton Woods conference, Keynes was able to push for at least some elements of his vision to be put into practice. It was at the Bretton Woods conference that provisions were made for the setting up of an International Monetary Fund (to attend to balance of payments problems amongst member countries), and for an International Bank of Reconstruction and Development (later the World Bank) whose purpose was apparent in its name. Keynes' main regret was that his proposal for a new international unit of account – the bancor – was not accepted by the Americans in 1944; they preferred that the US dollar should serve as the main means of international liquidity post-World War II. Keynes was also concerned about the internationalism which the US was embracing at Bretton Woods. Keynes favoured free trade between countries, but he was not in favour of greatly expanded capital flows within an internationalized world-economy; this would takes powers away from elected national governments.

All of this was soon to be of relevance to an ex-colonial Third World. The world-economy into which these countries were born was the Bretton Woods world-economy. At the level of ideology and accepted practice, it was also a Keynesian world-economy. State intervention in an economy was accepted in a way which it is not today. Even the socialist impulses of the Soviet Union and China were not entirely frowned upon. Although these countries were not trusted by the West in political terms – and were assumed to ride roughshod over individual political freedoms – there was not at this time a general presumption that socialist economies were mismanaged or inefficient. In some quarters there was a feeling that the socialist world might outproduce the capitalist world, as Khrushchev had promised (or threatened) in 1960.

In the Third World, this commitment to state intervention took a more specific form. In the socialist Third World it first took the form of Soviet-style Five Year Plans, with great efforts being made to transfer an economic surplus from the countryside to the city for the support of large-scale industrial projects. This project was abandoned in China after the Sino-Soviet rift of the late-1950s, with events (and absences: Soviet capital and know-how) then pushing China to make a virtue of 'balanced development' betwixt town and country (Blecher, 1986). In the non-socialist Third World a commitment to state intervention was most often expressed through a strategy of import-substitution industrialization. This strategy had the support of most Keynesians, and it enjoyed favour also within British Fabian circles, within the UN's Economic Commission on Latin America (fronted by Raul Prebisch) and in a nascent development studies community.

In Latin America the emphasis was placed mainly on consumption-goods based ISI. Foreign capital was often welcomed into the region, and the intention was to generate local industrial supplies of consumption goods by means of private- and public-sector investment and high effective rates of protection. Local industrial products would come to substitute for imported industrial goods and Latin America would begin both to develop and to escape an 'inevitable' decline in its commodity terms of trade.

In India a similar process of ISI was signalled by the Second and Third Five Year Plans (1956–66). The Nehru–Mahalanobis strategy differed from that found in most Latin American countries by virtue of its emphasis upon capital goods substitutions. A broader substitution of consumption goods would be delayed until a local industrial infrastructure had been constructed, and with it a more dynamic domestic market for cheap final goods. India also refused to open its borders to massive inward investment by foreign firms. It preferred to develop its communications and power systems, and heavy engineering industries, by means of joint ventures with partners as diverse as the United Kingdom, the United States, Czechoslovakia and the Soviet Union. The fact that all of these countries were willing to make capital and know-how available to India hints not only at the geopolitical significance of Nehru's India, but also at a wider international consensus on the nature of development and on the conditions necessary to secure its

promotion. Although Milton Friedman was opposing the inefficiencies of aid and nationalized industries in the late 1950s (Friedman, 1958), his voice was very much a voice from the periphery; the consensus on development at this time was firmly within a Keynesian mould. The main disagreements within the development community concerned the nature and the degree of state intervention in a developing economy; its desirability was rarely in question.

A second resource open to most developing countries was provided by the institutions set up at Bretton Woods and added to by the United Nations and the General Agreement on Tariffs and Trade. These institutions and channels are hardly democratic, for voting within them is usually in proportion to the size of a member country's subscription. (The United Nations is an exception to this rule, although the five permanent members of the Security Council have veto powers over most important issues.) Nevertheless, this quasi-internationalization of the political structures of the international economy has afforded some autonomy to some developing countries at some stages of their postcolonial histories (Jackson, 1990). Most obviously, the World Bank has provided official development assistance to most of its member countries in the Third World, and these disbursements of money and goods have been in addition to those transfers made by bilateral donors like the USA, the UK, the USSR and Japan. In the period between 1950 and 1970 an estimated $92 billion was transferred from the First World to the Third World as official development assistance, and these transfers provided funds, or counterpart funds, which supposedly would have contributed to local development initiatives. Aid was intended to plug a gap in the economies of the postcolonial world. If industrialization and development depended mainly upon savings and investment, as the Keynesians suggested, then transfers from abroad could make good any shortfall in a country's domestic revenue base (for example, as a result of crop failures). In this same period, a further $47 billion was transferred from the First World to the Third World as foreign direct investment, the majority of it ending up in Latin America and South East Asia (Wood, 1986). These transfers were themselves testimony to the increased openness of the international economy under the Pax Americana and to the new locational possibilities which faced global corporations. Not dissimilar flows were also under way in the socialist world, with the Soviet Union moving also in the 1960s to buy Cuban sugar at a premium above the world market price.

The institutional structure of the post-war global political economy also afforded the Third World an opportunity to act collectively in defence of certain common interests. In the 1950s, the UN's Economic Commission on Latin America had provided a structuralist critique of the post-war world-economy, and in the process provided a radical Keynesian prospectus both for *dirigiste* development and for changes which would be desirable at the level of the international economy. Prebisch wrote a withering critique of comparative advantage theories, claiming that the doctrine of free-trade condemned the developing countries to provide the dynamic manufacturing centres of the core (the First World) with cheap commodities from the periphery (the Third World). This was the idea of an unequal exchange of embodied labour times; it also led Prebisch to maintain that primary commodity exporters would face a secular decline in their income terms of trade (Prebisch, 1962).

In the 1960s these issues were aired more formally at the General Assembly of the United Nations. As Hoogvelt points out:

The first UN Conference on Trade and Development, held at Geneva in 1964, was a landmark in the history of Third World solidarity precisely because it was here for the first time that Third World countries confronted the rich countries *as a group*, demanding certain economic reforms in the world economic order (i.e. price stabilisation and improvement for primary products, market access for manufactures from developing countries and greater financial flows from the rich to the poor countries). (Hoogvelt, 1982, 75–6; emphasis in the original)

In subsequent years a caucus Group of 77 countries (later 120 countries) pressed these demands both within the UN and with regard to the International Labour Organization, the IMF and the World Bank. Their advocacy was 'rewarded' in 1973–4 when the Organization of Petroleum Exporting Countries (OPEC) was able to raise the price of oil more than three-fold. More generally, the activities of the Group of 77 led, in the 1970s, to Third World demands for a Charter of Economic Rights and Duties of States. This in turn led to proposals in the mid-1970s for a New International Economic Order (NIEO). The NIEO would be secured: (1) if and when actions were taken to ensure a global redistribution of wealth: the Lima Target of 1974 was for the developing countries to increase their share of world industrial production from 7 per cent to 25 per cent by the end of the century; (2) if and when actions were taken to reform the international trading system (by means of compensatory financing programmes, less First World protectionism, buffer stocks, etc); (3) if and when actions were taken for the reform of the international monetary system (by means of one country–one vote in the IMF, an increased allocation of Special Drawing Rights); and (4) if and when actions were taken to increase international flows of ODA and to regulate the transfer of technology between countries. The fate of these demands is something we will return to.

A final set of resources upon which the ex-colonial countries could draw can be dealt with more briefly; they concern 'local' conditions and the question of a 'national spirit' for development. Crudely put, the fact of independence created a window of opportunity in Third World countries whereby a degree of legitimacy was conferred on all postcolonial states as they mapped out a prospectus for development, and as they sought to define anew how best to make use of their physical and mental resources. India provides a good example of this. With the deaths of Gandhi and Sardar Patel in the period from 1947 to 1950, Jawaharlal Nehru was left in *de facto* charge of the Congress Party and, thereby, India itself. In so far as Nehru was able to personify the developmental ambitions of a unitary India, he was in turn empowered to push for this development by means of a technocratic state apparatus and a Planning Commission which would consist of the country's top economists and politicians. Decisions would then be taken by the Planning Commission in the long-run interest of India as a whole. Industrialization would be paid for by local taxation and ODA, and it would be directed to those parts of the country where it was most needed and/or where raw materials were available. The revenues generated by ISI would in turn allow the state to support its social projects. These projects would be focused on health care and education, but would also include projects for the 'uplifting' of India's Scheduled Castes and Scheduled Tribes. Growth would provide for development in a unitary and unified political economy.

Obstacles to development

The failure of India to meet all of its Plan targets is something we will take up later on. In the rest of this section I want to consider how the economic resources for development just identified could also become economic obstacles to development. There are three main sets of obstacles to describe.

Consider first the limitations of ISI as a strategy for development. These are not only economic, although these limitations are now well known. Despite not inconsiderable per capita rates of growth in countries as diverse as Brazil, India and Nigeria in the 1950s and 1960s, ISI can easily become sclerotic when protectionism leads to the over-valuation of the domestic currency, when domestic corporations are bailed out of the bankruptcies which otherwise would beset them, and when economic decisions are subjected to a time-consuming, and sometimes corrupt, bureaucracy (Little, 1982). The burdens of ISI can also be political. For ISI to work it is incumbent upon the state to administer a programme of economic sequencing which directs scarce resources to different sectors and regions of the economy on a priority basis. Inevitably, this can generate a measure of political resistance from the less advantaged sectors and regions *before* the assumed benefits of ISI begin to make themselves felt.

In the 'four little tigers' of East Asia (South Korea, Hong Kong, Taiwan & Singapore) there were movements in the 1960s to wean local economies away from the ISI model. ISI had provided a basis for the formation of local (infant) industries, and now these industries were expected to compete in global markets on the basis of their relatively low labour costs. ISI

made way for Export-Oriented Industrialization (EOI). In India and most other developing countries, this transition was less assured and was delayed until the mid-1980s. In political terms, the Indian government, from the mid-1960s, did recognize that certain costs were associated with ISI – notably, that it did little to generate employment and incomes for the majority of India's working population, *and* that an over-valued rupee penalized not only consumers in general, but the rural sector in particular. Given the power of the rural vote in India, and of the richer farmers in particular, this was bound to be a concern for the Congress government. In 1967, Charan Singh, the first major farmer's leader in post-independence India, won a famous by-election in the country's Hindi heartland, largely on the platform of standing for the true rural India (Bharat) against the colonial urban-industrial India which had been favoured by Congress. In the 1980s this same politics was to resurface in the mass farmers' movements which shook the north and the west of the country. 'Bharat versus India' became the rallying cry of Sharad Joshi's Shetkari Sanghatana, and his campaign for higher farm output prices, and lower farm input prices, was taken up by thousands of peasants who marched on New Delhi under his leadership and that of Mahendra Singh Tikait. India also ran into balance-of-payments problems in the 1960s, caused in part by its reliance on industrial imports in the short term to make possible industrial substitutions later on. Like most postcolonial countries, India faced a recurrent crisis with regard to its holdings of foreign currency reserves. These pressures combined to erode the political legitimacy which had accrued to Nehru and the Congress in the 1950s and early 1960s, as did India's disastrous war with China in 1962. The Brazilian government of Goulart suffered a not dissimilar loss of public support in the 1960s.

The practical difficulties of ISI (which should not disguise its several achievements: Chakravarty, 1987), were added to in the 1970s and 1980s by a reappraisal of the political and economy theory upon which ISI was based. The 1970s proved to be a pivotal decade in the history of the post-war world-economy, and in the history of development economics as a discipline (Toye,

185

1987). In 1971 President Nixon broke the link between the price of gold and the price of the US dollar which had been one cornerstone of the Bretton Woods system; in 1973 a system of fixed exchange rates (and of currency management) gave way to a system of floating exchange rates. The 1970s were also the years of the OPEC oil-price rises, of the privatization of international liquidity creation and transfer (through the Euromarkets), and of a worldwide rise in inflation. The old Philips Curve (suggesting that a country's level of inflation varied inversely over time with unemployment) broke down, and with it went the belief that countries could trade off unemployment against inflation. In short, the time was ripe for a re-evaluation of Keynesianism, of social democracy, and of development studies itself.

The demise of Keynesianism did not herald the 'end of history', as some might suppose, but the 1970s and 1980s did see some significant changes in the international economic climate in which Third World countries were compelled to do business. At the level of economic theory, this was evident in the rise of market economics in various guises (including monetarism and supply-side economics). It was also apparent in Lal's critique of the 'poverty of development economics' (Lal, 1983) and in the charge made by him and others that 'what the world's poor countries need most is less government' (*Economist*, 1989, 56). Markets could work in the Third World just as they could and did work in the First World. The fact of market failures could not alone justify a *dirigiste* approach to development. Government failures were at least as common as market failures, and probably were more damaging to a country's economic prospects (Beenstock, 1984).

At the level of economic practice, most developing countries were forced to adjust to a new economic climate based around a relative decline of ODA tranfers and a massive expansion of private bank loans to some middle-income countries. When these loans fell due in the early 1980s, it was amidst a steep recession induced by high interest rate policies in the United States and some other OECD economies. The recession in the North was also associated with a decline in world commodity markets and prices, which in

turn made the repayment of debts less assured. The 'debt crisis' then broke in August 1982 when Mexico defaulted on the repayments due on some of its publicly guaranteed debt; several other Latin American countries (including Brazil and Venezuela) followed Mexico into default later in 1982 or in 1983. This in turn brought many Third World countries into a new relationship with the World Bank, and with the IMF in particular (Corbridge, 1992). The IMF became notorious in Africa in the 1980s for 'Imposing Misery and Famine'. In return for arranging a local rollover of debts outstanding, the IMF imposed on its indebted member countries a series of structural adjustment programmes which entailed cuts in public expenditure and currency devaluations (to make a given economy more outward-looking). In short, the IMF now policed the world-economy – and the indebted world-economy most especially – in line with a new economic orthodoxy which favoured markets over states, and which warned that: 'There is no viable alternative to adjustment' (World Bank, 1987, 35). In not a few cases, the political latitude which was once granted to a developing country now disappeared, or was effectively closed down. Precisely at a time when dependency theory was in crisis in the academy (Booth, 1985), *de facto* dependency was being reinvented or reinforced in countries as diverse as Brazil, Nigeria and the Philippines. Political economy seemed to be losing out to the 'apolitical' claims and verities of the market.

Finally, there is the matter of the Third World itself and of Third Worldism. In the 1950s and 1960s the Third World was a recognizable entity in economic terms (which is not to say that the ex-colonial countries did not vary enormously in their economic programmes and capacities). In the 1970s this seeming unity seemed to come undone. Commentators now remarked upon the end of the Third World (Harris, 1986), and Mrs Thatcher chose to decry the demands of the South (as expressed in the first Brandt Report: Brandt, 1980) in just these terms. There was no South to voice a single set of demands; rather, there were well-managed, outward-looking (ex)-developing countries like South Korea and Taiwan, and there were poorly-managed, inefficient, inward-looking countries like India and

Nigeria. These countries could no longer sustain a coherent Third Worldist politics, and, in any case, demands for a New International Economic Order were easily discounted in the harsher economic climate of the late 1970s and 1980s.

The developed world, for its part, was less and less minded to accept its responsibilities for the plight of its once imperial world. The new world order of the 1980s was a world of individual countries, battling it out in global marketplaces which supposedly were less and less regulated by state authorities. The North acted with a surprising unity of purpose in both failing to dismantle its own structures of protectionism, and in resisting those demands which were still raised in favour of taking 'actions for development'. Some parts of the Third World no longer mattered very much to the real world-economy of which they were supposedly a part. The combined gross domestic products (GDP) of forty-one Low Income Economies in 1989 was less than 5 per cent of total world GDP; in 1965 it was closer to 10 per cent (World Bank, 1991).

Geopolitics

A second area where one can find both resources for, and obstacles to, purposeful Third World politics, concerns the field of international political relations, or geopolitics. It should be self-evident that a postcolonial country will be more able to effect policies in support of development and democracy to the extent that these policies are (a) supported from abroad, and/or (b) not undermined by hostile actions on the part of foreign governments and institutions. The likelihood of friendly/supportive/hostile inter-state relations will in turn depend on the capacity of Third World countries to prosecute a Third Worldist political agenda in the institutional arena. We can consider these points in turn, and with particular reference to US relations with India, South Korea and Nicaragua.

Superpower relations with the Third World

India provides an example of a country which was able to pursue a policy of ISI in the 1950s and early 1960s in part because of certain guaran-

tees which were offered to it by the USA, the USSR and other interested countries. This much has been explained already with regard to the provision of knowhow and capital to India in the form of ODA during this period. It was also apparent with respect to food policy and the countryside. Briefly stated, India was able to pursue its Second and Third Five Year Plans – its prospectus for 'heavy industrialization' – in part because the Indian government was able to treat the agricultural sector as a bargain basement (Ahluwalia, 1985). In the 1950s agricultural production was expected to increase in India on the basis of institutional reforms – the abolition of the zamindars, land ceilings legislation, and the Community Development Programme – rather than on the basis of massive public-sector investment in irrigation, fertilizers and pumpsets. Land extensification would also play its part. A common assumption at this time was that there was significant disguised unemployment in the countryside, and that 'feudal' styles of landlordism were acting as a 'depressor' upon India's agricultural performance (Thorner and Thorner, 1962). The main duty of the state was to relieve this deadweight by means of state-wide legislation to transfer land to those who tilled it. (This legislation was rarely effective in practice: Herring, 1983.) Labour could then be transferred to India's urban-industrial regions without a significant loss of food production in the countryside.

In the 1960s this diagnosis of India's food production needs and capacities was sharply challenged by India's richer farmers. It was also called into question in 1965–7 when famine returned to an India palpably not blessed by an efficient food production sector. By 1966 the Green Revolution was already being spoken of as India's likely saviour. In the 1950s, however, the poor performance of India's foodgrain sector was disguised to the extent that India's plans for ISI were underwritten by massive transfers of food aid from the USA under Public Law 480. These shipments of food aid, which served US geopolitical interests in South Asia at the time, furnished Nehru's governments with a degree of independence from India's richer farmers that later governments would not enjoy. It was this conjuncture which allowed Nehru to commit

India to an urban-industrial future, and which allowed the Planning Commission to seek to reform India's countryside by means of land reform and community development. By contrast, when the famine years returned to India in the mid-1960s, the USA proved itself unwilling to continue to supply food aid. More precisely, the USA now took advantage of India's precarious position (soon after the death of Nehru in 1964), to place India on a 'short-tether' with regard to food aid. The government of India would now be granted food aid from the USA on a month-by-month basis, and only in return for India agreeing to certain political and economic preconditions: for example, that India would not offer any support to North Vietnam; that India would devalue the rupee; that India would relax some of its prohibitions on foreign direct investment; and that India would allow foreign companies setting up in the Indian fertilizer industry to set their own prices, and run their own distribution systems, for seven years from March 1967 (Frankel, 1978).

The example of India illustrates very well the importance of the geopolitical dimension to our understanding of politics in postcolonial countries (which is not to say that most Indians were reluctant to embrace the Green Revolution). Most developing countries are in a dependent relationship with one or more superpowers, but their capacity to exploit such a relationship, or to be damaged by it, depends on various conjunctural factors and agendas which are rarely under their control. India has learned this lesson very well. In the 1970s and 1980s India rebuilt its food production capabilities to such an extent that it is now a net exporter of foodgrains (World Bank, 1989). Some of this rebuilding was paid for by the state (irrigation works, fertilizer subsidies) and involved a transfer of state incomes away from the urban-industrial project which had hitherto dominated in India. In this manner, India continued to define its future development not just in terms of economic growth, but with reference also to the dependence of its economy and polity upon the economies and polities of outsiders. It is only since the mid-1980s that India has been 'required' to embrace a more open strategy of development as the country began to face a fiscal crisis brought on by the budget deficits of the central government.

The case of South Korea illustrates a more contented (or at least more continuous) set of relationships between a developing postcolonial country and a global superpower. In large part, this has to do with the country's history immediately after the surrender of Japan – then the occupying power in Korea – in 1945. At this point: 'The southern two-thirds of the Korean peninsula was occupied by the US to receive that surrender and resist the Soviet Union' (Hawthorn, 1991, 36). The Americans in turn moved to establish the First Republic of Korea in Seoul in 1948, both to defeat a local movement for a Korean People's Republic, and to counter Soviet interests in the region. This First Republic lasted until 1960 when most of its corrupt and authoritarian leaders were ousted by popular demonstrations in Seoul. A prospective (and democratic) Second Republic was then sidelined by an Army coup, and from 1961 onwards 'one single (if intensely factionalised) party was in control' (*ibid.*, 36–7). Led by Syngman Rhee as a one-party–one-person state (more or less), South Korea continued to enjoy the support of the US military machine, while turning also to the Americans and the Japanese for support for the country's programme of ISI-EOI. This programme was in turn made easier to effect (and more successful in due course) by the absence in South Korea of large landowners opposed to the efficent production of local food supplies. The Americans had insisted on a radical land-to-the-tiller land reform in South Korea immediately at the end of World War II. They thus helped to secure a conservative, but efficient, countryside in South Korea, in which most farmers owned between one and three hectares of land which they and their families worked themselves. When the Green Revolution came to South Korea in the 1960s it was taken up by almost all of South Korea's farmers, and it proved to be less socially polarizing than elsewhere in south and east Asia. Finally, South Korea has enjoyed substantial transfers of money, capital and arms to its government from US aid programmes, US military spending, and the World Bank. It is also a favoured location for Japanese capital. A considerable number of the preconditions for take-off which Rostow identified in 1960 were present in South Korea by 1970, in no small part because of its privileged relationship with the United States (Amsden, 1989).

In Nicaragua, as in Chile under Allende (1970–73), the story was different again. In 1979, the longstanding dictatorship of the Somoza family was overthrown by the Sandinistas. The Sandinistas immediately began a process of land reform (which did not amount to a programme of collectivization), and the new government voted funds for improvements in the standards of literacy and health care of the majority of its citizens. These improvements were duly secured. Almost as quickly, the US government, especially under President Reagan from 1981, began to support a programme of destabilization in Nicaragua which focused upon US funding for the Contras. (At this time, also, the USSR was an occupying power in Afghanistan.) The USA argued that Nicaragua was a totalitarian state and thus quite unlike the authoritarian states to be found elsewhere in Latin America. (The distinction was probably first made by Jean Kirkpatrick, the US ambassador at the UN in the early to mid-1980s.) The 'communist' state of Nicaragua was a threat to democracy in the entire region of the Americas and as such it had to be confronted. The USA later chose to ignore a ruling of the International Court of Justice in The Hague which declared that US aid to the Contras was in violation of international law. Indeed, the US House of Representatives 'went ahead and approved President Reagan's request for $100 million-worth of new aid for the counter-revolution. In what looked like an act of retaliation, President Daniel Ortega of Nicaragua . . . announced the closure of the opposition newspaper, *La Prensa*, and the expulsion of two turbulent priests, Bishop Vega and Monsignor Bismark Carballo. The storm was brewing' (Rushdie, 1987, 13).

The storm came to a head with the decision of the US to mine the ports of Nicaragua and to subject the country to an economic blockade. The denouement, fortunately, was less violent. In free elections held in 'communist' Nicaragua in 1990, the Sandinistas were voted out of office and they promptly ceded power to Violeta Chamorro's UNO alliance. Whether the population was voting against Ortega, and/or whether it was voting against a civil war and associated depradations caused largely by the USA is a moot point. What is undeniable is that Nicaragua's room for manoeuvre – its potential for supporting both democratization and development – was narrowed from the outset by the aggressive actions of the USA, and by the willingness (and ability: cf. Iraq in 1990–91) of the USA to ignore the efforts of supranational organizations to make it abide by certain international conventions.

The use of Third World countries as sites for proxy wars between the superpowers is taken up elsewhere in this book, and it is by no means confined to Nicaragua and Chile. The Soviet Union contributed massively to the escalation of violence in the Horn of Africa in the 1980s (Clapham, 1990), just as the United States was active in Grenada, and just as China continued to impose its rule on Tibet. Three decades prior to this, the British and the French were still active in the affairs of colonies past and present, as their invasion of Egypt demonstrated at the time of the Suez crisis (1956). On that occasion, however, European attempts to revive old glories were quashed firmly by the West's new superpower, the United States.

Third World as 'Third Estate'

Mention of Nasser's Egypt also reminds us that the Third World has not been slow to defend a collective political and geopolitical interest. The concept of the Third World was first used in this context, and not as a euphemism for backward or developing economies. Alfred Sauvy argued that the Third World offered a third way between capitalism and communism which was akin to the 'way' or role played by the Third Estate in the French Revolution of 1789. The Third World would offer a new vision of the future; a new enlightenment which would be fleshed out politically in and through a politics of non-alignment (Pletsch, 1981). It was this definition of the Third World which was picked up at the 1955 Bandung Conference of African and Asian countries. As Hoogvelt points out, it was at this conference that 'the recently decolonised nations of the world met for the first time *as a group*, and recognized their communality of interests as both ex-colonial and as underdeveloped economies' (Hoogvelt, 1982, 74; emphasis in the original). The principal concern and significance of the Bandung Conference was political. Its energies were devoted to the political issues of anti-colonialism, the setting-up of a nonaligned movement (led by Nehru, Nasser, and Tito of Yugoslavia), the solidarity of the Third World, and

on getting a political say for the nonaligned countries in international fora. This preoccupation with political issues was also a distinctive characteristic of the nonaligned conferences in Belgrade in 1961, Cairo in 1964 and Lusaka in 1970. Latin American countries did not partake in the non-aligned movement until 1973, when Peru and Argentina became members. (Ibid., 74–5)

Whether the nonaligned movement could ever hope to achieve tangible results in the field of international relations is a question for debate. Certainly in the period since the 1960s the record has not been a happy one, with the two main global powers doing their best to detach 'friendly states' from the nonaligned movement. As argued above, Third World solidarity has also proved difficult to sustain, not least because the regional ambitions of some Third World countries have been encouraged by arms flows from the superpowers and other arms-exporting countries. Geopolitics are not only about proxy wars in the Third World; they are also about the regional conflicts which have seen India pitted against Pakistan and Sri Lanka, and which have found repeated expression in South East Asia and the Middle East (Chomsky, 1983; Halliday, 1990; Rowley, 1989).

All this accepted, the promise of Third Worldism, or non-alignment, should not be discounted lightly. As a resource for development, and for independence itself (if not always democratization), these ideologies worked to the advantage of some regional powers in the 1950s and 1960s, as the examples of India and Egypt make plain. Countries such as these could seek to play off the dominant superpowers in such a way that economic and military resources could be accessed from more than one source and in such a way that counterpart funds could in theory be found for development. In the geopolitical world

order of the 1990s even this option would seem to be denied the Third World.

Political institutions, state-building and the new states

Thus far we have focused upon those resources at a global scale which can be used both to foster and to hold back initiatives for development in postcolonial countries. A third set of resources has more to do with political institutions within a postcolonial country – resources which encourage and/or discourage the formation of democratic political parties and a developmental state. As Geoff Hawthorn points out:

Those aspiring to rule these countries have had first and fast and in unpropitious conditions to establish a claim to political authority. (This has not everywhere been necessary. It has been possible in one or two places for such people to deploy pure power and use it simply to cow their subjects and rob them. But this disposition, in the Duvaliers' Haiti, for instance, or in Ngeuma's Equatorial Guinea, has been the exception rather than the rule.) To make a claim to authority, the putative leaders have had to claim an identity for the political space, the territory, over which they hoped to exercise it; to claim an identity for the subjects in that space; and to suggest a plausible connection between those subjects and themselves. The first two claims have had to be particular, to a particular set of people in a particular place; the third has had to be more general, more 'universal', a claim to acceptable rules of rule. (Hawthorn, 1991, 25)

Hawthorn contends that: 'The only conceptual and practical possibility for making all three [claims] is now the nation' (*ibid.*, 25–6), and herein lies a problem. The fact is that most Third World countries came to independence as incomplete 'nations', if they were nations at all. Many Third World countries were also bereft of those institutions and technologies which have connected the state and civil society in the West: institutions which would include an elected and representative assembly, an independent judiciary and an autonomous bureaucracy or civil service. This was particularly evident in Pakistan after the trauma of its partition – or parturition – from India in August 1947. India was left with more than 90 per cent of the members of the old Indian Civil Service, whereas Pakistan was bequeathed a void, more or less, into which could easily step the substitute authority of the Army. Many African countries faced a similar 'loss' at independence – a similar 'history' of political and administrative absenteeism imposed on them by the presence of an occupying colonial power.

It would be unwise at this stage to suggest that there are but few resources for state-building in the Third World, even though it is necessary to signpost these 'local difficulties' before returning to them later on. Hawthorn's focus upon the politics of nationalism and national identity suggests one such resource, to which we might add the threat of external force, and the roles played by political leadership, skill and charisma.

Constructing a developmental state

The relationship between nationalism, charisma and development was vividly illustrated in India during the years when Pandit Nehru was Prime Minister (1947–64). India's main commitment was then (as now) to a politics of development, and to make this possible a strong central state was considered desirable. The central state would be strong, first, to the extent that it supported an executive agency (in this case the Planning Commission) which would be relatively independent of India's domestic proprietary elites (the rich farmers and big business); secondly, to the extent that Nehru's political party, the Congress Party, was not effectively challenged in the national and State elections; and thirdly, to the extent that Nehru could present himself as the embodiment of India's new capabilities and aspirations. Evidence that such a strong state was in place would come in turn from the fact of development (tautologically), and with reference to the capacity of the government of India to raise resources sufficient to finance its planned programmes of ISI. These resources would come from taxation, from ODA and from judicious government borrowing. A state which is unable to raise these resources, or which failed to allocate them efficiently, is a state lacking in infrastructural capacities and/or the will to development.

There is general agreement that India came close to meeting this model of a developmental state in the 1950s and early 1960s. The country inherited a system of government built around a capable bureaucracy into whose lower ranks many Indians had been inducted since World War I. India was also able to transfer resources from the countryside to the city for reasons already explained. Above and beyond this, Nehru and the Congress Party were able to define a sense of 'Indian-ness' which was inclusive rather than exclusive, which was secular in tone and intent, and which was connected to a vision of future greatness. More precisely, Nehru was able to present himself, almost subliminally, as the Father of the Nation. Nehru had been a leader of the anti-colonialism movement in India and he was untainted by hints of collaborationism or personal greed. He had also spent many years in the gaols of Empire. The fact that Mahatma Gandhi was assassinated in January 1948 only added to Nehru's standing in independent India. Nehru's position as the boss of the Congress Party was further strengthened by the death in 1950 of Deputy Prime Minister Patel, his only other rival for that position. A similar capacity for independent actions would not be enjoyed by Nehru's daughter and eventual successor (after Shastri), Indira Gandhi.

Beyond the simply personal, Nehru was able to part-invent a new nation of India which would be defined both as a developing postcolonial country and as a prominent member of the non-aligned Third World. The new myths by which India would live (myths in the sense of Barthes: Barthes, 1972) were the myths of socialism, secularism and federalism.

India's commitment to 'socialism' (or the mixed economy) was embodied in the Five Year Plans which began in 1951, and in the country's commitment to an ostensibly Soviet-style process of rapid industrialization. It was also apparent in Nehru's commitment to make sure that the fruits of India's growth would be shared by the poorest communities, the Scheduled Castes and Tribes, by means of programmes of positive discrimination. These programmes ran counter to the intention of the Indian state, through its Constitution of 1950, not to promote a politics of discrimination; they were justified, nonetheless, and promulgated, as Directive Principles of State Policy. The state could discriminate positively in the short term to prevent a private and more negative discrimination in later years (Galanter, 1984).

India's commitment to secularism was expressed most visibly in Nehru's unwillingness to allow new states to be formed in India on religious grounds. When the map of the country was redrawn in 1956, it was language which provided a key to where the new boundaries between new states would be drawn. Even when Punjab was created as a full state in 1966 (after Nehru's death), it was in theory as a state for Punjabi speakers, and not as a Sikh homeland. In this manner, Nehru and the Congress Party moved India away from the religious politics of divide and rule which the British had promoted, and which found its inevitable and tragic culmination in the communal massacres which scarred the partition of British India. The new voice of India was the soothing voice of reason itself. In the new and supposedly secular Republic of India, planning would become the godhead of a society geared to development and to a fabled 'unity in diversity'.

Finally, there was Nehru's commitment to a conception of India as a federal democracy. This vision was expressed in the Constitution of India, which married elements of the Westminster model of Parliamentary democracy with aspects of Washington-style federalism. Although federalism has never been as secure in India as it is in the United States – not least because the President of India, acting on orders from the Prime Minister, can with 'due cause' suspend state governments – the federal ideal in India was practised with some vigour in the 1950s and 1960s. Nehru tried hard not to involve himself in the affairs of state governments – or, rather, he tried hard not to be seen to be meddling in these affairs (Brass, 1990). Nehru preferred that Congress governments in the states should deal with local problems as they arose and without reference to New Delhi. Significantly, Nehru was less willing to maintain his distance with regard to the Communist Party of India (CPI) governments in Kerala.

India, then, under Nehru, provides one indication of what might be achieved in terms of nation-building after independence. It also offers a gloss upon the construction of a developmental state (albeit given several preconditions, most of which we have described). Still other postcolonial countries were able to bind together the three claims identified by Hawthorn in different, but no less successful, ways. In postrevolutionary China the binding process was focused upon Chairman Mao and made reference to a new and transformational model of men and women which emphasized the responsibility of an individual to a wider collective unit or community. The developmental state in China was also strongly girded by the Red Army, by a centralized system of extended state education, and by a system of urban–rural relations which allowed a massive transfer of resources from the latter to the former (until the late 1970s). In China, the tension between 'development' and the rights and liberties of the individual was more obvious than in India (in formal terms, if not always substantively: freedom begins with breakfast, some would say). In Cuba, too, the possibility of a development state was linked directly to the cult and persona of Castro, just as South Korean development was linked to the strongman Syngman Rhee, and just as Tanzania built its accounts of itself, its citizens and its future around Nyerere and all that was embodied in the 1967 Arushra Declaration.

Not constructing a developmental state

These early successes should caution us against too pessimistic a view of the political process in Third World countries. Strong states have been built in the postcolonial world and they can still be built there. The problem is that many more weak states are apparent in the developing world, together with too many strong states which are strong by virtue of their being opposed to a sympathetic relationship with civil society and representative political institutions. Why should this be?

Part of the answer to this question lies with the initial difficulties faced by many postcolonial countries. It bears repeating that the colonial legacy was rarely a happy one. Most African countries achieved independence at a time when civil society in the western sense was more or less absent from their territories. Whereas the common presumption of modernization theory was that 'a modern politics, a proper politics for moderns, follows from and cannot precede an already functioning civil society' (Hawthorn, 1991, 27), this luxury (this precondition) has not often been granted to the so-called latecomer societies. The state has then had to step into the void of civil society, with the intention of fostering development, and with at least half an eye to inventing the institutions of modern politics themselves. The problem is that such a state, however weak as a developmental state, can then become too strong – too overdeveloped – almost by definition and from the outset. The state can easily subsume civil society, with the result that the resources of the state become a major area of day-to-day politicking. What then tends to follow is a politics not so much of representation, as of colonization. States become the goal of politics for private interests, rather than serving as institutions which mediate between competing interest groups and different definitions of the common-wealth.

Some African countries would seem to offer an illustration of this thesis. Consider the case of Zaïre, or what was once the Belgian Congo. From 1908 this territory saw an intensive exploitation of its mineral deposits in the south by the Belgian Société Générale, while peasants elsewhere in the country were forced 'to produce food on pain of punishment' (Hawthorn, 1991, 37). Notwithstanding the attentions of some in the Catholic Church, the Belgian Congo came to independence in 1960 with not one trained native doctor, engineer or lawyer.

The new Congo was precipitated into independence with virtually no resources for rule, and with poorly articulated programmes of what to do with such rule . . . At the end of a prolonged civil war, in part a war simply between different contenders for power, in part a war about control of the abundant deposits of copper and other minerals in the south (and a war which was exacerbated by international involvements), Mobutu, who controlled the Army, took power and has retained it ever since. (Hawthorn, 1991, 37–8)

Despite support from the western powers and the Bretton Woods institutions since the mid-1960s, Mobutu's grip on Zaïre has never been quite secure. More to the point, the country under Mobutu has suffered what can only be described as a continuing underdevelopment – an underdevelopment which continued long after the collapse of international copper prices in the mid-1970s. Meanwhile, political institutions have withered away, or have been reinvented with brute force. In the north of the country Mobutu's rule is patchy, and in the south his control over the state (and the copper reserves) is maintained largely by a ruthless oppression of his political opponents, by his capacity to finance a palace guard, and by his ability to maintain continued support from certain foreign (western) backers. For the majority of Zaïre's population life is bleak indeed. Having contracted huge debts in the 1970s, Zaïre in the 1980s was required to tighten its belt still further (but not the belt of Mobutu) as IMF structural adjustment programmes were put in place to make the economy more 'efficient' (Korner, 1991). Politics in the western sense of that word was distinguished mainly by its absence. Large numbers of Zaïreans preferred to minimize their contacts with the state (which provided little in the way of health care or education), accepting that the largesse of the state was monopolized by Mobutu and his closest supporters and could not safely be challenged by ordinary people. The situation is only likely to change if Mobutu loses the support of the Army. Even then, Zaïre would be condemned to start from scratch in terms of constructing a developmental state and a functioning civil society.

Neo-patrimonialism

It might be argued that Zaïre is not broadly typical of African politics, let alone of the politics of the Third World. In one sense this is true, but in other respects Zaïre does point to a more general malaise in the postcolonial world. There is evidence from many countries in Africa (and not a few elsewhere) that the postcolonial state is regarded by its rulers as a private fiefdom or resource, rather than as a resource by which might be effected the common development of a country or a nation (where these two units overlap). Clapham takes up this point when he refers to the prevalence in developing countries of a politics of neo-patrimonialism.

This term is adapted from the work of Max Weber and it bears a little discussion. According to Weber, the political principles of a modern society are founded on a rational-legal conception of authority. This is an ideal-type, of course, and as such it is not realized in practice, but it does suggest that politics in the West are guided by the idea that individuals should only exercise power over their fellow-citizens in accordance with certain stable rules of law, and as mandated by the public office which they hold at the behest of an elected and representative government. In other words, the modern system of politics presupposes a strict division between the public and private lives of an individual. 'In office, the official acts simply as an official, exercising the powers which his offices gives him and accepting the restraint which it likewise places on him, while treating other individuals impersonally according to the criteria which the office lays down, whether they be his superiors, his subordinates, or the "public" with which he deals' (Clapham, 1985, 44).

In the postcolonial world this ideal-type of rational-legal authority is approached with still less regularity than it is in the West. Indeed, in a majority of the countries of Latin America and Africa (excepting the Caribbean), the constitution of a given country has been changed on at least one occasion in the past twenty-five years. The politics which then emerges is tied either to Weber's second ideal-type, charisma, or to his third ideal-type, patrimonialism (or neopatrimonialism). Charismatic rule has played some part in Third World politics as, for example, with the cults of Mao and Haile Selassie in China and Ethiopia respectively in the 1960 and 1970s. Far more common, however, is what Clapham calls neo-patrimonialism. 'The distinctive feature of patrimonialism,' he suggests, 'are that, in contrast to rational-legal relationships, authority is ascribed to a person rather than an office-holder, while in contrast to charisma, that person is firmly anchored in a social and political order' (Clapham, 1985, 47).

Patrimonial states – or neo-patrimonial states, where the rhetoric of government is anchored in the discourse of rational-legal authority – need not be unstable. Patrimonialism implies a concept of authority which is akin to that of a father over his children. Implicit in this relationship is a degree of loyalty, and also the notion that the

father-figure bears some responsibility to his supporters, or clients. Very often, allegiance will be given to the patron by virtue of kin relationships, lineage or ethnicity, with these values in turn being strengthened, and identified in part with those of the nation. More prosaically, a clientalist system of politics can secure for the ruling group of a country a political hierarchy based on local vote banks, and the capacity and willingness of clients to turn out this vote. V. S. Naipaul describes the process very well in his book, *The Suffrage of Elvira* (Naipaul, 1969). A measure of democracy is by no means excluded by neo-patrimonialist systems, any more than it was excluded in the patrimonial world of 'boss politics' which shaped some US cities in the early part of this century.

Neo-patrimonialism and patron–client relationships are not always a sound basis for statebuilding, however. We have noted already that such a system can be associated with a capricious use of the law and with a willingness to adopt new constitutions as new leaders come to assume new offices. Neo-patrimonialism has also been associated with an abiding corruption in some countries (in the sense of using public powers to achieve private goals), although some would doubt whether corruption is a meaningful term in a political system which eschews a strict public/private divide (Ward, 1989). More worrying, perhaps, is the tendency for local class and ethnic divisions to be reinforced in those postcolonial states which from the beginning adopted neo-patrimonialist systems of rule. Already privileged groups can monopolize the resources of the state, milk it for private gain, and thereby provoke unease in the rest of the country. The prospect then is one of possible military intervention in the politics of a postcolonial country. The army has long played a dominant role in the politics of many Third World countries, but its ability to enter the political arena legitimately is strengthened when and where neo-patrimonial systems of civilian government begin to break down and founder in a politics of violence and dislocation (Finer, 1962). Mobutu's regime simply brings the worst aspects of neopatrimonialism together in one place: endemic corruption and continuing personal rule define what some would call a kleptocracy. In such a country, the prospects for purposeful national politics remain a long way off.

It is worth noting, finally, that neo-patrimonial systems of rule are not only born at independence; even political systems founded on rational-legal authority can tend towards neo-patrimonialism when the local and international resources for state-building begin to change in shape and form. India provides a cautionary example in this regard. Nehru had been able to rule India according to what Rudolph and Rudolph call the tenets of 'command politics' (Rudolph and Rudolph, 1987). Nehru's commitment to rational-legal authority was underpinned by a favourable international economic and geo-political conjuncture, by the absence of an effective political opposition in India, and by his own personal prestige or charisma.

This set of circumstances was not bequeathed to his daughter, Indira Gandhi. During her years as Prime Minister (1966–77, 1980–84), the Indian model of ISI fell into disrepair as the country's dominant proprietary classes (through the Congress Party) refused to surrender to the central state the resources necessary for continued capital investment in India's development programme. The Indian economy also became encumbered by the bureaucracy which had grown up alongside the planning and licensing systems. The result was that the Indian economy no longer offered a 'goal' or a comfort to the majority of India's population in the 1970s. The political party in charge of the country, the Congress Party, was then condemned to join in the more decadent scramble for votes which characterized the post-Nehru years of 'demand politics'. In effect the state, or the Congress Party, was forced to borrow or print money in order to finance those social programmes which helped to keep its supporters happy (for example, increased student loans, fertilizer subsidies, aid for India's Scheduled Castes and Tribes). This in turn led to a fiscal crisis of the Indian state in the 1970s, the burdens of which New Delhi attempted to transfer to the states (and away from the centre). This did not play well in the states, as might be expected, and in the late 1970s there was a resurgence of regionalism in India which attached itself to ethnic, linguistic and religious affiliations (Kohli, 1990). Punjab

was the most visible sign of this centrifugalism, but the regional challenge was also evident in Assam, in Tamil Nadu and in Kashmir (Akbar, 1985). For her part, Mrs Gandhi tried to maintain the unity of India in two ways: by presenting herself as Mother India ('India is Indira, Indira is India' was one example of this narcissism), and by destabilizing state governments to which she and her party were antipathetic.

By the mid-1980s, the political system in India had moved a long way from the rational-legal model of authority mapped out by Nehru. In a multi-party democracy, parties were now mobilizing religious and ethnic mythologies (and refusing the myth of secularism), while party politicians began to change parties with a regularity which undermined their claim to be pursuing a meaningful ideological politics. The fact that one in four of India's MPs were changing parties in any given year in the 1980s led some commentators to remark on how 'banal' Indian politics had become under the Gandhi dynasty. More generally, it indicates how a change in economic and political circumstances can radically alter the resources which are available for state-building in a well-founded postcolonial polity.

State-building, civil society and oppositional movements

The seeming failure of state–society relations in many developing countries is a legitimate cause of concern for activists and scholars in both the western and the non-western worlds. There are good reasons for believing that economic growth and 'development' are desired by most of the populations of the ex-colonial world, and that development can be aided by purposeful state actions, and damaged by states which seek to pursue their own agendas with little regard for the long-term welfare of their citizens. When states support politics which deny human rights to a significant proportion of their populations (as in Myanmar, China, South Africa and El Salvador today), they deserve to be condemned. Equally, when states pursue policies which reproduce the power and wealth of a minority of a country's population (as in Brazil, and as in

India increasingly), it is right that critical attention should be drawn to the possible failure of institutional politics in such countries. (It is important to talk of possible institutional failures; it is also worth noting that the Indian economy grew rapidly in the second half of the 1980s notwithstanding the often precarious state of the Indian polity.) One might also raise questions about those states and countries which make use of a particular account of Islam and Islamic politics to deny the basic needs and rights of some of their female citizens. In all these cases, the comparative absence of rational-legal systems of rule in many postcolonial countries *is* a cause for concern; it is not simply a case of judging Europe's Others (its ex-colonies) against a grid of right and wrong which is appropriate only to the West, or to the ex-colonial powers.

This point made, there are still difficulties in approaching Third World politics through western eyes, and it is time now to signal some of these difficulties. More especially, it is important that we do not conclude this chapter before we have signalled three more positive developments within the field of Third World politics; three developments which in part offset the drift to a banal or corrupting politics which has so far concerned us. These developments concern: (i) the role of non-governmental organizations; (ii) the formation and success (in some cases) of new social movements in the periphery; and (iii) relatedly, the emergence in some postcolonial countries of popular movements which are critical of western models of development, democracy and modernization.

Non-governmental organizations

Non-governmental organizations (NGOs) vary in size and scope throughout the developing world and some are more obviously politicized than others. In the Philippines, according to Garilao, NGOs comprise an economic sector.

Sector is not used loosely here. It is used precisely to denote that NGOs as a group have a distinct socio-economic-political function. As such, they can be juxtaposed with the traditional sectors of the economy, the public and the private sectors . . . As NGOs expand and professionalize their services, and attempt to bring in more of their population from the margins

of society, they are in fact creating a new service industry – the social development industry. (Garilao, 1987, 115–16, quoted in Friedmann, 1992, 147)

This account of NGOs would also cover western NGOs like Oxfam, Cafod and Save the Children. These organizations are registered charities in the West and as such they are obliged not to promote 'political goals' or to engage directly with politicians (for party purposes) in the countries in which they are engaged. NGOs of this sort seek to empower people directly and not always in opposition to the local state. Their strength is their willingness to allow men and women to participate in the construction of their own 'local development initiatives', and to reflect on what these agendas should be.

Gandhian NGOs in India often serve a similar purpose. They provide education where the state in practice is failing to provide it; they also encourage in local people an attitude of self-reliance which is often at odds with state-centred definitions of development, but which has a strong mandate in the wider Indian political tradition. In the 'tribal areas' of India some Gandhian social workers are concerned to link these local strategies of empowerment to a wider social critique of 'development and deforestation'. A focal point for this critique is the area of medicine. Tribal people are reminded that their traditional herbal medicines come from the local forests. Deforestation threatens this stock of wealth and threatens a future dependence on expensive western medicines. (At this point, and at such times, the role of NGOs can become problematic for the state and for such western sponsors as they might have. Western governments have become increasingly supportive of NGOs in recent years, in part because of a general reaction in development theory and policy against 'state interventions', and in part because NGOs are expected to create systems of rational-legal authority where few currently exist. The problem is that many NGO workers do not quite share this vocabulary of development. Their emphasis upon empowerment and localism – on putting the last first (Chambers, 1983) – can then rub against the more mainstream and macroeconomic development intentions of the state and the international development industry.)

The line between political and non-political NGOs is always a thin one, but in Latin America NGOs tend to define their political purpose in more explicit terms than is common in Asia and Africa. Friedmann reports a meeting of thirty NGOs from nine Latin American countries in January 1987, at which the following statement was adopted:

In terms of relations with other social actors, the relationship with the state has been accorded priority, based on the assumption that the division between civil society and the state should not be seen as absolute. The Centres [NGOs] define as one of their functions permanent denunciation and criticism. But the Centres also consider that they are in a better position than state entities for developing creative proposals that respond to the most fundamental social problems. (Friedmann, 1992, 146, quoting Landim, 1987, 32–3)

In Latin America, then, NGOs stand both outside the state's domain and in opposition to the state. To the extent that they accept certain existing realities, 'they are also prepared to *work with* the state in developing creative proposals that will benefit the disempowered, for the division between civil society and the state should not be seen as absolute' (Friedmann, 1992, 147; emphasis in the original).

The rise of NGOs raises important questions about the sovereignty of some postcolonial states. On the one hand, the main purpose of the NGOs is to empower local peoples where the state is unable or unwilling to meet its rhetorical and statutory obligations to them and to their milieu. In India, the state of the environment is debated with some regularity in Parliament, and various declarations have been made in favour of environmental protection; yet little gets done. The NGOs provide a welcome line of defence against a definition of development which can easily reduce to an unconsidered blueprint for 'modernization', and non-governmental actors can provide resources in and through which oppositional movements can be mobilized. In Gramscian terms, the NGOs can function as a countervailing force against the hegemony of a centralizing state apparatus.

At the same time, some NGOs can pose a threat to the sovereignty of some postcolonial states, and this threat (or opportunity) extends also to quasi-state institutions like the IMF and the World Bank. There are examples already of governments which have ceded territory, in effect, to non-governmental organizations like Oxfam, in return for those organizations bearing the costs of the development which they choose to promote within that territory. Meanwhile, there are pressures within the donor community for ODA to be targeted to those countries which agree to abide by certain economic and political pre-commitments. Prime Minister Major of the UK has suggested that British aid should not usually be given to countries which abuse the human and civil rights of their citizens. Aid would also be made dependent upon signs that a country in receipt of ODA was moving in the direction of greater democratization (and was eschewing the politics of a one-party state).

It is significant that some in Africa, especially, are not minded to dismiss this development out of hand, or to condemn it as yet another instance of western interference in the legitimate concerns of a postcolonial country. Given the breakdown of rational-legal authority in some countries in Africa – indeed of government *per se* in some cases – some intellectuals do now support a greater involvement of quasi-state institutions in Africa, notwithstanding the affront that this poses to most accounts of sovereignty. (It should be pointed out that there is little support for aid programmes which are made conditional on structural adjustment programmes; nor is there an indication that support for quasi-state institutions and interventions is anything other than an interim measure – the expectation is that local state capacities will be rebuilt in the medium-term.) The rise of the NGOs can offer some hope of a politics of empowerment, even where the dominant political culture of a country is marked either by violence and dislocation and/or by an apparent absence of public meaning.

'New' social movements

A further reason for optimism with regard to Third World politics concerns the rise of the so-called 'new' social movements there. Many of these movements have roots which are anything but new (especially peasant movements based around a longstanding defence of the moral economy: Shah, 1990), but that is not our chief concern here. There is evidence that social movements are growing in significance in the developing world and that large numbers of men and women there are turning to these movements as vehicles in and through which they can express their political concerns and expectations. There is some evidence, too, that the new social movements are most successful where they resist attempts by formal political parties to incorporate them into the routinized politics of the postcolonial state (Routledge, 1991).

Social movements in the developing world defy easy categorization. In Latin America, some of the most important social movements have been movements which have pressed for the provision of basic social services and public utilities (Gilbert and Ward, 1985). (These demands are often raised in combination with the NGOs.) Land invasions by squatter communities provide one illustration of this tendency; related to these invasions are community movements which press for the provision of water and sanitation facilities, transportation connections and street lighting. These movements typically ask the state for 'equal treatment' with the already-settled urban populations. These movements are not directly challenging the authority of the state, except in relation to the use of small pockets of urban land. Indeed, one might argue that the institution of private property in land is legitimized by periodic invasions of this sort. One might also note that political parties are often to the fore in the careful planning of land invasions in Latin America.

Although urban land invasions are not unknown in Asia and Africa (Schuurman and van Naerssen, 1989), rural social movements are perhaps more common in these continents. In India the two most important social movements of the 1980s were the peasant movements referred to above, and various movements which sought to protect India's forests and forest communities. Significantly, the tactics of each of these movements have drawn liberally from Gandhian values and symbols – Gandhi (or the Mahatma) being a very difficult symbol for the Indian state to ignore or reject (Amin, 1988).

In the case of Sharad Joshi's farmers' movement, resistance to the state is avowedly non-

violent. The preferred theatres of confrontation are local road blocks (*rasta roko*) in the rural areas – to draw attention to the adverse rural–urban terms of trade – and long marches upon New Delhi (Omvedt, 1990). These marches often do not take the most direct route, but make contact with towns and villages of significance in the life-story of Mahatma Gandhi. Gandhian imagery is also to the fore in most of India's environmental movements, although not exclusively. In the Jharkhand region of India a struggle for the protection of India's forest resources is linked to a struggle for a Jharkhand state, and for a greater say for the *adivasi* (tribal) peoples who once dominated there. This struggle has several times had resort to violence (Corbridge, 1991a; Duyker, 1987). Caste and class violence is also not unknown in the Indian countryside, notwithstanding the image of India abroad as a 'pacific Hindustan' (Bonner, 1990).

The most successful of India's environmental movements, however, have been Gandhian in form and inspiration. I refer to the Chipko movements which began in their modern form in north-west India in the early 1970s. The Chipko movements started with a series of spontaneous movements to hug local trees (*chipko*: to hug or embrace), and to prevent their clear-felling by timber companies and contractors. These movements were catalysed by the extensive flooding which hit the Uttarakhand region of Uttar Pradesh in the early 1970s: floods which the local people, prompted by Gandhian social workers and communist activists, interpreted as social, rather than natural calamities, brought on by an excess of deforestation (Guha, 1989). Given that it is women who are mainly involved in the collection of minor forestry produce, it is perhaps not surprising that women have often taken a lead in the Chipko movements (Shiva, 1991): movements which take on a subtly different character in different Himalayan valleys in the same region. Women have also provided a substantial part of the leadership of the Chipko movements (one thinks of Mira Behn and Bimala Behn in particular), although the best known protagonist of Chipko continues to be the ascetic Gandhian leader, Sunderlal Bahugana. Indeed, Bahugana was in the news again in February–March 1992 when he went on hunger strike to protest the construction of a dam in the Tehri-Garhwal District of Uttar Pradesh.

Notwithstanding certain rivalries within Chipko – and splits between the Gandhians and the communists, led by Chandhi Prasad Bhatt – the success of the Chipko movements can be measured by 'the fifteen year ban on green felling in the Himalayan forests of [Uttar Pradesh] by order of India's then [1980] Prime Minister, Indira Gandhi' (Ekins, 1992, 143). The Chipko movements were also influential in ensuring that India's Draft Forest Bill of 1980 was not passed by Parliament. This Bill proposed that still greater powers should be given to India's forest guards to police the increasingly commercial forests of India against the intrusions of local people who, supposedly, were its main enemies. (This advocacy of 'scientific forestry' is a classic example of a colonial construction of subject positions continuing after independence: see Blaikie, 1985.) Chipko activists successfully promoted the idea that local people were the main custodians of India's forest wealth, and that the main threat to the forests came from unregulated clear-felling by the timber industry. This suggestion, together with Chipko-style politics, was later taken up by environmentalist groups in states as far-flung as Karnataka in the south, Rajasthan in the west and Bihar in the east. In each of these states there are now village-based Forest Protection Committees. Although legal and illegal clear-felling continues in India, further threatening the country's precarious forest ecosystems, the recent promotion of social and community forestry projects by the Indian government, the World Bank and some donor organizations suggests that the political struggle for India's forests is not yet lost. Notwithstanding the power of the state in India, that country's support for democratic politics still permits the emergence of oppositional social movements which can legitimately contest the official line on development. In some Latin American countries, that scope for an alternative politics of empowerment is not always so apparent. In Peru, the struggle for an alternative model of development has pitted the state against the Maoist Sendero Luminoso (Shining Path) in a bloody civil war.

Modernity and its discontents

Mention of the Sendero Luminoso returns us to certain broader questions about the nature of social movements in the Third World. It also brings into focus the relationship between politics, development and modernity in some post-colonial countries.

In the West, the social movements literature owes a particular debt to Jurgen Habermas. Habermas suggested that, in Western Europe,

modern attacks on . . . the 'organic foundations of the life-world' had triggered new forms of protest and action. The tangible destruction of the urban environment and of the countryside through unchecked economic growth, the detrimental effects of pollution on the health of the population, and the modern obsession with material wealth were seen as key causes of the emergence of the ecological and anti-nuclear movements. (As summarized by Slater, 1991, 33)

Chantal Mouffe later extended this body of work when she highlighted

the emergence of three new forms of subordination under late capitalism: (i) the 'commodification of social life', whereby the expansion and penetration of capitalist relations of production into an ever widening sphere of social life has created a situation in which culture, leisure, death and sexuality have all become a field of profit for capital; (ii) the increasing bureaucratization of society, or a further penetration of civil society by the state; and (iii) a marked tendency to a more standardized, homogenized way of life, or a so-called massification of social life, resulting from the growing power of the mass media. (Slater, 1991, 34–5; summarizing Mouffe, 1984)

These new forms of subordination were in turn associated with new arenas of resistance, as social and political movements began to challenge the powers of capital and the state away from the once-privileged arena of production. Accordingly, a Marxist or quasi-Marxist politics of revolution gave way to – or would give way to – a more diverse politics of resistance which would challenge the construction of subject positions in terms of the archetypal categories of gender, race, ethnicity, community and region. The question of identity was now to the forefront (Laclau and Mouffe, 1985).

One can see how this analysis might have some validity in the advanced industrial world, but what about its relevance in the Third World?

Slater points out that the context for a new politics of opposition is often quite different in the postcolonial world, as compared to the First and Second Worlds. In particular, in most Third World countries there is not the safety net of welfare which one finds in the western countries; nor is the state's ideological penetration of civil society so great that consent (as opposed to coercion) can generally be relied upon to secure its reproduction and continued territorial scope. In some Third World countries, too, the centralization of state powers is such that a politics of anti-centralism is common there. This politics can take the form of political regionalism; it can also take the form of the 'base movements' (*basismo*) that are linked to the Church in Latin America (Lehmann, 1990). Finally, it is significant that the modernity of some postcolonial countries is a fractured modernity in western terms (a point realized well enough by those 'Third World' novelists who explore the tropes and terrains of magical realism). In many Third World countries, the modernization of the economy is incomplete and the political system often remains wedded to private interests (as much as to the public good). At the same time, the penetration of some Third World countries by 'western' ideas and modes of thought is much more complete – at least within the intelligentsia – and herein lies a paradox. What is emerging now in some Third World countries is a critique of modernity itself empowered in part by the language *of* modernity (a reasoned critique of reason . . .), and with reference to the role played by modernization in the promotion of unhealthy Third World economies and polities (see Chakravorty Spivak, 1990).

If this sounds a little cryptic, consider what it might mean in the case of India (and with regard to western accounts of Third World politics). In a powerful account of the 'State, society and discourse in India', Sudipta Kaviraj has argued that the apparent degeneration of politics in India is rooted directly in the fact that:

The independent Indian state followed a programme of modernity which was not sought to be grounded in the political vocabulary of the nation, or at least of its major part. As a result, precisely those ideals – of a modern nationalism, industrial modernity, secular state, democracy and minority rights – came in the

long run to appear not as institutions won by a common national movement, but as ideals intelligible to and pursued by the modern elite which inherited power from the British. More than that: subtle and interesting things began to happen to this logic of 'modernisation' which have gone unnoticed in the works of its supporters or opponents. Precisely because the state continued to expand, precisely because it went in a frenetic search of alibis to control ever larger areas of social life, it had to find its personnel, especially at lower levels, from groups who did not inhabit the modernist discourse. . . . Since major government policies have their final point of implementation very low down in the bureaucracy, they are reinterpreted beyond recognition. (Kaviraj, 1991, 91)

This is an arresting observation. If Kaviraj is right, it suggests that the Nehruvian era of high or command politics in India was an aberration; the paradox of democracy in India is that the more real rights and opportunities are given to ordinary people, the more these same people come to challenge the formal political grammars of the state. Again:

The paradox . . . is that if Indian politics becomes genuinely democratic in the sense of coming into line with what the majority of ordinary Indians would consider reasonable, it will become less democratic in the sense of conforming to the principles of a secular, democratic state acceptable to the early nationalist elite. What seems to have begun in Indian politics is a conflict over intelligibility, a writing of the political world that is more fundamental than traditional ideological disputes. (Kaviraj, 1991, 93–4)

Kaviraj concludes that: 'the state and the ruling elite [In India] uncritically adopted an orientalist, externalist construction of their society and its destiny reflected in the wonderful and tragic symbolism of "the discovery of India". Its initiatives were bound to be one-sided. To the world of India's lower orders, it simply refused or merely forgot to explain itself' (Kaviraj, 1991, 95).

Now these lower orders are fighting back, but in so doing they are using the new-old languages of casteism and communalism which the post-independence Indian state tried to erase – witness the rise of the Hindu nationalist party, the BJU, in the late 1980s and the early 1990s. In this they have the support (or at least the understanding) of some elements of India's intelligentsia. Kaviraj himself works at the Centre for Political Studies at Jawaharlal Nehru University, New Delhi. The result, perchance, is an India at odds with itself; an India in which the modern pathologies are Punjab and Kashmir, the two most potent symbols of secession and fragmentation.

One should not be hasty in pursuing this line of thought – modernity and modernization, along with Hinduism, also help to bind India together (Vanaik, 1990) – but it is easy to see the sense of what is being said. In Said's terms, the Third World, or parts of it, is now engaged in a critique of the project of modernization, not least because it can be constructed as another western, colonial project. This critique is bound to be ambiguous, and it is surely made possible as much by the unevenness of economic modernization as by its successes, but it does signal the depth and plurality of politics in the contemporary Third World. The failure of western-style politics to take root in some parts of the Third World does not mean that politics in the ex-colonial world is simply banal, or without meaning (see also Rowe and Schelling, 1991). We should be wary of reading and writing Third World politics with reference mainly to what is 'absent'.

Conclusion

History has a habit of providing surprises for those intent on making it and for those who are obliged to write about it. Lenin said as much at the turn of the century and he would as surprised as anyone at the traps that history has prepared since 1900. In the most general terms, these traps have been sprung in relation to colonialism, development and socialism.

At the start of the twentieth century, the writing of history, and the mapping of political geography, were not especially problematic enterprises. At this time the West was confident in its colonial missions and the main threat to the 'progress' that reason and modernization enabled came from a dissenting voice within this broad tradition; I refer to the voice of socialism. Some nationalist movements had been set up in the late nineteenth century, but most of them were seeking a greater say for native populations within the structures of Empire. Even the threat

of socialism appeared to be a distant one. Russia may have been lost to the Bolsheviks in 1917, but the Great War of 1914–18 did not provoke the revolutions which some had expected in the heartlands of industrial capitalism. Indeed, the main legacy of the Great War seemed to be the eclipse of Islam as a political force. With the formal ending of the Ottoman Empire in 1923 it was assumed that reason had finally replaced unreason as the main discourse of politics – just as Marx and Mill (but not Weber) had predicted at various points in the preceding fifty years.

With the benefit of hindsight, such optimism can seem almost comic, if it were not also so tragic in its consequences. By the middle part of this century a politics of anti-colonialism was entrenched in large parts of what was soon to become the ex-colonial world. The claims to power of France and Britain had each been dented by World War II and by the rise to power of the United States, just as the assumptions of white supremacism had been called into question by the rise of Japan and by a new emphasis upon equalities of opportunity in an interdependent (that is, postcolonial) world-economy. The new code by which individual countries would live was the code of development itself, usually conceived as a process of capitalist modernization/westernization – less often as an 'anti-western' process of socialist modernization/westernization. The Third World was now expected to imitate its erstwhile oppressors, in the process bringing economic growth and pluralistic politics to its member countries and citizens. The Third World would be helped in this endeavour by an international economic climate characterized by the willingness of the developed world to transfer resources to the developing world.

Now even this promise seems to be hollow and unfulfilled (Chipman, 1990). At the end of the twentieth century the projects of development and socialism each stand in some disrepair, and some in the West are prepared to laud the 'end of history' and confrontational politics (a paean as blind to geography as it is to history). The socialist project has foundered in large part because it sought to transcend the contradictions of capitalism by means of an excess of the instrumental reason which characterizes modernity (Sayer, 1991). Men and women too often became victims of a state-centred model of development (and politics), which made a virtue of loyalty to the state, and which tended to ride roughshod over the 'bourgeois' liberties of individual men and women. When the economies of some (but not all) socialist countries started to fail, the comfort of wealth and security vanished too, and with it a wider faith in the transcendental project of socialism-communism. The project failed not so much because it did not improve the life chances of most people – in China and Cuba it clearly did (Sen, 1989) – but because it failed to fulfil that intimation of perfection with which it had announced its presence and with which it staked its claim to 'true' development.

In the Third World the spread of development has been uneven throughout, with the result that livelihoods remain precarious for a majority of the populations of the postcolonial world. The 'average' African family was poorer in real terms in 1990 than it was in 1960. Meantime, the vital connections which should bind together the state and civil society often refuse to be knitted together, except under duress. In large parts of the ex-colonial world the wounds inflicted by colonialism continue to be exposed to view, and rational-legal authority and western-style politics, unsurprisingly, continue to fight against the odds to survive at all. In still other parts of the Third World development has been so discontinuous that even its iconic value is now under attack. Thus we see in even the most vibrant of Third World democracies – in India for example – a reinvention of political languages which are opposed not just to particular state policies, but also to the state's account of what is desirable and what should count as development. An earlier faith in modernity as such is no longer so evident; it is rather the case that one vision of modernization is being forced to compete with vernacular, regional and religious accounts of what the future should be like, and whether, indeed, it is desirable that the future should be unlike the past and the present. Thus is the gospel of modernity challenged; thus are the subalterns speaking back. Seen in these terms, the rise of Islamic politics is no simple fundamentalism, as some in the West would like to believe. It

is rather indicative of the fractured nature of modernity in large parts of the the ex-colonial world – a modernity which promises much, but which is unable to fulfil its promises; a modernity which offers the sceptical voice of reason an opportunity to contest a more singular and destabilizing vision of modernity as growth, as change, as westernization.

Uncertainty, then, is both the essence of the modern condition, and, perhaps, the best note on which to end this summary and partial review of politics in the Third World. But uncertainty is not a recipe for inaction, nor is it a ploy by which one can avoid making judgements. At any rate, it should not be. Certain trends in Third World politics do seem fairly clear at the end of the twentieth century. We can note, for example, that Third World politics are rarely understandable except with reference to the wider economic and geopolitical contexts in which they are shaped. We can also note that the failure of rational-legal authority to take hold in some Third World countries is a direct consequence of colonialism: which is not to say that local political skills, resources and agendas are unimportant. Some countries have been better managed than others. We might also note that it is important – and legitimate – for scholars and activists throughout the world to bemoan the absence of human and civil rights in many Third World countries (as in some First World countries), just as they might continue to call for a western politics which is sensitive to the needs and rights of distant strangers. (Writing off the debts of some Third World countries would be one instance of how such a politics could be put into practice.) There is surely a danger that a critique of Orientalism – and of reason itself – is taken too far; to the extent, indeed, that these most basic appeals to a universal morality cannot be made.

And yet uncertainty remains. If a survey of the politics of the Third World teaches us anything, it is that such a politics defies simple characterization and simple explanatory schemas. The recent politics of the Third World indicates how a politics of modernization and antimodernization can co-exist in many developing countries, in the process making such countries richer by virtue of this ambiguity. Instead of condemning Third World countries for failing to live up to the West (which itself is ironic), or to the expectations imposed at independence, we should be more circumspect in our judgements (though no less critical). The historical geography of Third World politics is a historical geography of presences and successes, as well as of absences and failures – as indicated by the survival of the democratic process in India, and as witnessed by the emergence of oppositional social movements throughout the Third World. The fact that some of these social movements are now sceptical of western accounts of development is a source of concern, especially when an unreflective populism finds itself in league with an authoritarian politics of dissent (or even rule: cf. Khomeini's Iran). At the same time, these social movements at last begin to impress upon the West not just the differences which define the political process in the First and Third Worlds, but also the possibility that a new politics of postcolonialism might hold important lessons for the West. There is surely a lesson here for political geography.

References

Ahluwalia, I. J. 1985 *Industrial growth in India: stagnation since the mid-sixties*. Delhi: OUP.

Akbar, M. 1985 *India: the siege within*. Harmondsworth: Penguin.

Amin, S. 1988 Gandhi as Mahatma: Gorakhpur district, eastern UP, 1921–2. In R. Guha and G. Chakravorty Spivak (eds.) *Selected subaltern studies*, pp. 228–348. Oxford: OUP.

Amsden, A. 1990 Third World industrialization: 'global fordism' or new model? *New Left Review* 182: 3–21.

Arnold, D. 1988 Touching the body: perspectives on the Indian plague, 1896–1900. In R. Guha and G. Chakravorty Spivak (eds.) *Selected subaltern studies*, pp. 391–426. Oxford: OUP.

Barthes, R. 1972 *Mythologies*. London: Jonathan Cape.

Basu, K. 1957 The history of Singhbhum. *Journal of the Bihar Research Society* 42: 17–38.

Bayly, C. 1989 *Imperial meridian: the British Empire and the world, 1780–1830*. London: Longman.

Beenstock, M. 1984 *The world economy in transition* 2nd Edition. London: Allen and Unwin.

Blaikie, P. 1985 *The political economy of soil erosion in developing countries*, London: Longman.

Blecher, M. 1986 *China: Politics, economy and society – iconoclasm and innovation in a revolutionary socialist country*. London: Frances Pinter.

Bonner, A. 1990 *Averting the apocalypse: social movements in India today*. London: Duke University Press.

Booth, D. 1985 Marxism and development sociology: interpreting the impasse. *World Development* 13: 761–87.

Brandt, W. 1980 *North–South: a programme for survival*. London: Pan.

Brass, P. 1990 *The politics of India since independence*. Cambridge: CUP.

Chakravarty, S. 1987 *Development planning: the Indian experience*. Oxford: Clarendon.

Chakravorty Spivak, G. 1990 *The post-colonial critic: interviews, strategies, dialogues*. London: Routledge.

Chambers, R. 1983 *Rural development: putting the last first*. London: Longman.

Chaudhuri, K. 1990 *Asia before Europe: economy and civilisation of the Indian Ocean from the rise of Islam to 1750*. Cambridge: CUP.

Chipman, J. 1990 Third World politics and security in the 1990s: 'the world forgetting, by the world forgot'. *Washington Quarterly* 14: 151–68.

Chomsky, N. 1983 *The fateful triangle: the United States, Israel and the Palestinians*. London: Pluto Press.

Clapham, C. 1985 *Third World politics: an introduction*. London: Croom Helm.

Corbridge, S. 1991 The povety of planning or planning for poverty? An eye to economic liberalization in India. *Progress in Human Geography* 15: 467–76.

Corbridge, S. 1991a Ousting Singbonga: the struggle for India's Jharkhand. In C. Dixon and M. Heffernan (eds.) *Colonialism and development in the contemporary world*, pp. 153–82. London: Mansell.

Corbridge, S. 1992 *Debt and development*. Oxford: Blackwell.

Dunkerley, J. 1988 *Power in the isthmus: a political history of modern Central America*. London: Verso.

Duyker, E. 1987 *Tribal guerrillas: the Santals of West Bengal and the Naxalite movement*. Delhi: OUP.

Economist, the. 1989 A Survey of the Third World. 23–29 September.

Ekins, P. 1992 *A new world order: grassroots movements for global change*. London: Routledge.

Emmanuel, A. 1972 *Unequal exchange*. London: Monthly Review Press.

Fanon, F. 1967 *The wretched of the earth*.

Harmondsworth: Penguin.

Finer, S. 1962 *The man on horseback*. London: Pall Mall.

Frank, A. G. 1969 *Latin America: underdevelopment or revolution*. London: Monthly Review Press.

Frankel, F. 1978 *India's political economy, 1947–1977: the gradual revolution*. Princeton: Princeton UP.

Friedmann, J. 1992 *Empowerment: the politics of alternative development*. Oxford: Blackwell.

Furnivall, J. 1948 *Colonial policy and practice*. New York: New York University Press.

Furtado, C. 1963 *The economic growth of Brazil*. Berkeley: University of California Press.

Galanter, M. 1984 *Competing equalities: law and the backward classes in India*. Delhi: OUP.

Garilao, E. 1987 Indigenous NGOs as strategic institutions: managing the relationship with government and resource agencies. *World Development* 15: 113–20.

Gilbert, A. and Ward, P. 1985 *Housing, the state and the poor: policy and practice in three Latin American cities*. Cambridge: CUP.

Guha, R. 1989 *The unquiet woods: ecological change and peasant resistance in the Himalayas*. Delhi: OUP.

Haberler, G. 1987 Liberal and illiberal development policy. In G. Meier (ed.) *Pioneers in development*, II, pp. 51–83. Oxford: OUP.

Halliday, F. 1990 Iraq and its neighbours: the cycles of insecurity. *World Today* 46: 104–7.

Harris, N. 1986 *The end of the Third World: newly industrializing countries and the decline of an ideology*. Harmondsworth: Penguin.

Hawthorn, G. 1991 'Waiting for a text?': comparing Third World politics. In J. Manor (ed.) *Rethinking Third World politics*, pp. 24–50. London: Longman.

Heffernan, M. and Sutton, K. 1991 The landscape of colonialism: the impact of French colonial rule on the Algerian rural settlement pattern, 1830–1987. In C. Dixon and M. Heffernan (eds.) *Colonialism and development in the contemporary world*, pp. 121–52. London: Mansell.

Herring, R. 1983 *Land to the tiller: the political economy of agrarian reform in South Asia*. New Haven: Yale UP.

Hirschman, A. 1958 *The strategy of economic development*. New Haven: Yale UP.

Holland, R. 1985 *European decolonization, 1918–1981*. London: Macmillan.

Hoogvelt, A. 1982 *The Third World in global development*. London: Macmillan.

Hourani, A. 1991 *A history of the Arab peoples*. Cambridge, Mass.: Belknap.

Huntington, S. 1968 *Political order in changing societies*. New Haven: Yale UP.

Jackson, R. 1990 *Quasi-states: sovereignty, international relations and the Third World*. Cambridge: CUP.

Kaviraj, S. 1991 On state, society and discourse in India. In J. Manor (ed.) *Rethinking Third World politics*, pp. 72–99. London: Longman.

Keynes, J. M. 1936 *The general theory of employment, interest and money*. London: Macmillan.

King, A. 1990 *Urbanism, colonialism and the World-economy: cultural and spatial foundations of the world urban system*. London: Routledge.

Kohli, A. 1990 *Democracy and discontent: India's growing crisis of ungovernability*. Cambridge: CUP.

Korner, P. 1991 Zaïre: indebtedness and kleptocracy. In E. Altvater, K. Hubner, J. Lorentzen and R. Rojas (eds.) *The poverty of nations: a guide to the debt crisis from Argentina to Zaïre*, pp. 229–34. London: Zed.

Laclau, E. and Mouffe, C. 1985 *Hegemony and socialist strategy*. London: Verso.

Lal, D. 1983 *the Poverty of 'development economics'*. London: IEA.

Landim, L. 1987 Non-governmental organizations in Latin America. *World Development* 15: 29–38.

Lehmann, D. 1990 *Democracy and development in Latin America: economics, politics and religion in the post-war period*. Cambridge: Polity.

Little, I. 1982 *Economic development*. New York: Basic Books.

Marglin, S. and Schor, J. (eds), 1990 *The golden age of capitalism: re-interpreting the post-war experience*. Oxford: Clarendon.

Metcalf, T. 1989 *An imperial vision: Indian architecture and Britain's Raj*. Berkeley: University of California Press.

Mitchell. T. 1989 *Colonising Egypt*. Berkeley: University of California Press.

Moggridge, D. 1992 *Maynard Keynes: An economist's biography*. London: Routledge.

Mouffe, C. 1984 Towards a theoretical interpretation of 'new social movements'. In S. Hanninen and L. Paldan (eds.) *Rethinking Marx*, pp. 139–43. Argument Sonderbad AS 109, Berlin.

Naipaul, V. S. 1969 *The suffrage of Elvira*. Harmondsworth: Penguin.

Omvedt, G. 1990 The farmers' movement in Maharashtra. In I. Sen (ed.) *A space within the struggle: Women's participation in people's movements*, pp. 229–70. New Delhi: Kali for Women.

Pakenham, T. 1991 *The scramble for Africa, 1876–1912*. London: Weidenfeld and Nicolson.

Palma, J. 1989 Dependency. In J. Eatwell, M. Milgate and P. Newman (eds.) *The new Palgrave: economic development*, pp. 91–7. London: Macmillan.

Pletsch, C. 1981 The three worlds, or the division of social scientific labor, *circa* 1950–1975. *Comparative Studies in Society and History* 23: 565–90.

Pratt, M. 1992 *Imperial eyes: Travel writing and transculturation*. London: Routledge.

Prebisch, R. 1962 The economic development of Latin America and its principal problems. *Economic Bulletin for Latin America* 7: 1–22.

Rostow, W. 1960 *The stages of economic Growth: a non-communist manifesto*. London: CUP.

Rothermund, D. 1986 *A history of India*. London: Croom Helm.

Routledge, P. 1991 New social movements in India: four case-studies. Syracuse University: Unpublished PhD dissertation.

Rowley, G. 1989 Lebanon: from change and turmoil to cantonization. *Focus* 39: 9–16.

Rudolph, L. and Rudolph, S.H. 1987 *In pursuit of Lakshmi: The political economy of the Indian State*. Chicago: University of Chicago Press.

Rushdie, S. 1981 *Midnight's children*. London: Jonathan Cape.

Rushdie, S. 1987 *The jaguar smile: A Nicaraguan journey* London: Picador.

Said, E. 1979 *Orientalism*. New York: Vintage Books.

Sayer, D. 1991 *Capitalism and modernity: An excursus on Marx and Weber*. London: Routledge.

Schuurman, F. and van Naerssen (eds.) 1989 *Urban social movements in the Third World*. London: Routledge.

Shiva, V. 1991 *The violence of the Green Revolution: Third World agriculture, ecology and politics*. London: Zed.

Slater, D. 1991 New social movements and old political questions: rethinking state-society relations in Latin American development. *International Journal of Political Economy* 21: 32–65.

Stein, S. and Stein, B. 1970 *The colonial heritage of Latin America*. Oxford: OUP.

Taylor, J. 1979 *From modernization to modes of production*. London: Macmillan.

Thorner, D. and Thorner, A. 1962 *Land and labour in India*. Bombay: Asia Publishing House.

Tomlinson, B.R. 1979 *The political economy of the Raj*. London: Macmillan.

Toye, J. 1987 *Dilemmas of development: Reflections on the counter-revolution in development theory and Policy*. Oxford: Blackwell.

Vanaik, A. 1990 *The painful transition: Bourgeois democracy in India*. London: Verso.

Ward, P. (ed.) 1989 *Corruption, development and inequality*. London: Routledge.

Washbrook, D. 1990 South Asia, the world system and world capitalism. *Journal of Asian Studies* 49: 479–508.

Watts, M. 1989 The agrarian question in Africa: debating the crisis. *Progress in Human Geography* 13: 1–41.

Wolf, E. 1982 *Europe and the people without history*. Berkeley: University of California Press.

Wood, R. 1986 *From Marshall Plan to debt crisis: Foreign aid and development choices in the world economy*. Berkeley: University of California Press.

World Bank 1987 *World development report, 1987*. Oxford: OUP/World Bank

World Bank 1989 *India: An industrializing economy in transition*. Washington: World Bank.

World Bank 1991 *World development report, 1991*. Oxford: OUP/World Bank.

Yang, A. 1989 *The limited Raj: Agrarian relations in colonial India, Saran District, 1793–1920*. Delhi: OUP.

5

The United States and American hegemony

John Agnew

The United States has been the first territorial state in modern world history with significant economic and military power to see colonial empire as a burden rather than a prerequisite for political and economic success. In the nineteenth century Britain had adopted economic practices relating to free trade that prefigured those of a future United States, but it remained strongly attached to the pursuit of overseas empire. However, from its origins in the late eighteenth century the relatively 'weak', institutionally divided, and decentralized American system of government left territorial and economic expansion largely in the hands of private individuals and organizations except when their lobbying made military intervention or government diplomacy unavoidable. The Founding Fathers' fear of a strong state produced a political economy in which power was vested largely in private hands rather than in public institutions.

This archetypal liberal state came to have numerous advantages over more bureaucratic and centralized states. By the early twentieth century these included: a flexibility in identifying emergent economic possibilities; a lack of commitment to citizenship or welfare standards that would raise labour costs; low military costs; and a regulatory environment geared towards capital accumulation wherever it might occur, rather than organized around a limiting territorial definition (Meyer, 1982). These advantages laid the groundwork for the emergence of an American hegemony in the aftermath of World War II in which US-based institutions isolated and contained the Soviet Union while at the same time they engaged in a reconstruction of the world-economy based upon principles derived from American historical experience.

However, the new world-economy that was created under American influence after World War II increasingly has failed to deliver the levels of economic growth and national political autonomy that it once promised. One cause has been the globalization of American and other business. It is no longer news when a business executive declares that 'the United States does not have an automatic call on our resources. There is no mindset that puts the country first' (Uchitelle, 1989b, 1). Another cause has been the creation of a powerful but volatile global financial system beyond the boundaries of state regulation. Finally, and not to be underestimated, in containing the Soviet Union for forty years the United States turned itself into a major military power. While this maintained a global space for the construction of the new world-economy, it subjected the United States to one of the burdens the country had always previously managed to avoid.

The consequence is an increasingly *hollow* hegemony for many Americans, in which they continue to bear many of the costs of American centrality to the world-economy but without the material returns of an earlier period. The paradox of the American position in the contemporary geopolitical world order is that, having 'won' the Cold War with the Soviet Union, the country cannot reap the fruits of its victory because US hegemony no longer delivers a guaranteed dividend to the US territorial economy. The purpose of this chapter is to describe how this came about.

First of all I provide a short theoretical discussion of the nature of American hegemony and its connection to American historical experience. A second section offers an overview of America's rise to a central position within the geopolitical world order. The third and longest section is a narrative account of the course run by the United States during its half-century of hegemony from 1945 to the present. A fourth section reviews the evidence for the recent emergence of a 'new' geopolitical world order of 'transnational liberalism'. A final section identifies the impasse in which the United States now finds itself, and the barriers to its resolution.

American hegemony and American history

Conventional wisdom suggests that from 1945 until the late 1960s the geopolitical world order was based upon a bi-polar conflict between the United States and the Soviet Union in which the former was the hegemonic power and the latter the challenger. Seen in this perspective the two states were successors, respectively, to Britain's maritime empire and French and German attempts to assemble an overwhelming continental bloc against it. However, the dramatic collapse of the Soviet sphere of influence in Eastern Europe and revelations about the fragile condition of the Soviet economy suggest that the Cold War was never a conflict between equals. Rather, the United States has been the one true superpower of the post-World War II geopolitical world order along economic, military and cultural dimensions.

This does not mean that the United States is not a successor to previous hegemonic powers such as Britain in the nineteenth century, or that the United States was not challenged by the Soviet Union with its alternative vision of world order. It is to suggest that the United States has been *more* hegemonic than Britain ever was, and that it has been so in a distinctive *American* way. This is to go beyond accounts of hegemony which view it solely as equivalent to a historically recurring pattern of single-state economic and military dominance over the state-system or the world-economy, and to see it at its fullest extent as involving the international diffusion and acceptance of norms, institutions, and social structures from the hegemonic power that regulate not only international relations, but also the domestic social order of other states. Hegemony is a difficult and controversial concept but most definitions would accept the criterion that the leading state cannot be simply the strongest militarily, but also must have the economic and cultural power to set and enforce the rules of international conduct that it prefers (Arrighi 1990).

American hegemony, therefore, does not only signify a shift in the identity of the hegemonic power from a previous one (e.g., Britain in the nineteenth century) or a world without one (e.g., as in the period from the 1890s until 1945), but also the particular institutions and practices the United States has brought to the world by virtue of its dominant position (Rupert, 1990; Gill, 1990). What have been these institutions and practices, and how did they spread?

1945 represents both the end of World War II and also the closing of a period of open conflict and crisis in the geopolitical world order. In the wake of the defeat of the 'rival imperialisms' that had grown in the pre-war period, and paralleling the containment of the Soviet Union – the only other truly Great Power left at the time – the post-war era saw the creation of a hegemonic order centred upon the United States and its purposive reconstruction of the world along liberal democratic lines. This involved the projection at a global scale of institutions and practices that had already developed in the United States, such as: mass production/consumption industrial organization; limited state welfare policies; electoral democracy; and government economic policies directed towards stimulating private economic activities. Ruggie (1983) calls the normative content of these policies taken together 'embedded liberalism'. This was the inspiration behind the creation of such multilateral international institutions as the IMF, the World Bank and GATT, and the sponsorship of domestic economic growth policies such as American-style Keynesian demand management and tight monetary policy.

Liberal thought and practice are deeply rooted in American history. Some, such as Hartz (1955), have seen liberalism as an American tradition, akin to a self-evident fact of American life. Certainly, in historically liberal America the definition of community in terms of individual interests, the privileging of private property rights, and the limiting of the sphere of government were built into the political, legal, and social systems from the founding of the state. But the *corporate*-liberal form that became the basis for American hegemony after World War II had more immediate roots in the development of American capitalism in the late nineteenth and early twentieth centuries. This involved the conjunction of a corporate or managerial capitalist organization of the domestic economy, and the beginning of its movement overseas, with a liberal capitalist state that became more activist in the face of the Great Depression. It was this combination that provided the material basis for America's role in the victory of World War II and contributed to the revival of the world-economy after the war. But it also provided the political-economic *model* upon which post-war American hegemony was based.

As an outcome of the compact between social groups and business interests implicit in the New Deal policies of the 1930s and wartime exigencies, from 1945 until the 1960s a *growth coalition* of big business leaders, labour leaders and politicians reproduced this corporate-liberal model in the United States. There were three major elements to it (Wolfe, 1981). The first was economic concentration. Continuing an intermittent trend from the 1880s, in almost every American industry control over the market came to be exercised by ever fewer firms. Expanding concentration was accompanied and encouraged by the growth of government, especially at the federal level. Much of this was related to military expenditures designed to meet the long-term and, seemingly, imminent threat from the Soviet Union. These trends were reinforced by what was to become the major challenge to the continued reproduction of the model in the United States: the investments of American corporations overseas. Most of these were increasingly located in industrially advanced countries and were direct rather than portfolio in nature. In the short run this benefited the American territorial economy in the form of repatriated profits. But in the long run, by the late 1960s, as domestic technology and management followed capital investment, traditional exports were replaced by foreign production of US affiliates to the detriment of employment in the United States. This has come to define the impasse or crisis facing the American model in the United States since the late 1960s (Agnew 1987). What Arrighi (1990, 403) calls a Free Enterprise System – 'free, that is, from . . . vassalage to state power' – has come into existence to challenge the inter-state system as the major locus of power in the world-economy.

The key institutions and practices of the American corporate-liberal democratic order spread rapidly in the late 1940s and early 1950s. They were eventually accepted by political elites in the major industrialized countries either through processes of 'external inducement' and coercion, as in the British and French cases, or through direct intervention and reconstruction, as in the German and Japanese cases (Ikenberry and Kupchan, 1990, 300). In all of these cases there was considerable compromise with local elites over the relative balance of growth and welfare elements and national particularities. In the conditions of devastation and political up-

heaval of the period 1945–50, however, it is not surprising that the United States, with its massive advantages in industrial and financial resources, was able to create a widely-accepted commitment to its liberal democratic norms and institutions among its erstwhile enemies as well as some of its recent allies. As quickly became apparent, the Soviet Union did not share this commitment and it became by 1947 the object of a policy of isolation and containment. As the Cold War deepened in the 1950s, criticism of the new order was easily identified with pro-Soviet sympathies, and this identification became one of the key ideological elements of American hegemony.

In sum, after World War II the United States provided the framework for a liberal democratic world order by fostering a growth-oriented 'politics of productivity' (Maier, 1978) in major industrialized countries, and by sponsoring a liberalization of the world trade and financial systems in which it played the key roles. American hegemony, therefore, reflected a fusion of power and purpose at a global scale of what had been established previously in the territory of the United States. According to Hartz (1964, 118): 'From the time of Wilson, indeed even before then, if we take into account a stream of thought which accompanied our early imperial episodes at the turn of the century, . . . [the US] has actually sought to project its ethos abroad.' After 1945 it finally was able to do so.

America's rise to power

The young American republic was vigorously expansive within the North American continent. Even Jefferson, the early American political leader most suspicious of centralized government, had a clear vision of an American Empire. The War of 1812 with Britain had its origins in American attempts at expansion into Canada and Florida. Though these attempts failed, a national policy of continental conquest, settlement, and exploitation in the following years produced a resource base unsurpassed by all other of the nineteenth century's imperialisms except, perhaps, that of the Russian Empire. In 1800 the United States stretched only to the Mississippi

River in the west. Though its northern border was much as today, the country did not as yet extend to the Gulf of Mexico in the south. Florida, which then extended along the coast to the Mississippi, was still controlled by Spain. But within fifty-five years, that is more or less the lifetime of a contemporary male European settler, the contiguous United States was to attain its present size – an increase in area of some 300 per cent.

The widely-shared consciousness among the Founding Fathers that they were the creators of a continental empire did not extend to the question of means. In general they were opposed to military operations. The central figure in early American diplomacy was Thomas Jefferson, and he objected to the old European tradition of 'reason of state'. Though he pursued ambitious goals, above all territorial expansion and commercial reform, 'he was determined to dispense, so far as was possible, with the armies, navies, and diplomatic establishments that had badly compromised the prospects for political liberty and economic prosperity abroad and would do so at home if ever they became firmly entrenched' (Tucker and Hendrickson, 1990, ix). To 'conquer without war' – to pursue the objectives of American policy by economic and other peaceable means of coercion and consent – was Jefferson's bequest to the nation.

The Louisiana Purchase, as is well known, was accidental. With a few pen strokes, and for only $15 million, Jefferson bought from a cash-hungry Napoleon a huge part of the continental interior that was with later annexations and purchases to extend the United States to the Pacific in the west and into Mexico in the south-west. Often, settlement ran ahead of the boundaries of the expanding state. The War of 1812 was stimulated by frontier settlers wishing to advance into Canada and Florida. The Mexican War had similar origins in the revolt of American settlers in Texas against their Mexican government.

The beginning of effective incorporation of the newly-acquired territories coincided with the turndown of the world-economy in the late 1820s and the 1830s. Low rates of return at home attracted British capital to the United States, especially into railroad investment, and low wages and unemployment in Europe attracted new settlers over the Atlantic. State and local governments were important in providing a congenial fiscal environment for investment and settlement. Of particular importance was public assistance to transportation projects through the so-called mixed enterprise, in which private and public funds were pooled. Federal land grants to railroad companies after the Civil War, and even later interventions such as in the New Deal policies of the 1930s, had their historical precedent set at this time (Lively 1955).

One concomitant of the geographical extension of transportation, especially railroad construction, was the growth of manufacturing. By the 1830s factories were sufficiently widespread, especially in New England and the Middle Atlantic states of New York, New Jersey, and Pennsylvania, to allow talk of the onset of industrialization. The Tariff Act of 1816 was particularly important in providing the beginnings of governmental protection of American industry against foreign competition.

But the cotton trade was the single most important economic activity tying the United States directly to the world-economy. The cotton-exporting South imported increasing amounts of food from the West, and with the income received the West bought manufactured goods from the North East. Thus, a regional specialization with between-region links was established that both integrated the country economically and created three distinctive regional economies within it (North 1961).

Slowly, however, the regional balance came apart. Industrial growth in the North East accelerated to the point where the region was a manufacturing economy generating its own internal demand. This in turn stimulated a demand for agricultural products from the West. As the extension of railroad lines further integrated the North East and the West, the South was increasingly isolated as an agricultural export economy based on a system of economic organization – plantation slavery – that did not fit with the emerging northern economy. The fundamental question over which the Civil War was fought concerned *which* economy would be favoured by the federal government (Moore, 1966, 136).

The Civil War produced a victory for northern business. This was manifested in the establishment of federal control over the banking and currency structure of the United States; the passage of highly protective tariffs; the massive subsidization of railroads through federal land grants; the encouragement of private exploitation of the country's resources under the Homestead Act, and the renewed importation of 'cheaper' foreign workers under the Contract Labor Law.

During the twenty-five years that lay between the end of the Civil War and 1890 the American economy began to take on many of its modern characteristics. The most important change was the shift from an agricultural to an industrial economy. In 1870 agriculture was still the chief generator of incomes in the United States. By 1890 industry had surpassed agriculture, and in 1900 the annual value of manufactures was more than twice that of agriculture. Relative to the rest of the world, American growth in manufacturing output was phenomenal. By 1913 the United States was to account for fully one-third of the world's industrial production (Gallman and Howle, 1971).

Fundamental to this expansion was the American resource base. Who had need of overseas empire when all of the resources needed for industrialization were available at home? Superimposed on this foundation was a political unity that encouraged the mobility of capital and labour. There were none of the linguistic, currency or customs barriers that disrupted economic interaction in Europe. But there was also a friendly political environment for private economic activity. All levels of government regarded business enterprise in a beneficent light. Political clientelism, monopolistic practices and price-rigging were all tolerated if not encouraged by the *laissez-faire* (if hardly free-trade!) policy of government. The other side to this was the outright hostility displayed towards labour when organized in unions or engaged in strikes.

In a country chronically short of skilled labour in its early years of industrialization, businessmen easily turned to machines to save on their wage bills. Coming late in the century, American industrialization was in an advantageous position to benefit from the new inventions, the improvements in industrial processes, and even the new products that appeared in increasing numbers. American openness to foreign investment was also vital. Thanks to the general political tranquillity of North America after the Civil War, the interchangeability of currencies afforded by the British-sponsored gold standard, and the attractiveness of American investments during the long depression of the late nineteenth century, American industry experienced little difficulty in drawing vast sums of portfolio investment from Britain and other European countries. It has been estimated that Americans saved only between 11 and 14 per cent of their national income in the 1890s, compared to over 20 per cent saved by the British during their rise to eminence (Tarbell 1936). Americans thus enjoyed a double stimulus to industrial development; they could both spend to stimulate domestic production, and borrow abroad to finance capital investment.

From the 1870s on American economic growth was increasingly managed by large industrial firms and investment banks such as J. P. Morgan. The first phase of industrial concentration coincided with the beginning of the long depression in 1873. Many firms had overexpanded in the post-Civil War years to meet the demands of the new national market. With the realization that excess capacity was forcing prices below the costs of production, small businessmen engaged in a flurry of pooling and merging. The result was a massive consolidation and centralization in a wide range of industries, especially those in consumer goods (e.g. leather, whisky, sugar, kerosene, etc.). The most famous firm produced by this wave of concentration was the Standard Oil Company (kerosene production) of John D. Rockefeller.

The severe downturn of 1893–6 brought a temporary halt to concentration. But in American national politics the growth of the 'trusts' had already become a key issue, dividing the country between the north-eastern industrial 'core' where concentration was favoured or accepted, and the remainder (the 'periphery'), where concentration was seen as an instrument of north-eastern domination (Sanders, 1986). Anti-trust legislation such as the Sherman Act of 1890 continued to give the impression of a country committed to small-scale capitalism, but the reality was otherwise. The economic recovery of the period 1896–1905 marked the largest spate of mergers and firm consolidation in American history, larger in real terms than later spates in 1925–30, the late 1960s and the 1980s (*Economist*, 27 April 1991, 11). The firms created included General Electric, Eastman Kodak, International Harvester and US Steel. The original shareholders were often 'robber barons' such as Cornelius Vanderbilt, John D. Rockefeller and Andrew Carnegie. Many of the negotiations over the prices of buy-outs were handled by J. P. Morgan, the investment bank that then dominated American finance (Chernow, 1990). This relationship between banks and industry signified the rapid shift that had taken place in the American economy from family ownership of firms to stockmarket-listings and managerial capitalism. Other countries lagged behind (Chandler 1980).

By 1905 roughly two-fifths of the manufacturing capital of the United States was controlled by 300 corporations with an aggregate capitalization of over $7 billion (at 1992 prices). The stimulus to concentration came from various sources. One was the cost of mass production and the consequent need to exploit economies of scale that required large capital investments. Another, important in the early mergers, was the straightforward desire to eliminate competition and set monopoly prices. A major stimulus, however, came from the desire to expand overseas. Bigger firms could better handle the initial costs and political difficulties involved in foreign activity. More importantly, once big through domestic expansion, firms could achieve greater size and profits only through expansion overseas (Wilkins, 1970). By 1914 a minimum of forty-one American companies, mainly in the machinery and food industries, had constructed two or more plants abroad. That the greatest number of factories after Canada was built in Britain – a country still committed to free-trade – shows that transport costs and meeting the demands of local markets were more important considerations in American foreign direct investment than avoiding tariff barriers (Chandler, 1980, 399).

The involvement of American business abroad was facilitated and legitimized by the American Presidents who acquired office after 1896. If the sudden burst of European colonialism in the 1880s had been stimulated in part by the industrial challenge from the United States, it soon encouraged some Americans to follow suit. In the Spanish–American War pride in American achievements was conjoined with the more mundane desire for open access to previously closed colonial markets. American leaders could preach against European colonialism while large American firms, through their advantageous economies of scale and vertical integration, began to colonize the world.

The transition from American 'isolation' to 'involvement' was unsteady and politically contentious. To extend from ocean to ocean had seemed too geographically neat not to be the product of fate or Providence. Overseas expansion was more difficult to justify, in particular when it required explicit territorial control. From McKinley to Wilson, American administrations encouraged business activity abroad, but justified it in terms of spreading American culture and values. America itself was sold as an idea:

American traders would bring better products to greater numbers of people; American investors would assist in the development of native potentialities; American reformers – missionaries and philanthropists – would eradicate barbarous cultures and generate international understanding; American mass culture, bringing entertainment and information to the masses, would homogenize tasks and break down class and geographical barriers. A world open to the benevolence of American influence seemed a world on the path of progress. The three pillars – unrestricted trade and investment, free enterprise, and free flow of cultural exchange – became the intellectual rationale for American expansion. (Rosenberg, 1982, 37)

Yet there was a continuing hostility among many Americans to involvement in Great Power politics. In 1914 the United States was not immediately drawn into the war that had broken out in Europe both because of a lingering ideological isolationism, and because of a sense that American economic interests would not be well served by participation. Eventually the United States did go to war and tipped the balance in favour of Britain and its allies. This resulted from an increasing perception that Germany represented much more of a threat than Britain to the American position within the world-economy. Germany's new technologies and industries were more directly competitive to those of the United States and its 'style of operation'; Germany was much more tied to territorial imperialism and was antithetical to the American gospel of free enterprise, free trade (abroad if not at home) and free cultural exchange.

The United States was the real winner of World War I. Britain was crippled economically and became America's debtor. Germany, the country that had most threatened American aspirations, was defeated and exhausted. In 1920 the United States was 'a major creditor nation, and its bankers, merchants, shippers, and manufacturers filled the vacuum left by the decline of the international monetary and trading system built by the British' (Becker and Wells, 1984, 461).

Economic victory, however, did not translate into immediate hegemony. In the 1920s there was a retreat from international political and military involvement. A fear that 'alien' influences (connected perhaps to the incredible cultural diversity of the 'new' immigrants to America in the period 1890–1920) were threatening the workings of America's unique institutions led to a powerful resurgence in isolationism. The notorious Sacco and Vanzetti case, involving questionable armed robbery and murder charges against two immigrant-Italian anarchists (Dallek, 1983, 99), represents this era in American history in a way analogous to the Dreyfus affair in French history at the turn of the century. Foreigners were dangerous. In this context the prevalent view was that clashing interests and disharmony at home would only be exacerbated by involvement in foreign affairs. Two important expressions of this isolationist impulse were the refusal to recognize the new Soviet government in the old Russian Empire, and the rejection of membership in the League of Nations (Dallek, 1983, 103–9).

The path of American foreign policy in the 1920s was also a reflection of the delicate rebalancing between the legislative and executive branches of the federal government that periodically affects the 'decisional' power of American government (Cerny, 1989). The three Presidents (Harding, Coolidge and Hoover) of the period

were unable to put much of a personal stamp upon foreign policy. After the 'activism' of President Wilson, a Congress reflecting the interests of rural and small-town America retreated from the call for a Pax Americana. This constituency yearned nostalgically for a time in which America, even if mythically, was a world apart.

In general, however, American Presidents and economic leaders remained committed to earlier policies of informal economic empire. American governments were able to pressure foreign governments into allowing American business entry into their markets, and they undermined European efforts to monopolize basic industrial resources in world trade. But despite these efforts, much of the world, including the United States itself, remained organized on neomercantilist principles. American macroeconomic policy remained narrow and restrictive. Tariff protectionism and the refusal to reschedule European war debts restricted the capacity of trading partners to make or keep the dollars necessary to buy American exports. American economic power was not used to promote an expanding global economy (McCormick, 1989, 27).

In the four years between 1929 and 1933 the American economy simply collapsed and did not recover effectively until 1941–2 – the first years of American entry into World War II. The view that this disaster was in large part due to the competitive economic nationalism of the 1920s came to prevail in the Roosevelt administrations of 1934–44. From this perspective, it had been the fragmentation of the world-economy into national and imperial markets too small for efficient production that had brought on the Great Depression. Only an assertion of American leadership in the direction of low tariffs, free convertibility, minimal capital controls, and open trading could prevent it from happening again. Before international restructuring could occur, the American economy would have to be reformed to turn it into a model for others to emulate. This was the burden of the New Deal policies instituted in the 1930s.

Some other countries, in particular Germany and Japan, reacted to the Great Depression in the rather more classic form of attempting to restore prosperity through territorial expansion. The consequent chronicle of war, holocaust and social upheaval is all too familiar. Less examined has been the American relation to it. *Lebensraum* (living space), the great claim of German geopolitics to the lands to the east of Germany (Burleigh, 1988), represented the greatest ideological and material challenge yet mounted to the dominant American conception of a world-economy and America's place in it. As McCormick (1989, 31) eloquently expresses the nature of the challenge:

Living space for Germany and Japan . . . was dying space for American private enterprise and for capitalism as an integrated world-system. If the two Axis powers succeeded, the world would devolve into four classic empires organized around industrial cores: Western Europe, North America, Russia, and Japan – each empire rationalizing and protecting its own space and resources and pushing and probing that of its rivals.

American acceptance of this possibility would have required rejection of the entire trajectory of American development and the ideology most closely associated with it: the world-economy seen as a potentially global system run along lines analogous to those operative in the United States. The 'security' of the United States, in conventional military terms, was never seriously in doubt. Invasion and conquest were next to impossible, and economic strangulation through a global blockade was of 'indeterminate feasibility' (Art, 1991, 11). From this perspective, the narrow-minded isolationists who opposed American entry into World War II were correct. But they were wrong in thinking that the war was about the defeat of Nazi Germany, Fascist Italy and Imperial Japan. Rather, the American war aim was the prevention of a world divided into autarkic economic blocs dominated by regional hegemons, irrespective of their particular identities.

This goal inspired many of the details of American strategy during World War II, in particular the American concern with conquering the industrial Ruhr-Rhine region rather than investing resources in such geopolitical 'sideshows' as the British focus on Italy, the Balkans, and the 'rush' to Berlin. It also lay behind the official American policy of waiting to intervene

massively in Europe until both German *and* Soviet forces had dissipated their strength.

The problem with the American vision in the 1930s as a blueprint for national policy had been that the 'American system' itself was in crisis. It was rescued by a restructuring that was ultimately realized only in the American industrial war effort. This involved new and more powerful roles for the federal government and organized labour in relation to private business. Not only did the national government establish an overt commitment to regulation of the economy; it provided the beginnings of a social-welfare 'safety net' for the population. Labour unions were brought into the liberal mainstream by public recognition of their central position in wage negotiations and in maintaining 'order' in the workplace. It was this new corporate-liberal order based upon a growth coalition of big business leaders, labour leaders and internationalist politicians that generated the social conditions for the explosion of industrial energy that in World War II produced an end to the Great Depression and defeat for the Axis powers. It was also this model that the United States was to project on to the world in succeeding years.

Superpower years

The completeness of the victory over Nazi Germany and Imperial Japan had two major consequences for the post-war world. First, Soviet influence extended over Eastern Europe and into Germany. When the war ended, Soviet armies were as far west as the River Elbe. This encouraged both a continuing American military presence in Europe, and a direct confrontation with the Soviet Union as a military competitor and sponsor of an alternative image of world order. This was quickly to find expression in the geopolitical doctrine of 'containment', whereby through alliances and military presence the United States committed itself to maintaining the political status quo as of 1945. The American acquisition of nuclear weapons and a demonstrated willingness to use them meant that the security of the United States was itself beyond doubt (Art, 1991, 11). What was in doubt was the allegiance of other countries to the United States and its political-economic model.

Second, in economic and political terms the United States was without any serious competition in imposing its vision of world order on both its vanquished foes and most of its recent allies. Unlike after World War I, when the United States turned its back on hegemony, this time there seemed to be no alternative. Europe and Japan were devastated. The internationalist reading of the origins of the Great Depression that predominated in the Roosevelt and Truman administrations suggested that the continued health of the American economy depended upon increasing rather than decreasing international trade and investment. Europe and Japan had to be restored economically both to deny them to the Soviet Union and to maintain American prosperity.

The period from 1945 until 1990 has been one in which these two consequences of World War II have played themselves out. The United States government set out immediately in 1945 to be the sponsor of a liberal international order in which its military expenditures would provide a protective apparatus for increased investment and trade across international boundaries. These would, in turn, rebound to domestic American advantage. The logic behind this lay in a presumed transcendental identity between the American and world economies. Progressively, however, the globalization of the American economy and the costs imposed on the domestic economy in general by military expenditures have undermined the identity between the two. The very success of the policies instituted after World War II has led to an impasse for the country that sponsored them (this argument is developed at greater length in Agnew, 1987).

However, hegemony does not just happen; it is made. A base of economic and military power is necessary for any hegemony, but that is not sufficient in itself. A vision and the will to pursue it are vital ingredients. In the aftermath of World War II the United States had both. The vision came from the successful New Deal experience allied with economic internationalism. The will came from an elite of businessmen, labour leaders, and politicians closely associated with the last Roosevelt administration, some of whose members or protégés remained active in policy-making into the 1980s.

Although they were challenged from the left (for example, by Henry Wallace) and the right (sections of the Republican Party), and despite the fact that these challenges have recurred since, they put together a set of policy orientations that survived in the main until the late 1960s, but in many respects persist up to the present. The major elements were: (1) stimulating economic growth *indirectly* through fiscal and monetary policies; (2) commitment to a unitary global market based on producing the greatest volume of goods most cheaply for sale in the widest possible market by means of a global division of labour; (3) accepting the United States as the home of the world's reserve-currency and monetary overseer of the world-economy (Bretton Woods system); (4) unremitting hostility to 'communism' or any political-economic ideology that could be associated with the Soviet Union; and (5) the assumption of the burden of intervening militarily whenever changes in government or insurgencies could be construed as threatening to the political status quo established in 1945 (Truman Doctrine).

The spread, acceptance, and institutionalization of this model has been by no means a preordained or easy process. In the United States itself, approaches to regulating American interaction with the world-economy and co-ordinating economic policies with other countries have shifted over time. One straightforward simplification of the course taken by US governments shows the combinations of domestic policy choices with international co-ordinating mechanisms in a two-by-two matrix (Figure 5.1). The most important point here is that the balance between market and institutional mechanisms in managing international transactions has shifted backwards and forwards over time rather than being set in a single mould. Some of these shifts, such as that of the first Reagan administration, can be seen as ideologically inspired, or in the later Nixon years as resulting from the temporary ascendance of more nationalistic or protectionist forces within American national politics. The increased rapidity of shifts after the breakdown of the Bretton Woods system is indicative of the increased relative economic power of Germany and Japan, and the increasingly globalized world financial

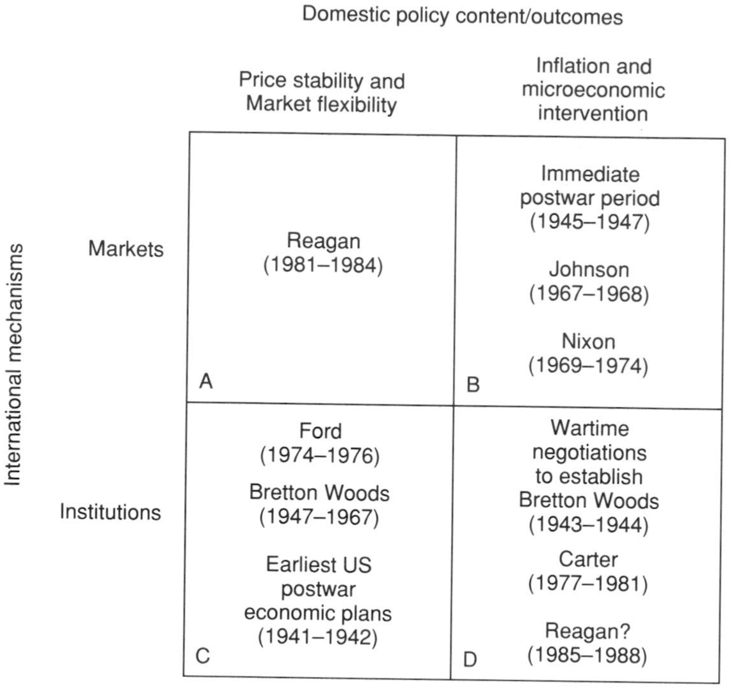

Domestic policy content/outcomes

	Price stability and Market flexibility	Inflation and microeconomic intervention
Markets	A Reagan (1981–1984)	B Immediate postwar period (1945–1947) Johnson (1967–1968) Nixon (1969–1974)
Institutions	C Ford (1974–1976) Bretton Woods (1947–1967) Earliest US postwar economic plans (1941–1942)	D Wartime negotiations to establish Bretton Woods (1943–1944) Carter (1977–1981) Reagan? (1985–1988)

(International mechanisms)

Figure 5.1 *Models of co-ordination of international and domestic economic policies.*

system. But there is also an interesting coincidence between confusion and disarray in US economic policy, especially in the early post-war period and during 1967–80, and the relative prevalence of left-of-centre governments in other industrialized countries (Nau, 1990, 48–9). This suggests, perhaps, the importance of at least a degree of elite solidarity across countries in the successful operation of American hegemony.

One fairly clear trend, however, signifies the

Table 5.1 *Total trade (imports + exports) of goods and services as a ratio of GDP for the seven major industrialized countries, 1955–89*

	1955	*1960*	*1965*	*1970*	*1975*	*1980*	*1985*	*1989*
Britain	48	43	39	45	54	52	57	52
Canada	39	35	38	43	47	54	54	50
France	27	27	25	31	37	45	47	43[a]
Germany	37	36	36	40	47	57	66	62
Italy	23	27	28	32	42	44	44	37[a]
Japan	24	21	20	21	26	31	29	27
United States	9	10	9	12	17	25	20	25

[a] Data are for 1988.

Sources: For 1955 data, Organization for Economic Cooperation and Development (OECD), *National Accounts Statistics, 1952–1981*, vol. 1 (Paris: OECD, 1983). For 1960–75 data, OECD, *National Accounts Statistics, 1960–1988*, vol. 1 (Paris: OECD, 1990). For 1980–89 data, OECD, *Quarterly National Accounts*, no. 1, 1990. See Webb (1991).

long-run success of the United States in opening up the world-economy irrespective of how it has been done: the increased importance of trade in the major industrialized economies (Table 5.1). Relative exposure to trade reflects one important way in which a national economy is integrated into the world-economy. The more integrated the economy is, the more likely it is that the balance of power between social groups will favour those with a stake in integration and, hence, favour further integration. The less exposed and the ones with powerful groups opposed to integration can be expected to resist the progressive removal of barriers to trade and investment (see Rogowski, 1989). American business in general reduced its attachment to protectionism after the 1890s. By 1945, and with the world-economy in a shambles, even American labour unions came to support lower tariffs and foreign markets. However, as other economies recovered, businesses and unions in sectors and regions facing intensive foreign competition have turned increasingly protectionist. This has been especially true of the more labour-intensive industries located predominantly in the North East and Mid-West (Rogowski, 1989, 120).

But irrespective of the level of trade relative to GDP, there can be important changes in a national economy's relation to the world-economy as barriers to trade and investment are reduced. The American economy has been transformed by the growth of free trade more than most others to the extent that some commentators now argue that conventional national economic accounts miss the major change that has occurred: that much trade, perhaps 30 to 40 per cent of total world trade, is now within multinational, or more accurately, transnational firms. As a consequence, therefore, 'ownership-based' rather than national accounting is more instructive. Thus, in terms of the national books, the US had a trade deficit of $144 billion (at current prices) in 1986. However, if the activities of US – owned corporations abroad and foreign firms in the United States are included, the huge deficit becomes a surplus of $14 billion (at current prices). Conventional accounts insist on nationalizing what are now totally internationalized transactions (Julius, 1990). The ultimate 'success'

of the American model would lie in a world where this pattern is repeated across all trading nations and all nations are equally exposed to trade.

The dissemination of the American model beyond the territorial boundaries of the United States has been constrained, however, by the vagaries of political-military conflict with the Soviet Union and the persistence of differences in political and economic institutions between the United States and other countries, most notably, perhaps, Japan. In particular, the stop–go arms competition between the United States and the Soviet Union and periodic American involvement in conflicts with states and groups seen as allied to the Soviet Union distorted the American economy and produced differences with allies who did not always share a simple bi-polar view of the world. In the 1980s, major policy differences on agricultural subsidies and the support of high-technology industries opened up between the United States on the one hand, and Japan and the countries of the European Community on the other. The political geography of the post-World War II world, therefore, is not a simple mapping of the triumph of American hegemony around the world, but the history of its uneven development.

Creating a 'Free World', 1945–50

Only one week after the surrender of Japan, the American Secretary of State James F. Byrnes laid out what were to become the guiding assumptions of American foreign policy. 'Our international policies and our domestic policies are inseparable,' he stated. 'Our foreign relations inevitably affect employment in the United States. Prosperity and depression in the United States just as inevitably affect our relations with the other nations of the world.' The Secretary of State then asserted his 'firm conviction that a durable peace cannot be built on an economic foundation of exclusive blocs . . . and economic warfare. . . . [A liberal trading system] imposes special responsibilities upon those who occupy a dominant position in world trade. Such is the position of the United States.' He then concluded by noting that: 'In many countries throughout the world our political and economic creed is in conflict with ideologies which reject both of these principles. To the extent that we are able to manage our domestic affairs successfully, we shall win converts to our creed in every land' (Dennett and Turner, 1948, 601–2).

Initially, American leaders hoped that this vision would be realized through such newly-established international institutions as the United Nations, the World Bank and the International Monetary Fund. But in 1946 they started to work more directly and quickly as a result of the economic crises that enveloped the countries emerging from the devastation of the war. The main concern was Europe, especially Britain and France. With pressure from their politically-organized working classes to increase incomes and provide social welfare programmes, European governments could ill afford to abandon economic protectionism and their captive imperial markets. The American weapon against a return to the pre-war order in Europe was its currency, the dollar.

The British loan of 1946 was the first sign of American willingness to use its superior economic strength to recast the world order. In this case the British need for dollars to buy American capital goods for its industrial reconstruction was exchanged for a promise that pounds sterling would be freely convertible to dollars within a year. The net result of this was the abolition of the mechanism of convertibility controls that British governments could use to manage their economy without American interference. Thereafter, decisions on government spending and how much of the nation's resources could be spent on American imports were vulnerable to 'international' (i.e. American) supervision. The only alternative for the Labour Party government of the time would have been the adoption of autarkic forced reindustrialization along Soviet lines without access to American goods and technology (Taylor, 1990).

The shortage of dollars or 'dollar gap' was not new; Britain and other European countries had run trade deficits with the United States since 1914. But previously the deficits had been covered by the foreign exchange earned by investments in Latin America, economic mono-

polization in colonies, and trade with Eastern Europe. These sources were now no longer available. Europe, and Britain in particular, had little choice but to accede to American demands. There were strict economic limits to the political choices possible in Europe in the immediate post-war years

As it turned out, the consequences of the British loan were not so dire. Because of a massive run on the pound sterling when it became convertible and the subsequent threat of financial disaster, the US allowed Britain to retain some controls. Not until 1958 did the British pound and other European currencies become freely convertible into dollars. In 1946 the world was not ready for free convertibility. What is clear, however, is that the United States was already planning for a world in which it would be.

This became obvious with the creation of the European Recovery Plan (ERP), or Marshall Plan in 1947–8. A multilateral response to the dollar gap crisis, the Plan involved supplying Europe with the dollars it needed over four years in return for European acceptance of American policies on reducing barriers to trade within Europe, integrating Germany back into the world-economy, and slowly adjusting to free convertibility in the medium term. European governments were required to match their American dollars in local currency, but their spending proposals were subject to veto by an American agency (the Economic Co-operation Administration). This agency often blocked efforts to use the non-dollar funds for social welfare purposes by the left-leaning governments in Britain, France and Italy, and instead, in the interest of increased price-competitiveness and fiscal conservatism, forced them to balance government budgets, suppress wages and reduce their dollar trade deficits. The message this sent to voters was unambiguous. The future arrival of American aid depended upon their electing governments that accepted the 'rules of the game' as written in Washington (Hogan, 1987).

There seems little doubt that the Marshall Plan was a major ideological success. It definitely helped to kill the hopes of the French and Italian Communist parties in the 1948 elections. It also provided a push to local investment. What is more contentious is whether it played the definitive role in re-establishing European economic growth that is often ascribed to it. In its first year, 1948–9, Marshall Plan aid exceeded 6 per cent of national income only in France among Western Europe's largest economies. Though this undoubtedly bolstered dollar accounts, there is reason to believe that growth rates would not have been much different *without* Marshall aid (Milward, 1984). Of course, this was not known at the time. The Marshall Plan provided an image of a giving America in the service of an open world-economy that even the most cynical Europeans found difficult to challenge.

The most difficult problem to be faced in creating a free world, however, was what to do about the Soviet Union and its dominance in Eastern Europe. The question was an old and, as we know today, a recurring one. Could the Soviet Union be admitted to a world-economy which it had left in 1918, without making fundamental changes in its economy and without acceptance of American pre-eminence? The answer was not long in coming, and it was no. By 1946 American leaders had decided that their plans for a new world order must exclude the Soviet Union as it was, and use it as *the* representative case of an alternative to the American model. Thereafter any opposition to American hegemony could be conveniently associated with support for the Soviet Union and defined as subversive and beyond the bounds of normal politics. Soviet hostility to the Marshall Plan and opposition to American attempts at 'normalizing' Germany reinforced American recalcitrance.

The two major founding actions of the policy of isolating and containing the Soviet Union were the Truman Doctrine of 1947 and the formation of NATO in 1949. The first grew out of Britain's inability to finance the campaign against a radical insurgency in Greece in the face of the dollar gap. The Truman administration literally leaped at the opportunity to replace Britain as the 'regional policeman' in the eastern Mediterranean. But as stated in the President's speech to Congress on 12 March 1947, it also amounted to the first official statement of the containment doctrine as formulated a year earlier by George Kennan, the US Ambassador to the Soviet Union. The regional context was little more than

a pretext for an essentially universal declaration of an American determination 'to support free peoples who are resisting attempted subjugation by armed minorities or by outside pressure'. This was a clearly-understood code for opposition to the Soviet Union and to those political movements, including some in Europe and the United States, that could be construed as allied with it.

The Truman Doctrine laid the groundwork for America's Cold War with the Soviet Union. It established the precedent of intervention on behalf of anti-communist regimes. It provided a way of taking over the global 'policeman' role from Britain by having the burden thrust upon it by Britain's penury. Most of all, it provided an ideological cover for a Democratic administration faced by a newly-elected Republican Congress suspicious both of an 'imperial presidency' in office since 1933, and of the internationalism that had become its official creed.

Truman's speech also provided the rationale for what became America's most important political–military measure in the sequestration of the Soviet Union: the NATO Pact of 1949. This was the security counterpart to the Marshall Plan's economic role in securing a free world order. The main purposes of NATO were: to anchor Germany in an entangling alliance; to commit American troops militarily to the conventional defence of Western Europe and, as a hostage force, to the use of American nuclear weapons; and to provide for a European ring of US military facilities encircling Soviet-controlled territory.

Yet, ironically, by clearly demarcating an American sphere of military influence in Western Europe, the NATO alliance institutionalized the division of Europe agreed to by Roosevelt, Churchill and Stalin at Yalta in 1945. Eastern Europe was thus tacitly understood by the US as a Soviet sphere of influence. There was an implicit recognition of two worlds of influence, in one of which the Soviet Union was viewed, at least in military terms, as the legitimately dominant power. The Soviet Union was seen as a necessary 'Other' against which the 'the Free World' could be contrasted empirically (O'Tuathail and Agnew, 1991; Dalby, 1989; Pletsch, 1981). The free world in the making was still one with definite geographical boundaries.

Militarization and 'struggle' in the periphery, 1950–6

The Soviet acquisition of the atomic bomb and the victory of the Chinese communists in the Chinese civil war reinforced the division of the world into two mutually hostile blocs. The idea that both were directly the result of the subversion of American government by communist sympathizers captured the public imagination in the United States. Domestic dissent and foreign policy failure became rhetorically connected as cause and effect in a gathering political consensus that had enveloped both political parties by 1950.

In addition to the enhanced external challenge and the collective paranoia this generated in the United States, there were recurrent problems in managing the new world-economy. Of primary importance was the failure in reviving the Japanese economy through production for export. Civil wars and revolutions effectively removed Japanese markets in China, South East Asia and North East Asia. Tariffs restricted Japanese access to American and European markets. A second problem was the slow growth in production of primary commodities, especially in Asia, relative to European demand. This kept prices high and limited the growth of markets for European manufactures. It was partly the result of the physical disruption associated with decolonization, for example, in India or Indonesia. But it also resulted from the adoption by governments in the 'periphery' of import substitution policies that did not accept the permanent division of the world into industrialized and primary-commodity producing parts. Third, the dollar gap, though perhaps less severe than before, still posed a threat to the continued growth of Western Europe.

The new conditions prevailing in the early 1950s could have been dealt with in a variety of ways. One possibility, never seriously entertained except by some State Department officials such as Kennan, was that the United States could negotiate with the Soviet Union over nuclear weapons, US German policy, and the possibilities of trade with both the Soviet Union and China. This was seen as a recognition of Soviet equality with the United States and not really open to consideration. The outbreak of the war

in Korea and the recent crisis over Berlin finally put it off limits. Another possibility was to further the opening of the world-economy by easing American tariffs on European and Japanese exports. The dilemma here was that this approach was not popular with powerful domestic interests concerned about American jobs and 'subsidizing' foreign producers, and it did nothing about the Soviet 'threat'.

The chosen response was to engage in a massive military build-up. In the years 1950–52 the US military budget increased from $14 billion to $53 billion. Much of this spending went for new military hardware (especially aeroplanes), the development of the H-bomb, and an expansion of the navy. But considerable amounts were also ploughed into increasing the US conventional presence in Germany, rebuilding a German army, and making military aid a major component of aid to peripheral countries. In purely economic terms this strategy made less sense than continuing at a faster pace the creation of a multilateral trading system begun previously. Its big advantage lay in its combination of economic stimulus with political–military response to the Soviet challenge. While overriding resistance from the American right in the name of national security, 'American military spending, with its insatiate demand for raw materials, fueled a worldwide triangular trade in which the periphery exported raw materials to the United States, which in turn exported capital goods and food to Europe and Japan, which exported finished consumer goods to the periphery' (McCormick, 1989, 95).

For this approach to work, the focus of American policy had to shift from Europe to the countries of what was now called 'the Third World', especially those where revolutions and nationalistic struggles threatened the process of integration into the world-economy. This was the zone of conflict between the competing visions of the 'First' and 'Second' Worlds (Pletsch, 1981). First of all, in the Asian 'rimlands' of Korea, Taiwan and Indo-China, in order to help Japan redevelop its East Asian regional economy and to counter Soviet 'success' in China and later in Latin America and Africa, and to integrate their extractive economies into the free-world economy, the US response to political 'instability' became a fixed formula of military aid and intervention. Sometimes, as in Indo-China, this involved taking over from the prior colonial regime, in this case France, even when direct American economic interests were almost non-existent. The American goals were more general: first, maintaining open economies and a specialized international division of labour; and second, preventing any one open economy from succumbing to closure in case adjacent ones would follow suit – the so-called domino theory. The great nightmare was if Japan, Western Europe and (who knows?) the United States itself was to 'fall' through a progressive closure of the world-economy.

Of course, Europe and Japan could not be entirely neglected. The military build-up had some of its major consequences in Western Europe. Germany was effectively rearmed within the NATO military alliance. The US commitment of large conventional forces to Europe, through its massive transfer of dollars, had an important economic-stimulative effect, especially in Germany. It also began a long-term drain on the US balance of payments. The Korean War was a major stimulus to Japanese economic growth. Quickly, the fear of Soviet intervention or subversion in Europe and Japan receded, but the effectiveness of the militarization strategy in overcoming the *economic* crisis of the early 1950s would not allow American leaders or their political allies elsewhere to admit it. The United States *needed* the Soviet Union (and vice versa) as a threat in order to justify the strategy of militarization (or 'military Keynesianism', as it is sometimes called) upon which economic stimulus was coming to depend (Kaldor, 1991).

The apex of hegemony, 1957–67

1956–7 was epochal for several reasons. In the Suez crisis the American attempt to manipulate the Egyptian President Nasser into abandoning 'non-alignment' in return for American financing of the Aswan Dam failed. However, the outcome of the crisis, with American pressure forcing Britain and France to withdraw their troops after their intervention against Nasser's nationaliza-

tion of the Suez Canal, both showed who was in charge of 'western' policy and suggested the importance of using aid more straightforwardly and in more fruitful settings than Nasser's Egypt. One major consequence of this was to be a search for 'moderate' regimes (most of which had previously been 'feudal' or 'backward' in the language of the day), rather than the 'nationalistic' ones such as Nasser's, to serve as regional examples and stabilizers in the Middle East, Africa, Latin America and South East Asia.

But the mid-1950s were also a turning-point in the creation of the international economic order that was at the heart of American hegemony. By 1958, the formation of the European Economic Community, the achievement of free convertibility among the main currencies, the closing of the dollar gap for both Europe and Japan, and the resurrection of the General Agreement on Tariffs and Trade (GATT) (originally established in 1948) to negotiate multilateral reductions in tariffs all signalled considerable success in converting the industrialized world to the virtues of free trade and free exchange under American auspices.

Even though these achievements were formidable by any standard, they were subject to serious constraint. In the first place, without Britain, which still had not decided whether it was part of Europe, the European Community was potentially a vehicle for French attempts at creating a European bloc *between* East and West. The return of de Gaulle to power in France in 1958 began a series of challenges to American dominance in Europe that culminated in French withdrawal from the military side of NATO in 1966. Second, the EEC was also a potential economic threat, particularly if it restricted American business and raised its external tariffs to exclude American goods and services. Third, and most significantly, the militarization strategy upon which the solution of the dollar gap had been based was increasingly costly to the United States. After 1957 the US trade balance was to be always in deficit, to a considerable extent because of a dollar drain produced by all the bases and troops overseas (Calleo and Strange, 1984, 104; Calleo, 1982). Yet the military budget was out of control. Military threats were seen everywhere, and an entire economic sector – what President

223

Eisenhower memorably referred to as 'the military-industrial complex' – was always available to respond. If Eisenhower at least noted the problem, his two successors were equally oblivious to it. Military spending leaped upwards during their administrations.

These costs spiralled upwards initially because of increased outlays to 'restore' American pre-eminence in nuclear weapons (which in fact was not seriously in doubt at this time) but, as time went on, mainly because of the strategy of military support for pro-American or 'moderate' regimes, especially that in South Vietnam. By 1967 US military involvement in Vietnam and surrounding countries became so intensive that not only was it imposing stress upon the American economy, particularly in the form of increased inflation, but it called into question the whole strategy of integrating the periphery into the world-economy by military means.

Decline and detente, 1968–76

Despite the drag of military spending, the American economy grew phenomenally between 1961 and 1967: in excess of 6 per cent annually. But in 1967 and 1968, inflation and unemployment went up together. A 'stagflation' set in that lasted until the early 1980s. The early 1970s were marked by large trade deficits, to be exceeded only by those of the 1980s, which in combination with currency speculation brought about the demise of the Bretton Woods system of international monetary regulation. Median American household incomes peaked in real terms in 1974. In the same year world oil prices were forced up at a rapid rate by the Organization of Oil Exporting Countries (OPEC), producing a major transfer of wealth within the world-economy from the United States and other industrialized countries to the oil producers. There was no longer a clear identity of interest between the American territorial economy and the working of the world-economy. What had gone wrong?

Under the militarization strategy, economic concentration paid dividends in the United States – expanding profits and providing higher incomes that stimulated demand for consumer goods. But big firms, protected from competition, failed to engage in the type of research and development that would fuel innovation in the *civilian* economy. Spin-off from military research was no longer enough. Productivity began to decline relative to foreign producers. Built-in wage increases for many workers in large firms also cut into profits. Corporate control over pricing prevented prices from falling as fast as general economic conditions mandated. This increased inflation.

The growth of government was an additional drain. The huge sums required to pay for the Vietnam War, to maintain the vast US military presence around the world, and to pay for the social policies enacted in the late 1960s to meet the demands of social groups not sharing in the general prosperity of the time, produced the bureaucratic state that the United States had always managed previously to avoid. Because tax increases were politically unpopular given the hostility of different organized groups to one or other variety of government spending, stimulation of the economy was possible only through increasing the supply of dollars and encouraging private indebtedness. The net effect was to increase inflation, begin a long-term process of massive public sector borrowing, and create instability in financial markets.

As first cause and then consequence of the economic slowdown within the United States, American transnational corporations found overseas investment more profitable than investment in their homeland. Not only did this increase unemployment, especially in the high-wage manufacturing sector, but it also reduced the availability of capital for re-tooling factories and investing in product innovation in the United States (Agnew, 1987).

The troubles of the American economy and dissent over the Vietnam War effectively ended the 'Cold War consensus' that had prevailed through the previous two decades. The increased exposure of industries in the north-east and midwest regions of the United States to competition in domestic markets from foreign producers (including, of course, the subsidiaries of US-based companies) led to an increase in protectionism on the part of those regions' political representatives. The economic and moral costs of policing the world were tangible and could be seen on the nightly news and in local communities.

The response of the Nixon administration was threefold. First, through cajoling and coercion it made Japan and Europe revalue their currencies against the dollar. In the short run this made American exports more competitive and imports less so. But its long-term effects were the demise of a system of stable rates of exchange and a major boost to a global financial system largely out of governmental control (Strange, 1986). Second, in terms of strategic nuclear weapons, the United States and the Soviet Union were almost at parity by the early 1970s. During the Brezhnev years, as the rest of the Soviet economy stagnated, a major emphasis was placed upon achieving equality with the United States in strategic weapon systems. This did not increase Soviet security, nor did it improve the generally low quality of Soviet forces as a whole (Edmonds, 1983; Cockburn, 1983). But it did increase American perceptions of the Soviet Union as a serious geopolitical threat to the territorial United States itself. In this context the rift between the Soviet Union and China made it possible for the United States both to accept the Soviet Union as a military equal, and to use China as a counterweight if Soviet demands became excessive. This was the essence of the Nixon policy of detente with the Soviet Union. But it also carried potential economic benefits in terms of American trade and investment if the Second World could be opened up. Third, as a consequence of the disastrous involvement in Vietnam, the Nixon administration substituted the arming and support of 'regional surrogates', such as the Shah's Iran, for direct American military intervention. This brought both the advantage of increased exports of military goods and the promise of 'no more Vietnams' in which Americans would be the victims of America's wars.

Whatever the impression given by Nixon or his major foreign policy adviser Henry Kissinger, these policies were the result of recognizing necessity rather than the product of design. The period 1967–74 marked a watershed in America's relations with the rest of the world. An administration that spoke openly of geo-politics was the first one since World War II so constrained domestically – by dissent and conspiratorial responses to it (as in the Watergate affair), and internationally, by an increasingly hostile economic environment – that it could not effectively practise it (Schell, 1989). The shifting course of American foreign policy in the early 1970s can be attributed in particular to the emergence of foreign competition to American business within the United States; the relative decline of American-based manufacturing industry; the collapse of an American-operated international financial system; the strategic parity between the US and the Soviet Union; and domestic division of opinion over both America's global policing and the virtue of an open world-economy.

Back on top? The 'second Cold War' and the Reagan era, 1978–88

By the mid-1970s the Nixon approach was increasingly in question. Its heavy emphasis on US diplomatic relations and unilateral economic action led to considerable unease among the most 'internationalized' sectors of American business (Gill, 1990). How to turn this unease into policy was another matter. In his first two years in office, President Carter tried policies which would make the US less dependent on foreign sources of oil, would stabilize arms competition with the Soviet Union, and would encourage a greater degree of respect for human rights among regimes allied to the United States. These policies foundered, however, because of three trends. The first was the influence of right-wing groups inside (in the person of Zbigniew Brzezinski) and outside the administration in identifying the Nixon policies of detente as nothing short of an appeasement of the Soviet Union. Although these groups differed in their proposals, all were agreed that the US needed to reassert itself militarily (Dalby, 1990) US military spending had out of fiscal necessity declined in the early 1970s. The claim, however, was that because of faulty policy the US was now being overtaken by the more relentlessly militaristic Soviet Union.

The second trend was that the recycling of the so-called petrodollars produced by the OPEC

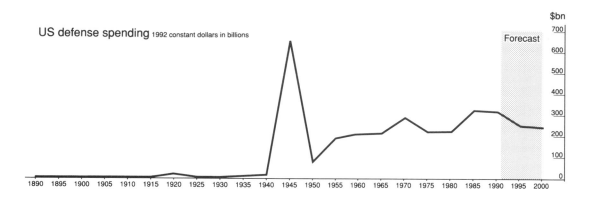

Figure 5.2 *US defence spending, fiscal years 1890–2000 (in 1992 $ billion).*

price increases into loans to the industrializing countries of the Third World through American, and some other, banks had by 1978 created the beginning of a global debt crisis. Servicing loans became difficult when American interest rates went up in order to keep attracting capital to pay for domestic consumption. It became impossible as oil prices again increased and demand for the borrowers' products went down in the global recession of 1979–80. The US was faced with the spectre of major bank failures if several of the largest debtors (e.g., Mexico and Brazil) were to default simultaneously. And, without the outlet of investment in the highly profitable newly-industrializing countries, there was a need to map out a strategy for investing in the reindustrialization of America.

A third trend was more by way of a collapse. In 1979 both the United States and the Soviet Union were faced with the disintegration of surrogate regimes in, respectively, Iran and Afghanistan. In other regions, surrogates proved to be unstable and unreliable (O'Loughlin, 1989). This was the American experience, for example, in the Horn of Africa and Latin America. Superpower competition could no longer be sublimated through the use of substitutes. The Carter administration was defeated in the 1980 election in large part because of its failure to maintain its surrogate (the regime of the Shah) in power in Iran and the shameful hostage-taking of the replacement.

In the absence of a viable nationalistic strategy

involving a national industrial policy, capital-export controls, and a decline in military spending, the only tried and true solution to the multiple dilemmas facing the country was a revival of militarization. Both the later Carter administration and the Reagan administrations of the 1980s chose this course (Figure 5.2). Increased military spending at home stimulated investment and employment, at least in those regions where it was concentrated, and increased military commitments abroad demonstrated an American resolve to reassert its centrality in a world it had created. For President Reagan at least, 'It was morning again in America.'

The first Reagan administration carried through the remilitarization strategy begun under Carter to a level not seen since World War II. Most emphasis was placed upon developing new weapons rather than upgrading existing conventional forces. The economic goal was to jump-start the American economy with a massive programme of government-subsidized industrialization. The political goals were to remind Europe and Japan of their reliance on an American military commitment that underwrote their economic development, to counter the Soviet attempt at military equality with the United States, and to encourage regimes friendly to American economic interests in the Third World.

In addition to massive military spending, the Reagan administration attempted to reverse America's competitive slide in the world-economy in three ways: recession, tax cuts and deregulation. In order to bring down inflation, cut wage rates and reduce the number of inefficient producers and thus to make US industries more competitive, the first order of business involved tightening monetary policy. This was done so drastically in 1981 that the US experienced its most severe downturn since 1937. Attention then was switched to the real economy. Income taxes, especially for those with high incomes, were cut dramatically on the theory that the extra money put into circulation would end up as new productive investment. Finally, deregulatory policies were introduced to allow 'the market' to choose between candidates for growth those heading for bankruptcy. The main path was to encourage mergers and acquisitions. But subsidiary policies included reductions in environmental and safety regulations, assaults on unions, and the elimination of many rules governing the conduct of financial institutions.

In the short run these policies were such a stunning success, particularly in reducing that traditional bane of the middle class – inflation – that 'Reaganomics' produced a landslide victory for President Reagan in 1984 and widespread imitation abroad. The US political economy's peculiar qualities, by the standards of the industrialized world as a whole – lower public spending on services, low levels of worker unionization, etc. (Rose, 1989; see Chapter 3 above) – all became standards of excellence in economic performance (Davis, 1985). The long-term consequences for the American economy, however, have been nothing short of disastrous. In the first place, the tax cuts did not produce a bonanza of capital investment in new plant or equipment in America. The money went largely into consumption, often of foreign-made goods which thus increased the trade deficit, into a frenzy of mergers and acquisitions among large firms and, in a world-economy with minimal capital controls, into overseas investment (Friedman, 1988, 233–70).

Second, the federal budget deficit exploded from $59.6 billion in 1980 to $202.8 billion in 1985. In lieu of the taxes forgone after 1981, the massive increases in military spending had to be financed by borrowing. Cuts on the expenditure side of the federal budget were insufficient to make up the difference. Because the US savings rate was so low, most of the borrowing had to be done overseas (Friedman, 1988, 209–32). Added to the trade deficit, this produced an explosion in the US, current-account imbalance (Figure 5.3). Whether this matters as much as it did in the past is now open to question. To one school of thought, as long as financing can be found, the nationality of its origin does not much matter; with variable exchange rates and few restrictions on capital mobility there is no longer the old problem of running down foreign-exchange reserves. However, trade and current-account balances do give a reasonable picture of the *relative* state of a particular national economy and are still used by governments and private investors as indicators of national economic health. They are not just 'tricks' of the economist's 'trade', even if with globalization they do not have the same significance they had when the world was less integrated economically (*Economist*, 30 March 1991, 61).

Third, only those groups and regions in the US, favoured by the tax cuts, military spending and deregulation have benefited from Reaganomics (Anderton and Isard, 1985; Corbridge and Agnew, 1991). Indeed, there was a huge redistribution of income in the United States in the 1980s in favour of already wealthier groups and regions (Phillips, 1990). Military spending pumped huge sums into Southern California, New England and Washington State. Other regions, such as the Mid-West, received relatively little. As a result of these and other government decisions, the US is now more socially and geographically polarized than it has been at any time since the 1930s (Figure 5.4).

The US trade and federal budget deficits could well be justified if they had given rise to a fundamental upward shift in American productivity. But the evidence is that they have not. Average annual rates of growth in GNP in the 1980s as a whole were no higher than those of the 1970s (Friedman, 1988, 187–208). They also pose dangers to the world-economy. To meet the interest payments on its current budget debt, the US will need to run a trade *surplus* in the 1990s that is equivalent to about 6 per cent of total world exports. How this can be done without undermining the delicate balance of the world trading system is far from clear. The most vital part of any reorientation will require that Japan and Germany in particular run trade deficits of a magnitude corresponding to the necessary US surplus (Friedman, 1988, 53–4; Corbridge and Agnew, 1991). With the erosion of the Soviet threat that has hitherto justified both the militarization strategy and European and Japanese subordination to it, there might be increasing doubt that those countries would do so, save for the fact of unprecedented global interdependence in which their fates and that of America are increasingly bound together. They may no longer be free to choose.

The new world order?

It has become a commonplace to observe that nation-states are less and less 'full societies' (Williams, 1983). They are at one and the same time too large and too small for a wide range of social and economic purposes. They are often too large to allow for full social identities and real, as opposed to artificial, national economic interests. This can be observed in the economic and political divisions between regions in the United States, often only papered over by the declaration of common 'national' interests at stake in some far-off corner of the globe. It can also be

seen in the claims to nationhood flowering among ethnic and linguistic groups in a wide range of states. But existing states are for many economic purposes also too small. They are increasingly 'market sectors' in an intensely competitive, integrated and unstable world-economy. There is, therefore, a profound contradiction between on the one hand the political claims and military pretensions of large states, and on the other, the economic realities they now face.

Yet in writing about international relations and political economy there has been a tendency to ignore this dual process of fragmentation and globalization. That *national* hegemony has been, perhaps, a temporary phenomenon of the nineteenth and twentieth centuries has not received much attention (Grunberg, 1990). Most images of world order retain a focus on territorial states and an assumption of 'dead space' around and within them that prevents the possibility of even seeing alternative spatial structures to the distribution of power. Rosecrance (1986, 67), a writer who does see alternatives, offers one plausible argument why this is so rare: 'One of the difficulties of most international theory is that it has been analytical rather than historical in character: it has been deterministic rather than contingent. Models have been offered that described one historical age in theoretical terms but failed to

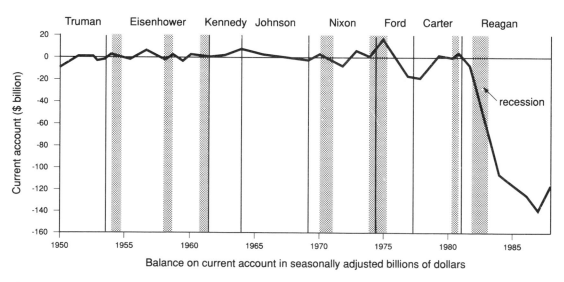

Figure 5.3 *US international accounts, 1950–88.*

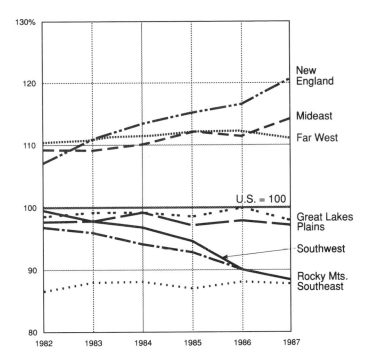

Figure 5.4 *Regional per capita incomes as percent of US average, 1982–7.*

account for others. The dynamics of historical development has in this way defeated any purely monistic approach.'

Globalization

If British hegemony in the nineteenth century made trade more free and interdependent, American hegemony has gone a step further in promoting the transnational movement of all the factors of production: capital, labour and technology. Free trade could always be limited as long as production was organized on a largely national basis. The possibility of a successful self-contained or autarkic national economy, however, has been reduced by the emergence of a world in which production as well as trade moves easily across national boundaries. This means that hegemony itself loses any clear national identity. Economic power, the material basis to military and political power, is no longer a simple attribute of states that have more or less of it.

What is the evidence for this view? First, since the late 1940s there has been an expansion of

world trade at rates well in excess of those of the nineteenth and early twentieth centuries. Between 1948 and 1980, the volume of world trade doubled about every eleven years (Table 5.2). In the worst period the average rate of expansion fell to 4.6 per cent, above the average rate of growth for the entire period 1840–1913 (4.2 per cent) (Rogowski, 1989, 88). The expansion of trade has been facilitated above all by massive decreases in the cost of transportation (between 1930 and 1990 the cost of sending cargo by sea halved in 1990 dollars (*Economist*, 20 July 1991, 117)), and by institutional innovations such as GATT and the European Community.

Different world regions have shared unequally in the post-World War II growth of trade. Expansion of trade in manufactures has been greater than expansion of primary products. By and large the more industrialized economies have experienced the greatest growth in trade. The heretofore isolated and insulated American economy has been a major participant in this as a node in a finely balanced *system* of trade. In the 1970s the main circuits of this system were as follows: (1) a US manufacturing trade deficit

Table 5.2 *World trade, 1948–80*

Year	Volume index[a]	Annualized growth (per cent)
1948	103	–
1953	142	6.63
1958	187	5.66
1963	269	7.40
1968	407	8.63
1970	490	9.72
1975	613 (640)	4.58 (5.49)
1980	817 (875)	5.92 (6.45)

[a] 1913 = 100.

Sources: 1948–1970, Rostow 1978, 67 and 669; values for 1975 and 1980 are extrapolated on the basis of United Nations 1984, 1225–26, and are not fully comparable since they cover only 'market economiest. United Nations 1986, 35, provides values, in current dollars, for *total* world trade in 1975 and 1980; deflating these by price indexes for US imports (Organization for Economic Co-operation and Development 1982, 76) yields the indexes and growth rates shown in parentheses, which may comport more closely with the earlier entries. See Rogowski (1989).

with East Asia (cars, electronics, clothing) offset by a surplus with Europe (computers, aerospace); (2) a net US trade surplus with Latin America balanced by an outflow of US direct investment, especially to Brazil and Mexico; (3) European manufacturing deficits with the US and East Asia balanced by surpluses with Africa; (4) Japan's surpluses with Europe and the US balanced by payments for energy and raw materials and the export of capital to East Asia and elsewhere (Davis, 1985).

In the early 1980s this system destabilized because of the debt crisis and an overexpansion of the US trade deficit with East Asia. But within this system the long-term growth of one party is determined by the growth of others. Hence there is an incentive to resolve trade conflicts and stimulate depressed circuits. By the early 1990s the US–East Asia link was rebalancing partly because of Japanese response to American political pressure, but also because of Japanese foreign direct investment (FDI) in the United States sub-

stituting products made in the US for ones previously imported (*Economist*, 8 June 1991, 67). In a world of large-scale trade there is a premium placed upon maintaining openness and balance rather than territorial expansion and military superiority (Rosecrance, 1986).

Second, American transnational firms have been major agents in stimulating a more open world-economy (Brett, 1985; Agnew, 1987). For example, about 30 per cent of the US trade deficit with East Asia (including Japan) is due to American firms' making goods there and selling them in the United States. Of the four largest exporters from Taiwan, one is RCA and another is AT&T: the largest exporter of computers from Japan is IBM. In 1984 US firms accounted for the same proportion of world exports (18 per cent) that they did in 1966. But at the same time the US territorial economy's share of world exports shrank by a quarter. About half the total exports of nominally American firms originates outside of the United States (Lipsey and Kravis, 1987).

Even small American firms have been 'going global'. As a first step they usually rely on joint ventures, partnerships and licensing agreements. The expansion abroad is often reluctant, designed to diversify markets and gain access to knowledge of foreign demand rather than to conquer global market share. But it illustrates the general pressure to 'compete internationally to be successful' (Uchitelle, 1989a), irrespective of firm size.

In the 1980s American firms were joined by others at an increasing rate. In 1985 European firms accounted for nearly 50 per cent of total world foreign direct investment (FDI), most of this flowing into the United States and other European countries. With the exception of flows into off-shore financial centres, such as Hong Kong or the Cayman Islands, there has been a rapid diminution of American and European FDI in developing countries (Balcet and Monforte, 1991). Overall, however, whereas in the past rates of growth of trade and FDI were comparable, between 1983 and 1989 FDI by transnational companies grew three times faster than world trade (Figure 5.5). Four-fifths of the total FDI came from the US, Europe and Japan (*Economist*, 27 July 1991, 58).

Third, even the relatively protectionist

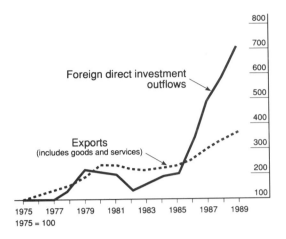

800
700
600
500
400
300
200
100

Foreign direct investment outflows

Exports
(includes goods and services)

1975   1977   1979   1981   1983   1985   1987   1989
1975 = 100

Figure 5.5 *The growth of foreign direct investment relative to the growth of exports, 1975–89.*

Japanese economy, sometimes touted as a successor to the US as a national hegemon, is increasingly internationalized and subject to stresses generated abroad (Higashi and Lauter, 1987). In 1987 Japanese foreign assets totalled close to $1 trillion compared to the $1.1 trillion of the United States (Uchitelle, 1988). Much of this is now FDI rather than bonds or portfolio investment (Thrift and Leyshon, 1988). The increased openness of the Japanese economy is also evident in the growing role of the yen in international transactions; in the importance of Japanese banks in global finance; and in the ties being forged by Japanese business with the economies of the European Community, the United States, East Asia, the Pacific Rim, and the Soviet Union (Fujita and Ishigaki, 1985; De Cecco, 1987; *Economist*, 13 July 1991, 83).

Fourth, there is an increasingly globalized world financial system (Wachtel, 1986; Strange, 1986; Thrift, 1989). The demands of institutional investors, such as pension funds and insurance companies, for more diversified portfolios has led to a transnationalization of many stock markets. Deregulation in the 1980s, particularly in London and New York, encouraged the development of new 'products', such as 'junk bonds'. To serve their worldwide clienteles, many financial markets now operate around the clock. These markets operate with a mix of currencies;

the dollar now shares the spotlight with the D-mark, the yen and the European Currency Unit. These trends have encouraged interdependence between national financial systems. They have also redefined the space for national sovereignty with regard to macro-economic management. In the 1980s fear of capital flight constrained governments to tight counter-inflationary policies. The absence of much international financial regulation has also increased the vulnerability of national banking systems to the looser practices and often outright criminality of banks operating outside regulated national channels (as, for example, in the case of the Bank of Credit and Commerce International which came to international attention in 1991 as a major conduit for the funds of terrorists and laundry for the cash of drug dealers).

For the United States, the shrinking role of the dollar poses a particularly difficult problem. In the 1980s manipulating the value of the dollar constituted the main policy instrument of successive governments in trying to stimulate exports and discourage imports (Agnew, 1987; Parboni, 1987). But this can only work if the dollar is the dominant metric of world trade and there is no alternative to it. In 1976 80 per cent of world foreign exchange reserves were held in US dollars. By 1990 this figure had shrunk to 51 per cent. The D-mark, yen, and other currencies then accounted for almost half of world reserves and, in the first half of 1991, for 69 per cent of international bond issues (*Economist*, 13 July 1991, 83).

Fifth, and finally, the transnationalization of production and exchange has produced various institutions and new social groups involved in managing the world-economy. The IMF and the World Bank have existed since the 1940s. Some, such as the annual summits of government leaders from the major industrialized countries, have encouraged multilateral macro-economic policy co-ordination, though the United States is still the dominant actor in them (Putnam and Bayne, 1987; Webb, 1991). Others, such as the Trilateral Commission, are 'unofficial' organizations that attempt to build an internationalist consensus among powerful businessmen, influential journalists and academics from the United States, Europe and Japan (Gill, 1990). In

231

a more sociological vein some commentators have noted the growth of a 'transnational bourgeoisie' or class of the managerial employees of transnational firms whose loyalties are as much to the firms for whom they work as to the nations from which they come (e.g., Sklar, 1976). Whether such nascent groups will become important actors in the world-economy and adopt an ideology of 'transnational liberalism', however, will depend in large part upon the successful activities of transnational organizations such as the Trilateral Commission in maintaining and legitimizing an open world-economy.

None of this evidence for globalization should be confused with an endorsement of it (a confusion apparent, for example, in Gordon, 1988). Neither should the global economy be seen as an inherently stable or an irreversible system. In the early 1990s, for example, its stability is threatened by a global capital shortage brought on by an explosion of demand, especially from Eastern Europe and the Soviet Union, in the face of a shrinking supply engendered by the US federal budget deficit and the collapse of the Japanese stock market (Senner, 1991). With respect to reversibility, at the turn of the last century the world-economy was in some ways more integrated than it is today. Take, for example, the world's capital markets. Over the period 1880–1913, investment and saving (both as a percentage of GNP) were much less correlated for a range of countries than over the period 1965–86. This suggests that investment in recent years has depended more on domestic saving, and therefore that capital is relatively more immobile now than it was ninety years ago – perhaps because of increased exchange-rate volatility in the absence of the gold standard (as prevailed at the turn of the century) or of a 'semi-fixed' monetary system such as Bretton Woods (Bayoumi, 1989). Examining the ratio of total trade to GNP is also instructive. Only in the 1970s did US and British ratios return to the levels experienced before 1913. World trade declined dramatically after World War I (Krugman, 1990, 195). It could well do so again if global markets fractured into closed regional trading blocs such as North America (the US, Canada, Mexico), the European Community and Japan–East Asia.

Regional trading blocs will tend to reduce world trade to the extent that as their 'domestic' markets increase in size, external tariffs will be increased to take advantage of the larger size in improved terms of trade (Krugman, 1991). If global trade and investment do collapse, then American global hegemony will be well and truly dead rather than, as it is now, 'ghosting' transnationally.

Fragmentation

Before the acceleration of globalization in the late 1960s and 1970s and under the rigid military blocs of the Cold War, improvements in incomes, regional economic policy, and state repression produced a world order in which state and society were mutually defining. However, an increasingly uneven process of economic development, the retreat from regional economic policy in the face of pressures to increase 'national' competitiveness in an integrated world-economy, and the decline of state socialism, have called this equation into question. Sectionalism, localism, regionalism and ethnic separatism have been on the rise all over the world.

There are, perhaps, two major aspects to this fragmentation: (1) a redefinition of economic interests from the national to regional or local scales; and (2) a questioning of political identity as a singularly national-state phenomenon. The first of these is a direct result of the breakdown of the national economy as the basic building-block of the world-economy. As Doreen Massey (1984, 295) summarizes her argument for an emerging spatial division of labour:

The old spatial division of labour based on sector, on contrasts between industries, has gone into accelerated decline and in its place has arisen to dominance a spatial division of labour in which a more important component is the interregional spatial structuring of production within individual industries. Relations between economic activity in different parts of the country [Britain] are now a function rather less of market relations between firms and rather more of planned relations within them.

This describes the breakdown of a long-established pattern of economic development based on regional-sectoral specialization (cars in Detroit, steel in Pittsburgh, etc.), and its replacement by a decentralization of production facilities to multiple locations.

The geography of this process can be seen in the deindustrialization of many older industrial areas and in the polarization between cities and their suburbs. The economics of it lie in the pressures firms face under globalization. Increased openness to foreign competition produces an enhanced concern with local conditions for doing business, such as labour skills and costs, taxes, financial services, availability of sub-contractors and research facilities. Lowered transportation costs and new information technologies have made the choice of decentralized operations a feasible solution (Agnew, 1988).

Economic restructuring of this kind has had important consequences in defining economic interests. Above all it ties local areas directly into global markets. They are 'communities of fate' in which local actions count much more than national economic policies geared towards either some fictive national 'average' or the interests of a particularly well-represented region. This explains the recourse of many local governments in the United States and elsewhere to their own local economic development policies (Logan and Molotoch, 1987).

The second aspect of fragmentation has also been encouraged by the crumbling of national economies, but relates more to the emergence of new political identities often based upon old ethnic divisions. Cases of secession from established states have been relatively rare in recent world history (Diehl and Goertz, 1991). But the last twenty years have seen a proliferation of political movements with secessionist or autonomist objectives (Williams, 1989). In Western Europe this trend can be related to the growing redundancy of national governments as the new supranational institutions of the European Community take shape, as much as to an increase in regionally-defined relative deprivation. In Eastern Europe and the Soviet Union the assertion of ethnic identities probably has more to do with the demise of strong central governments, the exhaustion of state socialism or communism as an ideology that transcended old divisions by incorporating ethnic elites, and the settling of old

political scores. If successful, political fragmentation of this type is much more threatening to the transnational liberal order than the new economic fragmentation or localization. Statehood must be justified economically; a national interest must be defined and defended against the danger of foreign domination. The ultimate irony for American hegemony would be a shattered Second World (Eastern Europe and Soviet Union) engaging in a competitive protectionism that spreads elsewhere and undermines the liberal world order the Cold War was fought to create.

America's impasse

The United States is itself caught between these two processes of globalization and fragmentation. The policy levers available in the past, such as military spending and national macroeconomic and tariff measures, are no longer appropriate to the problems at hand. Military spending, for example, can help some local economies, but it penalizes the competitiveness of the *national* economy. A world largely made by American design and energy can no longer be relied on to deliver growth to America as a whole. America is becoming just another country.

There is controversy over the extent to which the American economy and the economic welfare of Americans have declined. During the 1980s, the United States consistently accounted for around 40 per cent of the aggregate GDP of the OECD (the major industrialized countries). This was a 10 per cent decline from the 1960s, but the American economy was still two-and-a-half times the size of the second largest economy, that of Japan (Webb, 1991, 341). However, this kind of arithmetic masks important qualitative changes in the relative autonomy of the American economy within the world-economy in the post-war era as a whole. For example, Rupert and Rapkin (1985) have shown that US shares of global resources and capabilities (measured by the ratio of US GNP to global GNP) have declined in a more or less linear fashion since 1950, but that an indicator of interdependence or susceptibility to 'external shocks'

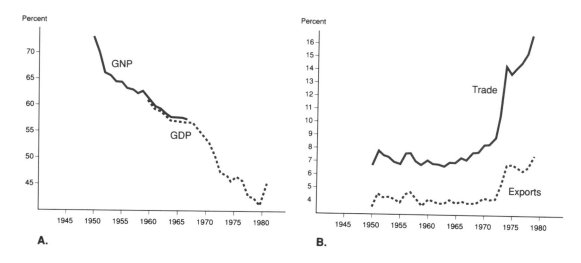

Figure 5.6 *Ratios of (A) US to total core GNP (1950–67) and GDP (1960–81) and (B) US trade and exports to GDP (1950–79).*

(the ratio of US trade to the GNP) has increased dramatically since 1970 (Figure 5.6). They borrow the phrase 'scissors effect' from Bergsten to describe the joint consequences of a declining capability share and increased interdependence. According to Bergsten (1982, 13): 'The United States has simultaneously become much more dependent on the world-economy and much less able to dictate the course of international events. The global economic environment is more critical for the United States and is less susceptible to its influence.'

I have provided a detailed analysis elsewhere of how capability and autonomy declined (Agnew, 1987). With respect to capability, the main problem lies in the decline of GNP potential or the total capacity to produce. This is partly a function of a changing labour force (more dependants, lower skills), but it is largely a product of declining national productivity. In 1990 this was indexed at 2.5, compared to 4.3 in 1965. As a consequence, real GNP growth has slumped from 4.3 per cent in the 1940s through 3.2 in the 1950s, 2.9 in the 1960s, 2.8 in the 1970s to 2.8 in the 1980s. One leading economic forecaster sees US GNP growth heading below 2.0 per cent per annum in the 1990s (Straszheim 1991).

Recent trends underscore the decline in relative autonomy. The deficits of the 1980s have

made the US the world's largest debtor and Japan its largest creditor. Above all, the increased interdependence and velocity of the world financial system affect the United States as much as any other country. Capital has become much more mobile in both time and space. Before 1973, currency exchange rates changed once every four years on average, interest rates moved twice a year, and companies made price and investment decisions no more than once or twice a year. This has all changed. Exchange rates now change several times a day, interest rates move weekly, and firms make price and investment decisions monthly or quarterly.

The most important information concerning the condition of the United States, however, involves the economic welfare of Americans. Median household income peaked in 1974. Since then the major feature of American incomes has not been their growth, but their redistribution from poor and middle class to rich. The pace of this shift was especially marked in the mid-1980s with the decline in the number of well-paying manufacturing jobs, an increase in part-time and temporary employment, and tax cuts that disproportionately benefited the affluent. An associated trend has been the re-emergence of interregional disparities and the deepening of central city–suburban income inequalities (Phillips, 1990). A

234

resurgence of protectionism and hostility to liberal internationalism in the areas most affected by job loss and economic decline (in particular, the old Northeast–Midwest Manufacturing Belt) is one not surprising consequence (Trubovitz, 1992).

None of this should be read as indicating an imminent political-economic apocalypse for the United States. The world-economy has changed so much that there is little possibility of the United States falling victim to a single competitor nation-state. Even if this were not so, there is no one other state in the 1990s with both competitive military and economic assets. Technological innovation, in the past often closely associated with a particular country, is now less 'contained' geographically (Thompson, 1990, 232). The development of new technologies is increasingly firm, rather than country-specific. Firms 'carry' them with them when they locate their facilities. Most importantly, a country so large, rich, and welcoming of capital and immigrants as the United States is assured of a key role in the developing global economy. The Gulf War of 1991 demonstrated the continuing military role of the US, even as allies were asked to pay for it. Some of America's recent economic problems, the deficits for example, can be traced in part to domestic policy choices, such as those detailed by Stockman (1986) and Greider (1987), rather than to foreign competition. In sum, the US territorial economy is at the very least still *primus inter pares* among the world's national economies (Strange, 1987; Parboni, 1987).

The undoubted success of American hegemony and its continuing power in the form of transnational liberalism, however, are increasingly challenged by those who see them as hollow victories. From one point of view, what really matters is whether the US territorial economy can provide rising living standards through enhancing the strength and competitiveness of the economic activities located within its borders irrespective of the nationality of ownership (e.g. Reich, 1991a). In this perspective, 'Money is unpatriotic; these days, investment dollars are speeding to wherever on earth they can get the higest return. People, however, are relatively immobile, and they belong to societies with particular cultures and histories and hopes.

It is up to governments to represent people, to respond to their needs and fulfill their hopes – not to represent global money. (Reich, 1991b, 53).

From another point of view, what really matters is that the United States restore its status as Number One by substituting 'geoeconomic' conflict with states that do not trade 'fairly' (usually Japan) for the geopolitical conflict of the Cold War (e.g. Luttwak, 1990). 'Japan-bashing' has become a major growth industry among some journalists and academics. Titles such as *The Coming War with Japan* guarantee high sales (see Fallows, 1991). Here, the major issues involve the relative aggregate economic gain of Japan (or another economic challenger) over the United States, and America's downward mobility within the world-economy (see Mastanduno, 1991). Except for some interest in the increased dependence of the US on foreign suppliers of weapon components (see Moran, 1990), there is little or no attention given to the changed nature of the world-economy.

Given the lack of 'decisional power' within the American system of government, a throwback to the Madisonian formula for a divided government of checks and balances (Cerny, 1989), and the power of an entrenched internationalist business establishment, neither viewpoint is likely to prevail. This is doubly so when national stagnation or low growth produces conflict over relative group or regional gains, rather than a national consensus over an 'internationalist' policy such as characterized the country during the years of high national growth from the 1940s to the 1960s. In these more straitened circumstances, 'muddling through' will merely highlight the difficulties now involved in making any *national* choice.

In the past there was no such dilemma. America's national economic health and its global status were mutually reinforcing. Today, globalization and fragmentation have cut the connection between them. Serving one no longer guarantees the other. As the country faces a new century, this is America's impasse.

References

Agnew, J.A. (1987). *The United States in the World Economy*. Cambridge: Cambridge University Press.

Agnew, J.A. (1988). 'Beyond core and periphery: the myth of political–economic restructuring and sectionalism in contemporary American politics'. *Political Geography Quarterly*, 7: 127–39.

Anderton, C. H. and Isard, W. (1985). 'The geography of arms manufacture', pp. 90–104 in D. Pepper and A. Jenkins (eds.) *The Geography of Peace and War*. Oxford: Blackwell.

Arrighi, G. (1990). 'The three hegemonies of historical capitalism'. *Review*, 13: 365–408.

Art, R. J. (1991). 'A defensible defense: America's grand strategy after the Cold War'. *International Security*, 15: 5–53.

Balcet, G. and Monforte, S. (1991). 'Investimenti diretti e multinazionali europe'. *Relazioni Internazionali*, 15: 70–9.

Bayoumi, T. (1989). *Saving–Investment Correlations*. Washington, D.C.: IMF Working Paper 89/66.

Becker, W. H. and Wells, S. F. (eds.) (1984). *Economics and World Power: An Assessment of American Diplomacy since 1789*. New York: Columbia University Press.

Bergsten, C. F. (1982). 'The United States and the world economy'. *Annals of the American Academy of Political and Social Science*, 460: 11–20.

Brett, E. (1985) *The World Economy Since the War: The Politics of Uneven Development*. London: Macmillan.

Burleigh, M. (1988) *Germany Turns Eastward: A Study of Ostforschung in the Third Reich*. Cambridge: Cambridge University Press.

Calleo, D. P. (1982). *The Imperious Economy*. Cambridge, Mass.: Harvard University Press.

Calleo, D. P. and Strange, S. (1984). 'Money and world politics', pp. 91–125 in S. Strange (ed.) *Paths to International Political Economy*. London: Allen and Unwin.

Cerny, P. G. (1989). 'Political entropy and American decline'. *Millennium*, 18: 47–63.

Chandler, A. D. (1980). 'The growth of the transnational industrial firm in the United States and the United Kingdom: a comparative analysis'. *Economic History Review*, 33: 396–410.

Chernow, R. (1990). *The House of Morgan: An American Banking Dynasty and the Rise of Modern Finance*. New York: Atlantic Press.

Cockburn, A. (1983). *The Threat: Inside the Soviet Military Machine*. New York: Random House.

Corbridge, S. and Agnew, J. (1991). 'The US trade and budget deficits in global perspective: an essay in geopolitical-economy'. *Society and Space*, 9: 71–90.

Dalby, S. (1989). 'Geopolitical discourse: the Soviet Union as other'. *Alternatives*, 13: 415–42.

Dalby, S. (1990). *Creating the Second Cold War*, London: Pinter.

Dallek, R. (1983). *The American Style of Foreign Policy: Cultural Politics and Foreign Affairs*. New York: Mentor.

Davis, M. (1985). 'Reaganomics magical mystery tour'. *New Left Review*, 149: 45–65.

De Cecco, M. (1987). 'Le relazioni finanziarie tra internazionalismo e transnazionalismo', pp. 17–30 in A. Graziani (ed.) *Il dollaro e l'economia italiana*. Bologna: Il Mulino.

Dennett, R. and Turner, R. K. (eds.) (1948) *Documents on American Foreign Relations*, IX (1947). Princeton, NJ: Princeton University Press.

Diehl, P. F. and Goertz, G. (1991). 'Cambiamenti territoriali e conflitti del futuro'. *Relazioni Internazionali*, 15: 22–33.

Economist (30 March 1991). 'Tricks of the trade', 61.

Economist (27 April 1991). 'From Morgan's nose to Milken's wig', 8–12, Survey of International Finance.

Economist (8 June 1991). 'Vanishing red ink', 67.

Economist (13 July 1991). 'A fistful of ecus', 83.

Economist (20 July 1991). 'Going cheap', 117.

Economist (27 July 1991). 'Globetrotting', 58.

Edmonds, R. (1983). *Soviet Foreign Policy: The Brezhnev Years*. New York: Oxford University Press.

Fallows, J. (1991). 'Is Japan the enemy?' *New York Review of Books*, 30 May: 31–7.

Friedman, B. (1988). *Day of Reckoning: The Consequences of American Economic Policy*. New York: Random House.

Fujita, M. and Ishigaki, K. (1985). 'The internationalisation of Japanese banking', in M. J. Taylor and N. J. Thrift (eds.) *Multinationals and the Restructuring of the World Economy*. London: Croom Helm.

Gallman, R. E. and Howle, E. S. (1971). 'Trends in the structure of the US economy since 1840', in R. W. Fogel and S. L. Engerman (eds.) *The Reinterpretation of American Economic History*. New York: Harper and Row.

Gill, S. (1990). *American Hegemony and the Trilateral Commission*. Cambridge: Cambridge University Press.

Gordon, D. (1988). 'The global economy: new edifice or crumbling foundations?' *New Left Review*, 168: 24–64.

Greider, W. (1987). *Secrets of the Temple: How the Federal Reserve Runs the Country*. New York:

Simon and Schuster.

Grunberg, I. (1990). 'Exploring the "myth" of hegemonic stability'. *International Organization*, 44: 431–77.

Hartz, L. (1955). *The Liberal Tradition in America*. New York: Harcourt Brace and World.

Hartz, L. (ed.) (1964). *The Founding of New Societies*. New York: Harcourt Brace.

Higashi, C. and Lauter, G. P. (1987). *The Internationalization of the Japanese Economy*. Boston: Kluwer.

Hogan, M. (1987). *The Marshall Plan: America, Britain, and the Reconstruction of Western Europe, 1947–52*. Cambridge: Cambridge University Press.

Ikenberry, G. J. and Kupchan, C. A. (1990). 'Socialization and hegemonic power'. *International Organization*, 44: 283–315.

Julius, D. (1990). *Global Companies and Public Policy: The Growing Challenge of Foreign Direct Investment*. New York: Council on Foreign Relations.

Kaldor, M. (1991). *The Imaginary War*. Oxford: Blackwell.

Krugman, P. R. (1990). *The Age of Diminished Expectations: US Economic Policy in the 1990s*. Cambridge, Mass.: MIT Press.

Krugman, P. R. (1991). 'Is bilateralism bad?', pp. 9–23 in E. Helpman and A. Razin (eds.) *International Trade and Trade Policy*. Cambridge, Mass.: MIT Press.

Lipsey, R. and Kravis, I. (1987). 'The competitiveness and comparative advantage of US multinationals, 1957–1984'. *Banca Nazionale del Lavoro Quarterly Review*, 161: 147–65.

Lively, R. A. (1955). 'The American system: a review article'. *Business History Review*, 29: 81–96.

Logan, J. and Molotch, H. (1987). *Urban Fortunes: Towards a Political Economy of Place*. Berkeley: University of California Press.

Luttwak, E. N. (1990). 'From geopolitics to geoeconomics'. *The National Interest*, 20: 17–24.

Maier, C. S. (1978). 'The politics of productivity: foundations of American international economic policy after World War II', pp. 23–49 in P. J. Katzenstein (ed.) *Between Power and Plenty*. Madison: University of Wisconsin Press.

Massey, D. (1984). *Spatial Divisions of Labour*. London: Methuen.

Mastanduno, M. (1991). 'Do relative gains matter? America's response to Japanese industrial policy'. *International Security*, 16: 73–113.

McCormick, T. J. (1989). *America's Half-Century: United States Foreign Policy in the Cold War*. Baltimore: Johns Hopkins University Press.

Meyer, J. W. (1982). 'Political structure and the world-economy'. *Contemporary Sociology*, 11: 263–6.

Milward, A. S. (1984). *The Reconstruction of Europe, 1945–51*. Berkeley: University of California Press.

Moore, B. (1966). *Social Origins of Dictatorship and Democracy*. Boston: Beacon Press.

Moran, T. H. (1990). 'The globalization of America's defense industries: managing the threat of foreign dependence'. *International Security*, 15: 57–99.

Nau, H. R. (1990). *The Myth of America's Decline: Leading the World Economy into the 1990s*. New York: Oxford University Press.

North, D. C. (1961) *The Economic Growth of the United States, 1790–1860*. Englewood Cliffs, NJ: Prentice-Hall.

O'Loughlin, J. (1989). 'World-power competition and local conflicts in the Third World', pp. 289–332 in R. J. Johnston and P. J. Taylor (eds.) *A World in Crisis? Geographical Perspectives*. Oxford: Blackwell.

O'Tuathail, G. and Agnew, J. A. (1992). 'Geopolitics and discourse'. *Political Geography* 11, 190–204.

Parboni, R. (1987). 'Il dollaro e l'economia mondiale', pp. 31–116 in A. Graziani (ed.) *Il dollaro e l'economia italiana*. Bologna: Il Mulino.

Phillips, K. (1990). *The Politics of Rich and Poor*. New York: Harper Collins.

Pletsch, C. E. (1981). 'The three worlds, or the social scientific division of labor, *circa* 1950–1975'. *Comparative Studies in Society and History*, 23: 565–90.

Putnam, R. D. and Bayne, N. (1987). *Hanging Together: Cooperation and Conflict in the Seven-Power Summits*. Cambridge, Mass.: Harvard University Press.

Reich, R. B. (1991a). *The Work of Nations: Preparing Ourselves for 21st-Century Capitalism*. New York: Knopf.

Reich, R. B. (1991b). 'Who do we think they are?' *The American Prospect*, 4: 49–53.

Rogowski, R. (1989). *Commerce and Coalitions: How Trade Affects Domestic Political Alignments*. Princeton: Princeton University Press.

Rose, R. (1989). 'How exceptional is the American political economy?' *Political Science Quarterly*, 104: 91–115.

Rosecrance, R. (1986). *The Rise of the Trading State: Commerce and Conquest in the Modern World*. New York: Basic Books.

Rosenberg, E. S. (1982). *Spreading the American Dream: American Economic and Cultural Expansion, 1890–1945*. New York: Hill and Wang.

Ruggie, J. G. (1983). 'International regimes, transactions and change: embedded liberalism in the postwar economic order', pp. 195–231 in S. D. Krasner (ed.) *International Regimes*. Ithaca, NY: Cornell University Press.

Rupert, M. E. (1990). 'Producing hegemony: state/society relations and the politics of productivity in the United States'. *International Studies Quarterly*, 34: 427–56.

Rupert, M. E. and Rapkin, D. P. (1985). 'The erosion of US leadership capabilities', in P. M. Johnson and W. R. Thompson (eds.) *Rhythms in Politics and Economics*. New York: Praeger.

Sanders, E. (1986) 'Industrial concentration, sectional competition, and antitrust politics in America, 1880–1980'. *Studies in American Political Development*, 1: 142–214.

Schell, J. (1989). *Observing the Nixon Years*. New York: Random House.

Senner, M. (1991) 'The grim shortage of global capital'. *New York Times*, Forum, 13 July: 13.

Sklar, R. L. (1976). 'Postimperialism: a class analysis of multinational corporate expansion'. *Comparative Politics*, 9: 75–92.

Stockman, D. (1986). *The Triumph of Politics: The Inside Story of the Reagan Revolution*. New York: Harper and Row.

Strange, S. (1986). *Casino Capitalism*. Oxford: Blackwell.

Strange, S. (1987). 'The persistent myth of lost hegemony'. *International Organization*, 41: 551–74.

Straszheim, D. H. (1991). 'Statement by Donald H. Straszheim, Chief Economist, Merrill Lynch & Co. Before the Joint Economic Committee, US Congress'. 26 July. Hearings on Midyear Economic Outlook.

Tarbell, I. (1936). *Nationalizing of Business, 1878–1898*. New York: Macmillan.

Taylor, P. J. (1990). *Britain and the Cold War: 1945 as Geopolitical Transition*. London: Pinter.

Thompson, W. R. (1990). 'Long waves, technological innovation, and relative decline'. *International Organization*, 44: 201–33.

Thrift, N. (1989) 'The geography of international economic disorder', pp. 16–78 in R. J. Johnston and P. J. Taylor (eds.) *A World in Crisis? Geographical Perspectives*. Oxford: Blackwell.

Thrift, N. and Leyshon, A. (1988). ' "The gambling propensity": banks, developing country debt exposures, and the new international financial system'. *Geoforum*, 19: 55–69.

Tucker, R. W. and Hendrickson, D. C. (1990). *Empire of Liberty: The Statecraft of Thomas Jefferson*. New York: Oxford University Press.

Trubovitz, P. (1993). 'Sectionalism and American foreign policy: the political geography of consensus and conflict'. *Political Geography Quarterly*, forthcoming.

Uchitelle, L. (1988). 'When the world lacks a leader'. *New York Times*, 31 January, Business Section, 1–6.

Uchitelle, L. (1989a). 'Small companies going global'. *New York Times*, 29 November, Business Section, 1–5.

Uchitelle, L. (1989b). 'US, businesses loosen link to mother country'. *New York Times*, 21 May, 1.

Watchtel, H. (1986). *The Money Mandarins: The Making of a Supranational Economic Order*. New York: Pantheon.

Webb, M. C. (1991). 'International economic structures, government interests, and international coordination of macroeconomic adjustment policies'. *International Organization*, 45: 309–42.

Wilkins, M. (1970). *The Emergence of Multinational Enterprise: American Business Abroad from the Colonial Era to 1914*. Cambridge, Mass.: Harvard University Press.

Williams, C. H. (1989). 'The question of national congruence', pp. 229–65 in R. J. Johnston and P. J. Taylor (eds.) *A World in Crisis? Geographical Perspectives*. Oxford: Blackwell.

Williams, R. (1983). *The Year 2000: A Radical Look at the Future and What we can Do to Change It*. New York: Pantheon.

Wolfe, A. (1981). *America's Impasse: The Rise and Fall of the Politics of Growth*. Boston: South End Press.

Epilogue

Fin de siècle geopolitics: towards a geographical dialogue

Nationalism versus 'world society': a view from Russia

Vladimir Kolossov

On the eve of the third millennium, history speeds up its pace. The victory of the western coalition in the Gulf War, the collapse of totalitarian regimes in Eastern Europe and the reunification of Germany gave rise to a certain euphoria among a large part of western public opinion, which interpreted the events of the last three years as the triumph of liberal democracy and the new international world order. But these estimations are too optimistic and based on errors of Eurocentrism.

In fact, the world centre won the battle of values. Such basic western values as human rights, the civic society, the legal state, individual and social mobility, cosmopolitanism and consumerism became truly universal and widespread across the globe. The formation of the world-economy with its global communications networks and accelerating innovation rates proved that under certain conditions the wider inclusion of national economies in the system of international productive and financial relations contributed to the rise of well-being of all the workers, thus removing the grounds of the ideological struggle against multinational capital. At the same time, the strengthening of the world-economy made senseless any attempt to achieve economic and political aims by military means.

In long-term plans the semi-periphery countries have no alternative to integration in global economic and political systems. Recent years have revealed a number of reassuring signs of 'world-society' creation. It needs three principal elements: economic integration, the formation of a global cultural community, and a political identity. Among these signs were the unprecedented unity of the world community in face of Iraqi aggression against Kuwait; the *rapprochement* between the West and the former socialist countries, and the readiness of the core countries to help them in solving hard problems of economic restructuring; attempts to renegotiate Third World indebtedness; the increasing number of multilateral regional and global environmental agreements, including 'the green swap'; widening international solidarity in the counteroffensive to such global threats as AIDS, nuclear disasters, the weapons trade, drug trafficking, etc. (Lévy, 1990; Durand, Lévy and Retaillé, 1991; Brown, 1989–91).

But the global core–periphery structure interacts with cultural patterns in rather contradictory ways. Global economic actors need facilities for spatial manoeuvre and substantial socio-spatial differentiation, but at the same time they create a uniform economic and social space. In destroying local social structures of the periphery and semi-periphery, they can cause additional tensions and destabilization. New cultural and social elements born by internationalization

239

come into conflict with forces claiming to pre-serve traditional values. The notion of social pro-gress varies in different parts of the world, and standards imposed by core countries are rejected or they seriously destabilize the political situ-ation, for instance, in replacing authoritarian rule by competition between political parties on issues that are not clear for the people. That is why the core countries' attempts to evaluate regimes in the Third World which are too auth-oritarian and do not correspond to modern stan-dards could be destabilizing and, from the long-term point of view, wrong.

Western political culture is very often in oppo-sition to cultures of the world beyond the core according to four key dimensions: it is individua-list, and not collectivist; it is pluralist, decentra-lized, and without sharp social stratification; it requires readiness for uncertainty, and not the will to avoid it; and it is 'low-masculine' (in accepting more or less wide participation of women in social and political life), not 'high-masculine' like in the major part of the Third World (Berry, 1989). These oppositions result in three typical political cleavages in semi-periphery countries: (1) between 'occidentalists' sharing values of western liberal democracy and human rights, and 'traditionalists' looking for a specific development route based on national grounds or 'fundamentalists' aspiring to self-isolation; (2) between 'liberals' more or less attached to the principles of a market economy, and supporters of the state's wide involvement in all the spheres of economic and social life (Vanlaer, 1991; Kolossov, 1991); (3) between adherents of the unlimited application of people's self-determination principle, often interpreted as the most important element of human rights, and supporters of existing centralized states which in the major part of the world are usually multi-ethnic. Sometimes these cleavages temporarily overlap, as in the Soviet Union before and after the coup of 1991. Such a coincidence can give rise to strong mass social movements – progressive as well as conservative.

The rate of change in the semi-periphery largely exceeds the ability of the core to integrate culturally and economically this part of the former 'Second World' and the Third World. It evokes a new wave of nationalism challenging

liberal democracy. It seems that at least in the present decade, *until 2000*, the interactions be-tween these two powerful ideological streams will determine the world geopolitical order. The failure of communism leads to a loss of identity, not only in multinational East European states, but also in a large number of socio-political movements and governments in developing countries which saw in the communist ideology their *raison d'être*, a means to maintain stability and social cohesiveness under conditions where there are no objective economic and cultural premises for liberal democracy. Nationalism, in combination with religious orthodoxy, is often the only substitute for communist ideology. It confirms once more the naivety of develop-mentalism: liberal democracy is not a logical culmination of social evolution (Taylor, 1989, 1990).

The political evolution of nationalist move-ments in the semi-periphery usually follows two stages: first, the escalation of claims as far as they can be satisfied within existing structures; second, at a certain moment this escalation acquires its own logic and national leaders must follow it if they want to keep their positions. Thus, in a few months during 1990–91 the for-mer First Secretary of the Communist Party of the Ukraine, and actually head of the Ukrainian state, moved to the active support of the inde-pendence declaration in August 1991 and to the organization of the successful referendum in December 1991 for independence, though only in March 75 per cent of Ukrainian voters backed the union with the other republics. Even in Belgium, every next crisis in relations between Flemish and Walloon communities compels party leaders to take another step towards the federalization of the country not only of social security and agriculture regulations, but also of foreign policy. Some observers believe that, taking into account the accelerating rate of feder-alization, the time when there will be nothing left to federalize is not far off.

The geopolitical consequences of the explosion of nationalism are largely counter to the positive trends in world development. *First*, the realiza-tion of the nation-state idea by the republics of the former USSR and Yugoslavia have created a very dangerous precedent. The nation-state emerged in the core countries under certain cultural and historical conditions, but now the stability of state boundaries is the basis of the world geopolitical order, which was confirmed by the Organization of African Unity at the very beginning of its existence as well as by the Helsinki Act of 1965. Being aware of it, the existing states traditionally protected this order from secessionism (Knight, 1986). To realize this danger, it is enough to remember the examples of Nigeria with its 250 peoples, or India where ethnographers have counted about 800 ethnic groups. Theoretically, secessionist processes have no logical end. In the USSR no fewer than thirty-five peoples had no territorial autonomy; the major part of the existing territories are similar to Russian dolls, for almost every possible new unit will have a further compact ethnic minority after each potential partition. For instance, in secessionist Moldova, the Gagauzes claim their autonomy or secession, but in Gagauzia itself there is a region populated by Bulgarians who would like to have their territorial autonomy too, and so on.

Second each nationalist movement and each new nation-state inevitably is based on the 'them–us' dichotomy, and they all aspire to legitimate themselves by looking for their *raisons d'être*, their 'vital interests', their particular views of potential political and military coalitions, by strengthening the barrier functions of their territorial units – their boundary. This logic obliges them to create their own armies and defence industries, which in developing countries are often considered a basis of economic modernization. Thus, it complicates international relations patterns and aggravates boundary conflicts.

Third, nationalism, with its stress on collective identity, is in principle radically distinguished from liberal democracy which accentuates common human values and human rights. In semi-periphery countries, nationalism is tied to the formation of unbalanced political systems without a political 'centre'; it revives the traditional ethic of social and political promotion based on the loyalty of each person to national, tribal or clan leaders in exchange for their support; it cultivates the disbelief of people in the ability of an individual to influence politics, hence the suspect attitude towards all authorities. Nationalist politics usually usurps the role of representative

institutions and elections as a cover for the struggle between elite clans, as the means to legitimize power in the eyes of international public opinion. People in a lot of semi-periphery and periphery countries consider elections only as a test of loyalty, and are convinced that this procedure has nothing in common with the real struggle for power. So, in Uzbekistan there are three true 'parties': the cultural/tribal clans economically and politically controlling Tashkent, Samarkand and the Fergana valley. These political clans happily existed under the one-party communist system and will exist further, if necessary, under the labels of 'democratic', 'liberal' or 'muslim' and 'people's' parties. That is why it can happen that in just a few days after a peaceful election, mass bloody pogroms can shake the whole republic. That is why the Uzbek declaration of independence seems to be no more than simply an attempt by its leadership to avoid the export of counterrevolution from Russia. But in the West, even some political scientists seriously interpret the organization of a multiparty system in such a country as the victory of one more state for liberal democracy and the global cultural community.

Fourth nationalism and secessionism, which in periphery and semi-periphery countries usually start in relatively rich regions, mean the elimination of the territorial redistribution system: 'we' do not want pay for 'them'. In the USSR, the rise of nationalism was preceded by false speculations in the local mass media that each republic subsidized the remaining part of the country: there were no winners, only losers – all the more because it was very difficult to check these calculations owing to the artificial price system. Selfish nationalist ideology is hardly compatible with the creation of a global and macroregional political identity, a cultural community, economic integration and an awareness of global problems.

But populations of multi-ethnic semi-periphery countries which are now especially affected by nationalism have no historical experience of life in small isolated 'independent' states. The cost of it could be rather high, and these expenses are inevitable. Until 2000 and even later, the world probably will see the painful births of new nation-states or at least radical readjustment of inter-ethnic relations and state organization in Eastern Europe and the former USSR, India, the Middle East and, perhaps, in several regions in Africa – in the most advanced ones, such as the South African republic, and in the poorest ones like the Sahel countries. These processes will seriously slow down and could even push back the global movement to the 'world-society'. Moreover, the awareness of the population about global environmental and other problems and their direct impact on everyday life will still be too weak to counteract tendencies to separatism and nationalism and attempts at self-isolation. So notions about people's absolute rights to self-determination and excessive enthusiasm about the disintegration of empires are illusions.

To reduce the consequences of this 'disease' of nationalism, our civilization has to invent radical and quite new remedies. The old ones of national-territorial autonomy and federalism institutionalizing ethnic boundaries can be only symptomatic and palliative. Among 103 states with more than 5 million people, thirty-five may be considered to be multi-ethnic, but only seven states have adopted a federal organization, and in only four states (Czechoslovakia, India, Belgium and the former USSR) is the federal system related to national and linguistic boundaries (Glézer, Kolossov, Petrov and Smirnyagin, 1991). Their experience shows that the existence of national-territorial autonomies does not by any means exclude cultural and economic decline. The degree of people's social, cultural and spatial interpenetration and mixture is often so high that it is impossible to draw federal boundaries without creating new minorities and conflicts. People living alongside other peoples for centuries cannot have exclusive rights to their respective territories. As the experience of the former USSR, Yugoslavia, Belgium and the other multi-ethnic countries shows, national boundaries, once officially delimited and institutionalized, tend to acquire more and more dividing functions.

One of the possible means of avoiding extremities of nationalism is a return to ideas of non-territorial forms of cultural autonomy. It existed in the USSR until 1934 and was applied at different territorial scales which allowed (at least

in theory) spatially dispersed peoples to have a kind of cultural self-government. Similarly, so-called national councils at the central level were suggested before the political crisis in August 1991 (Zubov and Salmine, 1991).

The other possibility for making political-territorial organization at all scales more flexible in general is that there should be enough networks of political units that every citizen can realize different social and political identities: national, regional and cultural. At the same time, it is necessary to avoid rigid hierarchies of territorial governments. Each political unit should fit into various political units at a higher level in order to assure stability and flexibility in the whole system. Local and international *ad hoc* authorities at intermediate levels could be gradually organized; with time, existing territorial governments should delegate them additional functions. New forms of political-territorial organization could be based on the integration not only of areas, but also of networks and their key points, for instance, of large cities (Durand *et al.*, 1991): the social distance between the central cities of core countries and those of peripheral countries can be much less than that between the core and the periphery of a separate political unit. Radical, arbitrary reforms should be avoided at all the levels; it is much better to combine old and new territorial governments and to maintain the continuity of political boundaries, as has been done, for example, with boundaries in metropolitan areas in France and some other western countries.

From the second half of the next decade, dramatic consequences of global environmental degradation will begin to be more and more felt. A couple of large-scale environmental and/or technological disasters will lead to an awareness of the priority of common human values everywhere. Owing to the further internationalization of the world-economy and the gradual integration of former socialist semi-periphery countries into its system, along with large migration flows, the creation of new communications networks and the strengthening of the modern middle class, the conviction that the most radical separatist and isolationalist forms of nationalism do not help to achieve economic and political aims will become generally accepted. By 2025, the world will probably have a kind of a global government; the global geopolitical structure will include a number of large overlapping macroregional units connected by close economic, cultural, ecological and political interdependencies. For instance, in the northern hemisphere it could be the West European, East European and Mediterranean communities making up a larger All-European community; a North-American community might include the US, Canada, Mexico and some countries of the Caribbean; an Atlantic–Pacific community could emerge instead of NATO and the OECD; and a Eurasian community could replace the USSR and the CIS and perhaps include some of its neighbours. But the way to a new flexible organization will be difficult and twisty, abundant in lateral paths leading our civilization to self-destruction via environmental or social catastrophes.

References

Berry, B. J. L. (1989) Comparative geography of the global economy: cultures, corporations and the Nation-State, *Economic Geography*, 65, 1–18.

Brown, L. (ed.) (1989–91) *The State of the World*, New York, Norton.

Durand, M. F. Lévy, J. and Retaillé, D. (1991) *Le monde, espace et systèmes*, Paris.

Glézer, O., Kolossov, V. Petrov, N. and Smirnyagin, L. (1991) *Rossiiskaia Gazeta* (special issue), 28 March 1991 (in Russian), Paris, Presses de la Fondation Nationale des Sciences Politique.

Knight, D. B. (1988) Self-determination for indigenous peoples. In: *Nationalism and self-determination*. D. B. Knight and E. Kofman, eds; London, Croom Helm.

Kolossov, V. (1991) Electoral geography of the USSR, 1989–1991 (retrospective comparisons and theoretical issues). A paper for the conference of the IGU Committee on the World Political Map, Prague, 1991.

Lévy, J. (1990) Espace politique et changement sociale, *Espace Temps*, 43/44, 112–29.

Taylor, P. J. (1989) *Political Geography*, Second edition, London, Longman.

Taylor, P. J. (1990) Extending the world of political geography. In *Developments in Electoral Geography*, R. J. Johnson, F. M. Shelley & P. J. Taylor (eds.), London and New York, Routledge, 257–71.

Vanlaer, J. (1991) Les premières élections libres en Europe de l'Est: systèmes de partis et clivages régionaux, *Révue Géographique Belge*, 115.

Zubov, A. and Salmin, A. (1991) The Union treaty and the mechanism of the elaboration of the new national-territorial organization of the USSR, *Polis (Political Studies)*, 42–57 (in Russian).

From hegemony to co-operation: a view from Japan

Akihiko Takagi

According to Wallerstein's world-system analysis, we are experiencing a Kondratieff B phase economic recession, coinciding with the decline in the hegemony of the United States (Wallerstein, 1991). Extrapolating this cyclical model, we will enter a new growth phase and yet another hegemonic system will begin. Will this really occur and if so, what kind of system will it be? In the extrapolation, it is thought that Pax Japanica (or Pax Nipponica) may follow Pax Americana, as Japan is now undoubtedly a great economic power second only to the United States, having made a miraculous recovery after its defeat in World War II. The author maintains that Japan, in all likelihood, will not replace the United States as the next hegemonic power.

I state my reasons below and I refer to the creation of a co-operative world-system over a short and medium time-scale.

The hegemonic system and the post-hegemonic system

Japan, defeated in World War II, now proclaims itself to be a peaceful nation. Article 9 of the Japanese Constitution states its renunciation of war and forbids the maintenance of any armaments and the right of belligerency of the state. In reality, however, Japan has a self-defence force (Jieitai), but Japanese people are very sensitive to an increase in the defence budget. A bill calling for the dispatch of a Japanese contingent to participate in the United Nations Peace-Keeping Force in the Gulf War was submitted to the Diet, but it was not passed. This bill is still under deliberation and was carried over to a subsequent dietry session.

Neighbouring countries such as China and

Korea are also extremely sensitive to any enlargement of a Japanese force. The extent of this sensitive concern can be illustrated, for example, by the level of consternation in deciding what terms to use to describe Japan's World War II military strategy in Japanese high-school textbooks. Their attitude is understandable given that it is said now, based upon the well-used former metaphor to describe US/Japanese relations, that if Japan sneezes, its neighbouring countries catch a cold.

Since World War II Japanese diplomacy focused on co-operation with free nations, limiting itself to playing a role in Asia and adhering to the guidelines laid down by the UN. During the Cold War period, priority was given to co-operation with the US under the security pact between Japan and the US, and hence co-operation with free nations. However, this policy in itself would seem to contradict policy regarding its role in Asia. Due to Japan's increasing status in the world, it is understandable that other Asian countries are dissatisfied and regard Japan as a member of the advanced countries rather than an Asian counterpart. On the other hand, most Japanese find it difficult to come to terms with Japan being a great power, in the light of its diplomatic subordination to the US. Yet it would seem Japan is a greater power than it recognizes itself to be. Even trivial words or actions can produce far-reaching effects on neighbouring countries. They, in turn, fear Japan's remilitarization. However, Japan wishes to avoid any recurrence of the former 'bad dream of invasive war', and its only course would seem to be to observe its peace constitution as a peace-loving country.

However, the US will demand military expenditure by Japan corresponding to its economic power. Then what system will be desirable? The answer would seem to be the post-hegemonic system. Inoguchi (1987) maintains that a co-operative management system with mutual regulation of cost-sharing by countries whose economy is above a certain level, as opposed to a hegemonic intiative, can better achieve the long-term stability of the international economic order. She calls for an international order based on a policy of co-operation to suceed the hegemonic system. Evidence to support this construction and operation of this kind of post-hegemonic system can be shown by the partici-pation of various countries in summits and the formation of G7. In this case, the post-hegemonic system can be taken to mean a new world order maintained by a process of consortium and regulation, and could be termed 'Pax Consortis' or 'Pax Diplomatica'.

What efforts should be made to establish the post-hegemonic system? With the collapse of the geopolitical world order of the Cold War, the first problem to be highlighted will be the north and south conflict. An increase in the 'ethnic' problem will also be inevitable. This will lead to an increase in the role of the UN as it was originally established to deal with affairs after World War II, but as half a century has elapsed since then, it may be necessary to make some organizational reforms in future.

Short-term perspective to 2000

As Taylor (1991) states, the year 1989 was an epoch-making one. The democratization of Eastern European countries brought an end to the political system under Communist Party rule. The Berlin Wall, which symbolized the Cold War, was broken down. In December that year, a top-level conference of the US and the Soviet Union was held in Malta to put an end to the Cold War. The wave of democratization did not stop there, for it was followed by the break-up of the Soviet Communist Party and three Baltic States gaining their independence. These changes will be further accelerated with the market unification of the EC, which may enlarge by the addition of Eastern European countries and Russia as its members. However, it will take time to see the realization of the actual unification.

On the other hand, regimes indicative of the Cold War remain in Asia, especially in the Far East. Despite this, however, both Koreas, North and South, have been admitted to membership to the UN, gaining worldwide acknowledgement of their status both in name and in reality. Further, as movements to promote exchange have begun, the possibility of unification is for the first time seriously being considered. However, unification would seem beyond realization, at least before the year 2000. Radical democratization in China may well be realized in the future, depending on how the post-Deng regime develops by the year 2000. Hong Kong will be returned to China in 1997, and China will hope to promote the opening of its economy, centring on Hong Kong and its coastal regions. There are also signs for a solution to the Cambodian conflict.

If this is the case, strained political relations throughout Asia should be seen to ease in the 1990s. Given the continued economic growth of Japan and the Asian newly-industrialized countries, we can expect a formation of East and South East Asian economic blocs. With the advance of these trends, the political and military role of the US will be seen to decrease in this region thereafter. Therefore, tri-polarization of the world – United Europe, the US and East and South East Asia – will proceed slowly. There are some structural impediments between the US and Japan, so it could take a little more time for the formation of a Pacific Rim economic bloc. Possible outbreaks of local conflicts like the Gulf War will not disappear altogether, and consequently measures are needed to prevent and deal with these conflicts. The next decade will be a period for establishing these measures.

Medium-term perspective to 2025

By the year 2025, the world-system will have shifted to the post-hegemonic system. In such a system, the role of the UN in resolving various world problems will be important, so the necessary reforms of the UN should be completed by that time.

In the Pacific region, structural impediments will be removed gradually as the Asian bloc increases, and a Pan Pacific economic bloc will be formed. Thereafter, a multilateral system in which the US, United Europe and Japan are three poles will shift towards a two-pole system, consisting of the Atlantic bloc and the Pacific bloc. As Funabashi (1991) states in his discussion of 'geoeconomics' – a situation where international politics is prescribed by economic, technical and geographical conditions rather than geopolitical conditions – these major elements in international relations are becoming more and more important to this system.

In this way, while the world-system shifts from a tri-polar to bi-polar system, economic factors will strengthen interdependence between each bloc. Solutions to various problems by co-operative policies centring on the UN will occur and will give political shape to the post-hegemonic system. However, in the cyclical theory of hegemony, as in world-systems analysis, the emergence of the pax era of hegemonic power will not occur until the middle of the next century. Given this, the post-hegemonic system mentioned above could be understood as part of the process of hegemonic cycles.

References

Funabashi, Y. (1991): *Post the Cold War*, Iwanami-Shoten, Tokyo (in Japanese).

Inoguchi, K. (1987): *The Emerging Post-Hegemonic System: Choices for Japan*, Chikumashobo, Tokyo (in Japanese).

Taylor, P. J. (1991): A theory and practice of regions: the case of Europes, *Environment and Planning D: Society and Space*, 9, 183–95.

Wallerstein, I. (1991): *Geopolitics and Geoculture – Essays on the changing world system*, Cambridge, Cambridge University Press.

Demography and division: a view from the Middle East

Ghazi Falah

Introduction

At a time when democracy and liberalization has been sweeping the globe from Latin America to the former Soviet bloc, the Middle East with its Arab core has not remained stagnant. Similar trends have been occurring in most countries of the region (Hudson, 1991) with important implications for both inter-state and international world relationships. The recent Gulf crisis highlighted these trends; it cannot be regarded as a purely local or regional issue, but should provide a 'testing ground for the *rapprochement* between East and West as applied to north–south relations' (Sid-Ahmed, 1991, 16). Iraq's August 1990 invasion of Kuwait and the widespread western anger against Saddam Hussein should not obfuscate trends or divert emphases. Nor should the rise of Islamic forces (Sharabi, 1988) necessarily be viewed as a retreat from pluralism or incompatible with democracy. Political reforms today are on the agenda of key Arab states (The Editors, 1922, 3). Some authors may regard these reforms as a tactic to cope with economic and social crises, as in the reforms that followed political unrest in Algeria, Egypt, Jordan, Morocco, Sudan and Tunisia (Esposito and Piscatori, 1991, 429; Amawi, 1992, 26–9). Orientalist stereotyping (Hudson, 1991, 407) and traditional disengagement from the Middle East and Arabs (Said, 1991, 15) are likely to be the main factors influencing western writers to underestimate contemporary Middle Eastern changes.

The Middle East and the new world order

In August 1990, Saddam Hussein confronted both Islamic and Arab worlds with two dilemmas that went to the very heart of its masses and intelligentsia.

First, Iraq's invasion of Kuwait was aimed at obliterating a sovereign Arab country, thus violating the sacred goal of self-determination that all Arabs had struggled to 'earn' throughout the long decolonization process. The invasion not only 'wrest[ed] the right of self-determination of Kuwait from those to whom it naturally belongs – the people of Kuwait' (Khalidi, 1991, 13), but also undermined the remarkable achievement that had just crystallized in the mind of the Arab people of focusing on the *state* as a viable place freed of the burden of pan Arabism. From these perspectives, Saddam was not supported in his invasion of Kuwait.

The second dilemma was associated with spill-over effect of the invasion and the US-led response at it related to the long history of foreign interference in the region especially for the exploitation and control of the 'God-given gift to the Arabs' – oil. Because of such interference, the main feature of Arab society was the incongruence between the geography of wealth and the population distribution. The distributive injustice is shown by the following: '8 percent of the 200 million Arabs who lived in the oil countries owned more than 50 percent of the

aggregate gross national product of the Arab world and . . . the per capita income of the native population of the oil countries ranged between \$15,000 and \$20,000 while that of the vast majority of the 200 million Arabs was below \$1,000' (Khalidi, 1991, 19).

In his appeals to the Arab and Muslim masses (outside the oil countries) to support Iraq against US-led forces, Saddam managed to convey a clear message calling for a new Arab order and for the breaking of the existing socio-economic status quo. He accused the Saudis and the Kuwaitis of being responsible for this injustice and of inviting in foreign troops to perpetuate the situation. In this particular perspective, Saddam gained considerable support. Even before the war erupted, there were mass demonstrations in the *mashrig* and *Maghrib* countries calling for the immediate withdrawal of foreign troops from the Arabian peninsula. Nationalist and Islamist forces spontaneously coalesced in the streets of many Arab capitals. This suggests that a 'latent power' within the masses exists, and it is possible that this power will take the form pan-Arabian combined with Islamism to confront western and foreign domination.

The future political map of the Middle East will be shaped by two political realities: the degree to which the United States translates its current military presence into permanent bases in the region, and the reaction of nationalist and Islamic voices in the Arab countries. These two voices are likely to be unified against the anti-American presence if it becomes permanent. In such a situation, Saddam's accusations against the West will fuel the national forces, while the Islamic movement will see 'a new crusade led by the world's great superpower, coming to gobble up our oil resources, humiliate our people and so forth' (Quandt, 1991, 12). I would submit that the US military arsenal will remain in the Gulf area, and in Arabia in particular, at least to the end of the present century if not longer. The two immediate justifications for that will be: (1) Saddam remaining in power in Iraq, and (2) the war's strengthening of Iran, and Iran's Islamic activities in the region – two factors that are perceived by the West as a source of instability. However, there might be a simpler, more logical reason for the continuing presence of the US

Army in Arabia: that America has acquired a political position in the region and will remain there to preserve it. Even more logical would be the transfer of the American military arsenal from the former West Germany to Arabia as part of a new strategic distribution of American forces across the globe.

Assuming that the US Army will remain in Arabia as a form of neo-colonialism, I will discuss in the following two sections the political geography of the Middle East in the years 2000 and 2025. For the purposes of this discussion, the term 'Middle East' comprises the Levant countries, the Arabian peninsula countries, as well as Iran and Egypt. Libya and Sudan, located on the periphery of the Middle East, are excluded from the discussion. The two non-Arab countries, Cyprus and Turkey, that belong geographically to the Middle East are likely to be more European-oriented in their future development, and are excluded from the present discussion.

Towards the year 2000

In the coming four or five years, most Middle Eastern countries will still live in a post-Gulf War atmosphere. At the same time, internal forces will continue to pressure leaders of states to respond positively to democracy and liberalization. Islamic forces will reinforce their position and influence in countries where the economic situation is most uncertain. The Palestinian–Israeli conflict will surface in the context of the general Arab–American relationship, but is likely not to influence much of the inter-state Arab relationships, nor the internal processes of liberalization in a given state. In other words, since the war, Arab countries have been forced to set new priorities. Support for the Palestinian struggle for statehood ranked low in the list. Those Arab states who opposed Saddam Hussein and joined the US-led coalition are likely to push for the peaceful solution of the Palestinian problem, at least to justify for their people their decision to join the western coalition against a fellow Arab country. But this enthusiasm will probably vanish after a time due to their inability to influence United States foreign policy or reduce US support of Israel.

A short review of priorities for the major countries of the region provides a useful indication of their domestic orientations. Starting with the Gulf states, Kuwait's priority is rebuilding the country. Estimates of the cost for rebuilding generally start at $60 billion. It may take 5–10 years to restore the emirate's oil production to pre-war levels (Sadowski, 1991, 7). This situation has a direct impact on the flow of capital from wealthy Arab oil states to non-oil countries. What is probably more fundamental is the emerging patterns of labour migration. Kuwait and the rest of the Gulf countries (Qatar, Bahrain and the United Arab Emirates) will probably restrict future labour migration from Arab countries and give preference to workers from the South Asian countries who neither speak Arabic nor care about the political affairs of the host countries. The rate of workers' remittances from the Gulf states will be decreased correspondingly, and this will add to the burden of the non-oil countries.

Iraq has been economically devasted by the war. Allied air strikes rapidly erased much of its economic infrastructure: bridges, power plants, factories and refineries. The post-war reconstruction in Iraq is estimated roughly at $100–200 billion (Brynen and Noble, 1991, 137, note 19). Iraq's main priorities towards the end of the present century are to build the country and rebuild its army. Much effort will be given to suppress internal dissent (e.g. Kurds and Shii Muslims). Iraq will also seek to establish good neighbourly relationships with Iran – a new emergent regional power following the war and a potential ally for Iraq against an American presence in Arabia. Iran as a new political power will lead the Islamic forces and will probably play an important role in determining the fate of some Arab regimes. At the same time, Iran will devote much effort to building its technological capability and improving its diplomatic relationships with European countries, and probably with East Asia (Japan and China). After the war, Saudi Arabia gained new international respect and self-confidence at home. The country emerged from the war relatively unscathed. Although the Saudis invited the foreign troops, they paid the major cost of the war. It should be noted here that a substantial portion of this cost will be recovered through extra oil production and the rise of oil prices during the crises. Saudi Arabia remains the only viable source of wealth to provide monetary aid to non-oil Arab countries, although its primary efforts will probably go towards building its military capability and increasing its influence in the Gulf. Its main weakness derives from its internal political system wherein an elite based on kinship forms what amounts to a state within a state. The masses are too wealthy to push for change, and the rulers, who have accumulated both power and tremendous wealth, feel most vulnerable to any external ideas, particularly new ideologies from Iran.

As for the non-oil countries of the region, Egypt and Syria joined the US-led coalition against Iraq and were rewarded with some relief from the economic woes that had plagued them in the 1980s. Jordan, Yemen, Sudan and the Palestinians, which the Saudis felt had shown too much sympathy for Iraq, are faced with enormous economic problems. Since the flow of monetary aid from the oil-rich states will dramatically drop, these countries will have to generate some aid from western countries, particularly the United States. They will allow a certain extent of democratization, and show a keen interest in resolving the Palestinian Israeli conflict peacefully. In doing so, they might be credible for western support.

In summing up this period, the Middle East by the year 2000 will be much poorer than it was in 1992. Inequalities between Arab oil and non-oil states are likely to become even more pronounced. In addition, three possible changes that could alter the Middle East map and bring a new Palestinian state will be mentioned here. First, if the United States, considering the strategic advantages of the Arabian peninsula (if only because of its large area), changed its foreign policy towards Israel, the Israelis would as a result probably withdraw from the 1967 Palestinian-occupied territory after appropriate pressures, including an ultimatum. The second possibility could occur within the political sphere of Israel itself, if the new government of Israel, elected in 1992, is ready to agree to the formula 'peace for land'. A new era of economic and political transformation may follow. The third but bitter solution to Palestinian–Israeli conflict is found in

the spirit of a recent article by Stephen Green (1991) referring to the old world order: 'first the mutual fear and mutual respect, and then the discovery of common interests and peace' (Green, 1991, 51). According to Green, the recent negotiations will not bring any results 'if Israel feels no fear, there is going to be no motivation to compromise, and the negotiations are unlikely to progress toward respect or toward any meaningful resolution' (Green, 1991, 50). Unfortunately, this third possibility is closer to the reality than the other two options. If this is the only way to resolve conflict between Israel and the Palestinians, it also may imply the rationale for Israel's conflict with both Syria and Lebanon. If Green's thesis proves to be correct, the Middle East will witness another war before the close of the century.

Towards the year 2025

Before discussing the political geography of the Middle East in the year 2025, it is necessary to present the demography and resources of the region. According to the Population Reference Bureau (1989, 1991), by the year 2025 the Middle East population (excluding Cyprus and Turkey) will increase in relation to its present size (mid-1991) by almost 2.4 times. If today the population of the United States exceeds that of the Middle East (in the study area) by some 63 million, in the year 2025 the latter will exceed the former by 120 million. Iran and Iraq's population alone will equal 74 per cent of the US population in that year. Regionally, between the years 1991 and 2025 the population of Iran (82.8 million) alone will exceed by almost 22 million the total population of all the Arabian peninsula's oil countries combined (Saudi Arabia, Oman, Bahrain, Qatar and the United Arab Emirates). Since there is no evidence that resources in the Middle East will increase at a similar rate to that of the expanded population, the rate of per capita resources will decrease as time progresses. The unequal distribution of wealth among oil and non-oil states will continue and will even become greater.

The balance between demography and resources in the year 2025 is likely to form the major pillar for crystallizing the region's political geography. It will create a new order of power and eventually result in partitioning the region into sub-areas. For some states, demography will create serious disadvantages and reduce sharply its regional political role, while for others demography will turn out to be an advantage. Egypt and Iran are illustrations of this phenomenon. Egypt's demography has for a long time been an advantage over other Arab states. It has the largest Arab army that could stand against Israel, so it acquired a unique political position in the Arab world (before 1979) because of its larger population. But the year 2025 will turn Egypt's large population (projected to be 105.4 million) into its great disadvantage in view of its limited resources. Only 2 per cent of its land is suitable for agriculture. Egypt is a non-oil country with a GNP of $630 per capita in 1989 and with an external public debt of $49,970 million in the same year (Sadowski, 1991, 9). This country is likely to join the ranks of the Third World and gradually lose influence in Arab world politics. Its future technology and democracy will not solve its economic problems. On the other hand, the West may not need Egypt any more as an intermediary between itself and the rest of the Arab world, since by then the Palestinian–Israeli conflict will be resolved in one way or another. Equally, the Arab world will see Egyptians as people in search of food and may not be able to help much. In such a situation, Egypt will be more African than Middle Eastern.

Iran in 2025, on the other hand, with a projected population of 141.4 millions, will not have problems supporting its population. It is a relatively rich oil country which possesses other natural resources and 9 per cent of its land is suitable for agriculture. Its population in 2025 will comprise 31 per cent of the study area total. Its prospects of becoming a dominant political power in the region are most promising.

Iraq and Syria, with projected populations of 43.8 and 41.4 millions respectively, will also emerge as important political powers in the year 2025. Syria was often considered the cradle of pan-Arabism. Although a non-oil country, over 30 per cent of its land is suitable for agriculture, which will be crucial in overcoming its economic problems. Iraq, on the other hand, is a relatively rich oil country, with 13 per cent of its total land suitable for agriculture, putting it in a better position than Syria to support its projected 44 million population. Iraq and Syria are likely to lead the Arab World in the year 2025. Pan Arabism will revive once again with the turn of the 21st century. This assumption is based on the manifestation of the changes that will occur during the present decade (up to 2000) within most of the Arab states.

Democracy and liberalization will introduce to the Arab world what Edward Said sees as most needed: 'a critical language and a full-scale critical culture' (Said, 1991, 18). The struggle of groups for democracy, economic justice, women's rights and human rights within state territories is likely to have a spill-over effect on the region as a whole. In such a situation, regional co-operation will emerge on a non-official basis. A good example of such co-operation is the 'gathering of intellectuals in some transnational Arab institutes, like the Institute of Arab Unity and the Arab Human Rights organization. Various lawyers' groups, university and intellectual groups that collaborate on small projects . . .' (Said, 1991, 18). I would submit that such co-operative ventures will continue after the year 2000 and eventually redefine and revive pan-Arabism. A new modification of pan-Arabism will be created 'from below' which will be different from the 1960s pan-Arabism which was created 'from above'. This kind of pan-Arabism comprises the secular nationalist forces which will co-operate with the Islamic forces to protect the scarce resources that remain in the region.

The core area for pan-Arabism will be Syria and Iraq. These two countries are likely to overcome the undeclared war between their present leaders, Assad and Saddam. Equally, by the year 2025, the two leaders will be replaced by new leaders who will unite the two countries. If Saudi Arabia and the Gulf states link their future to the United States through alliance treaties, the Middle East will be divided into two major alignments (see Figure Epil. 1).

Syria, Iraq and Iran will form the first alignment here called SYRAQIAN, while Saudi Arabia with the Gulf states and appropriate presence of a US arsenal will form the second alignment, here called AMERABIA. The political

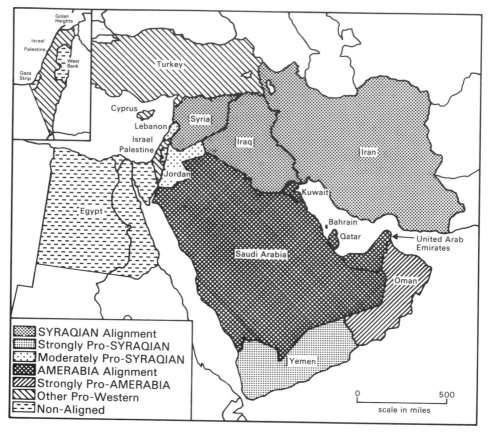

Figure Epil.1 *Political map of the Middle East in the year 2025.*

behaviour of the other Arab states will be divided along the lines of these two alliances. Egypt will be neutral because it will be absorbed by its economic problems. It probably will enter some co-operation with Sudan (on agriculture) and Libya (for oil). Oman probably will not join AMERABIA but may support it or seek the support of another western power, such as Britain. Yemen, a poor country with strong nationalist tendencies, will be likely to support SYRAQIAN and derive some economic benefit. Lebanon and Jordan may support SYRAQIAN morally, but at the same time both countries will maintain a balanced relationship with the West. The new Palestinian state (if it is realized) may sympathize with SYRAQIAN but remain neutral, it being assumed that the new entity would be devoting its efforts to building a viable state and hence steer clear of regional disputes.

Palestine would like to be a centre of science, computing and technology for the whole Middle East, and would probably be competing with Israel.

Israel's position in the Middle East in the year 2025 will be determined by how the Palestinian–Israeli conflict and the territorial conflict with other Arab states is resolved. If a peaceful solution is reached between the two countries resulting in the establishment of a Palestinian state, then Israel is likely to be in a better position in the Middle East from both economic and political standpoints. Israel's projected population for the year 2025 is 7.2 million, including at least 1.4 million Palestinian citizens who will form 20 per cent of the total state population. If we add the 1.4 million Palestinians in Israel to the 3.6 Palestinians living in the West Bank and the Gaza Strip, we would reach a figure of 5 million

Palestinians inside what was mandatory Palestine in the year 2025 (bearing in mind that an additional 7 million Palestinians would be outside). This figure should be compared with that of 5.8 million Jews occupying the same territory if the present status quo remains until 2025. Here is another example of the role of demography in shaping the Middle East political map in the year 2025. Assuming that an additional 1 or 2 million Jews immigrate into Israel by the year 2025, Israel could not annexe the West Bank and Gaza Strip and claim to be a democratic state. If Israel intends to remain a Jewish state with a Jewish majority, it will not achieve that without the partition of Palestine between the two nations (Arabs and Jews).

Conclusions

This essay suggests that recent political changes (if they continue at the level of individual states), combined with democratization trends, will lead to a strengthening of the state as a political institution. The participation of men and women in building a viable nation within the framework of state boundaries will eventually spill over and 'invent' a new form of pan-Arabism. This new form, which will develop 'from below', will find an agreeable co-existence formula with Islamism. It will not weaken the *raison d'être* of any state of the region, and will act to unify states with the aim of protecting resources and holy places and of resolving internal conflicts.

No viable state will be able to exist in the modern world if half of the state's population – the women – remain passive economically or even politically. Already there are great disparities in the Arab world. Whilst a Palestinian woman leads a delegation that discusses crucial issues for the fate of her people – the possibility of solving the Arab–Israeli conflict politically – women in Saudi Arabia are prevented from 'driving cars or working in offices and factories with men' (Hooglund, 1992, 21). It is not Islam or Arab culture that has decided that the women cannot drive cars; it is the Saudi man (Doumato, 1991).

References

Amawi, A. (1992). Democracy dilemmas in Jordan, *Middle East Report*, No. 174 (January–February), 26–9.

Brynen, R. and Noble, P. (1991). The Gulf conflict and the Arab state system: a new regional order?, *Arab Studies Quarterly*, Vol. 13 (1 & 2), 117–40.

Doumato, A. E. (1991). Women and the stability of Saudi Arabia, *Middle East Report*, No. 171 (July–August), 34–7.

Esposito, L. J. and Piscatori, P. J. (1991). Democratization and Islam, *Middle East Journal*, Vol. 45(3), 427–40.

Green, S. (1991). Middle East peace process: the old world order again, *American Arab Affairs*, No. 37 (Summer), 45–51.

Hooglund, E. (1992). Iranian populism and political change in the Gulf, *Middle East Report*, No. 174 (January–February), 19–21.

Hudson, M. (1991). After the Gulf War: prospects for democratization in the Arab world, *Middle East Journal*, Vol. 45(3), 407–26.

Khalidi, W. (1978). Thinking the unthinkable: a sovereign Palestinian state, *Foreign Affairs*, Vol. 56(4), 645–714.

Krämer, G. (1992). Liberalization and democracy in the Arab world, *Middle East Report*, No. 174 (January–February), 22–25, 35.

Population Reference Bureau (1989). *1989 World Population Data Sheet*. Washington, D.C.

Population Reference Bureau (1991). *1991 World Population Data Sheet*. Washington, D.C.

Quandt, B. W. (1990–91). After the Gulf crisis: challenges for American policy, *American Arab Affairs*, No. 35 (Winter), 11–19.

Sadowski, Y. (1991). Arab economies after the Gulf War: power, poverty and petrodollars, *Middle East Report*, No. 170 (May–June), 4–10.

Said, W. E. (1991). The intellectuals and the war-interview, *Middle East Report*, No. 171 (July–August), 15–20.

Sharabi, H. (1988). *Neopatriarchy: A Theory of Distorted Change in Arab Society*, New York: Oxford University Press.

Sid-Ahmed, M. (1991). The Gulf crisis and the new world order, *Middle East Report*, No. 168 (January–February), 16–17.

The Editors (1992). The democracy agenda in the Arab world, *Middle East Report*, No. 174 (January–February), 3–5.

Globalization and the semi-periphery: a view from Brazil

Bertha K. Becker

In the face of simultaneous movements towards spatial globalization and fragmentation, a multi-dimensional power is challenging the state as the only scale of power, making its future the central question for political geography at the end of the century (Becker, 1982). In other words, the question is raised of the continuation or the unmaking of the inter-state system. Since the inter-state system and the centre–periphery structure constitute the spatiality of the capitalist system, and as such are crucial to its operation, the unmaking of the inter-state system associated with the end of US hegemony would indicate the beginning of the demise of the capitalist world-economy itself (Taylor, 1993).

Nevertheless, at this moment any discussion of the new global political geography must in a large part remain speculative. Accepting that the recognition of otherness makes it impossible to speak in the name of 'the rest', this particular speculation will reflect a Brazilian perspective, Brazil being understood as an integral part of the semi-peripheral sector of the capitalist world-economy.

Technological innovation, political movements and the new world (dis)order

In an era of uncertainty, indications both that the inter-state system will continue and that it is no longer ideal for the operation of the capitalist system are visible in the practices associated as much with the logic of domination as with the logic of post-modern culture which generate new territorialities 'above' and 'below' the state scale.

The logic of domination indicates that global political geography becomes incomprehensible if it does not consider the moulding of the planet by the modern scientific-technological vector now intertwined in the social power structures. This movement contains state-subversive elements, accentuating trans-stateness as much as territorial fragmentation.

The last years of the twentieth century correspond to the transition between crisis/ restructuring of the world-economy and the implementation of the new regime of accumulation which, developed in the 1970s, will tend to dominate in the beginning of the twenty-first century. The techno-economic pattern and the forms of political control dominant since World War II have exhausted themselves. The essence of the new regime is the increasing internationalization of the world-economy associated with a new form of production introduced by the technological revolution, and based on scientific knowledge and information.

The specific historical significance of the introduction of new technologies is the creation of new space–time relations. Velocity becomes the key element, capable of altering not only the civil and military techno-productive complex, but also social and power relations (Virilio, 1976). A new dialectic establishes itself between the space of flows and decisions made according to a global logic and the space of places, i.e. of historical experience (Castells, 1985).

The space of flows tends to go beyond states and borders, altering geopolitics and the concept of sovereignty. The single market extends itself through the selective valuing of differences, of resources and spaces; territorial dominion is derived from the possession of circulation and communication networks, especially with regard to telecommunications. Through these networks it is possible to establish a direct relationship between firms, and between local and trans-national spaces, bypassing the state. Economic and power advantages of a territory, on all geographical scales, in a great part derive from the velocity with which it is able to adopt the new form of production and from its position in the networks, this becoming an object of competition. Such a flexibilization of the capitalist system may result in a more democratic tendency, but equally may lead to a more authoritarian one.

Globalization driven by the transnational corporations intensifies from the fusion and centralization of the firms which, relating themselves directly and selectively to regions and places, remove from the state its control over the productive process and fragment the national territory. Moreover, the proliferation of modern arms prevents the leading states from making war, and the instantaneity of information stimulates autonomous movements and permits an immediate reaction to internal intervention, as seen in the USSR.

If, however, sovereignty involves both the internal and the external capacity of the state, multiple reasons negate the premise of the end of the inter-state system. First, the new scheme of accumulation is not the result of the free inter-action of market forces, nor is it predetermined by technological advance; it is a social and political process. Corporations command the process, but the velocity of certain territories in adopting new technologies is also the product of policies managed by states. It is in this way that in the three economies which most rapidly recuperated from the crisis – Japan, Germany and the United States – though with different models, the state has a central role. Second, national territory is one of the legal bases of the states which guarantee property rights and administrate currency and the labour market, necessary for productive reconversion. Third, war assumes new expressions which remain linked to national strategies, government spending decisions and to the national survival imperative. Finally, the inter-state system assures the necessary distinction between productive and consumptive states, i.e., assures the reproduction of the centre–periphery relationship.

However, the inter-state system is not just being challenged by the logic of domination. For the first time in centuries, separatist nationalisms and social movements threaten the territorial integrity of the states. In the early eighties, Wallerstein maintained that the logic of the 'civilizing project' challenging the universalization of western culture was the great unknown variable of the century's end. This could contribute towards the creation of a socialist order or provide an outer wrapper for the logic of domination. In other words, the transitional structures of the states would constitute the decisive political battlefield in the restructuring process of the capitalist world-economy (Wallerstein, 1983).

The tie between nation and territory through the state which made the inter-state system so powerful in the modern world is being affected by the divergence between the space of flows, of economic organization, and the space of places, of cultural identity. However, nationalisms and regionalisms are not defining themselves in an anti-systemic sense: to the contrary. Although with differences deriving from historic specificities, what is dominant is their search for greater autonomy to rapidly insert themselves into the world-economy. They are movements against authoritarian modernization and state capitalism in favour of the 'freedom' to compete; they reproduce the system, and even strengthen it through the eventual creation of new forms of power in the case of regions with greater economic and political capacity, while the less able cannot live without state support.

On the other hand, Non Governmental Organizations (NGOs) and global social movements not centred on the state are important state-subversive elements. A doubt still remains, however, as to their capacity to confront co-operation by the states in order to strengthen liberalism: a justifiable doubt when ecology becomes an ideology and when NGOs act mainly in the periphery but are headquartered in the central countries.

One must certainly recognize the end of the welfare state, of the pretensions of centrally-planned economies, and of regimes which neglected otherness. The new form of production and the demands for autonomy require a new kind of state: a flexible economic and social organization, open to internationalization, which favours and stimulates competition. Liberal ideology, with its strategy for the modernization of institutional apparatuses, including as central components debureaucratization, privatization and decentralization, seeks exactly to liberate the state from its charges with respect to the nation in order to guarantee its survival (Becker, 1991).

A new territoriality also becomes necessary. According to Taylor, two alternative tendencies are defined regarding the inter-state system (Taylor, 1993). First, there is emergence of a new spatiality which would indicate the beginning of the unmaking of the inter-state system, such as is being configured in Europe: three levels of political power, represented by regions, states and the community. Second, there is the temporary reinforcement of the system by a new, higher level of one-scale authoritarianism.

A third alternative which also seems viable is

the simultaneous existence of superstates and multi-scale power represented by regions and/or states which act as regions. In other words, the accentuation of the opposite yet complementary tendencies of globalization and fragmentation as a form of spatial adjustment for the survival of the world-economy, and not as the end of the system – at least until the first quarter of the twenty-first century.

The new public–private relationship is sustained by a new multilateralism. Not only the IMF and the World Bank, but also the UN, under the command of the Group of Seven, become collective instruments of international administration. The multilateral agenda has a global scope and a limited number of themes and relevant actors, and it defines norms which are imposed through collective actions on national societies.

The centre–periphery structure: the intensification of inequalities

One of the most important political questions at the end of the century is the accentuation of inequalities between the centres and the peripheries. As the East–West dispute disappears, the world becomes divided into the fast and the slow through the possession, or non-possession, of scientific knowledge and communication networks. We are witnessing the age of technological apartheid.

World commerce is regionalizing itself. Globalization is forcing each nation to direct its energies towards international competition for markets and gains. To avoid war and to achieve a broader scale and greater speed in the production of new techonologies, the central economies create supra-national markets even though the basis for these coalitions will for a long time be the nation-state and the defence of national interest. Integrated by the space of flows and by networks, these markets are strongly exclusive.

The reduction of the volume and types of raw materials used with the creation of new materials signifies the crisis of the principal markets of traditional raw materials and of mass production. Cheap labour ceases to be a market advantage for the peripheral countries, and the disequilibrium

in the global distribution of telecommunications infrastructure tends to exclude them from international transactions. As well, the elevated cost of technological warfare excludes them from the possibility of making it.

The accelerated rhythm of the decision-making processes of the new multilateralism prevents the participation of a great number of countries. The great novelty in terms of alliances, as an expression of the new space–time relationship, will be the 'shifting coalitions' with specific and limited military and/or political objectives with an eye for geo-strategic rearrangements, religious and ethnic questions, and national questions which affect 'global security'. These multinational forces will be formed in order to act on the interfaces of the interests of the power blocks and, consequentially, will have different compositions for each case. Global security and the new concept of the 'duty of interference' which divides the world into the 'responsible' and 'irresponsible' will justify extra-jurisdictional and territorial action and the implementation of a system of limited sovereignties in the peripheral and semi-peripheral sectors of the world-economy.

However, the new tendencies affect the semi-peripheries of the world-economy more intensely which, directly hit during the eighties by the end of the growth cycle sustained by the external debt and state intervention, are being threatened by the prospect of a return to peripheral status. The velocity of their recuperation, however, is not predetermined, but depends on internal and external factors. It is especially the semi-peripheries of great territorial extension, the 'whales', which have the greatest difficulty in recovering (Becker and Egler, 1992).

For the 'whales', productive reconversion signifies the rapid obsolescence of the industrial park already created and difficulty in transferring to the new productive form. Furthermore, they suffer all kinds of external pressures to submit to the new economic rules and disassemble the market slices and areas of influence conquered by them, their centralized states and their national development strategies, all of which reduces the speed of their recuperation. This is the case of China and the USSR itself; it is the case of Iraq, an emerging regional power cut short by a war which was also the result of a belligerent nationalism adopted as a development strategy by the country's leaders; it is the case of Brazil, also pressured by the environmental question of the Amazon (Becker, 1991).

Together with the conventional pressures, new parameters in world geopolitics also emerge: first, the discourse of 'destatization', when in the central countries the state is modernized but not dismantled; second, technological curtailment, under the justification of nuclear non-proliferation and the restriction of arms sales to 'irresponsible governments'; third, the GATT Uruguay round as to service instalments, intellectual property, and investments; fourth, the limiting of sovereignty in vast areas of the planet under a variety of pretexts such as drug trafficking, access to energy resources, and ecological preservation. The new technologies alter the concept of value (until now associated with goods obtained through work), and nature is now attributed value: use value, value as a living warehouse, and value as capital for future realization, i.e. natural capital.

In the semi-peripheries, the external pressures add up to internal instability as a result of the fiscal and political crises of the state, the unattended demands of poor mass societies, and the regional attempts to directly relate with the world market. The question to be raised is with regard to which new pattern of insertion into the world-economy these countries will have. While the USSR, India and China open themselves to the international market under the resurgence of ethnic and cultural conflicts, in Brazil it is the absence of a plain nation which undermines the necessary support for the state in order to negotiate in its own favour.

Nevertheless, globalization has its price. Besides containing the largest share of the assets of the international financial system, the semi-periphery contributes to the stirring up of political instability in the planetary 'order' affecting the direction, the nature and the velocity of transformation in historic capitalism.

Who commands? The question of hegemony

The geopolitics determined by the ideological and military rivalry between the two great powers being spent, the USA ceases to be a superpower. The force of the so-called superpowers resulted from their rivalry. With one in collapse, the antipode loses its role.

With the power vacuum left by the USSR, a situation of competition and systemic chaos is being configured, similar to the periods which characterize the end of the hegemony of one state and the emergence of a new player within various contestants: in this case, the United States, Japan and Germany.

The potential for conflicts and instability is amplified with the formation of supra-national markets due to competition, to cultural confrontation and to the excess population not absorbed into the new form of production. Differentiated cultures and velocities of transformation resuscitate border conflicts, which are sharpenend by vast migrations. Claims to citizenship and to scarce resources on the periphery contribute to the increasing mobility of the world population affecting the dominant powers, who also become vulnerable as a result of commercial, financial and technological rivalry.

The ability to respond to these tensions will influence the future lead position which will reside with the state most capable of utilizing all its managerial capacity to confront three great challenges: regulation, i.e., the administration of interdependency in the world-economy; distribution, breaking the vicious circle of inequalities; and the recognition of the 'other'.

Three alternatives may be proposed regarding hegemony at the beginning of the twenty-first century and with these, diverse alternatives for the semi-peripheries. One alternative would be the end of US hegemony with the division of power amongst various superstates as is being witnessed today. The United States is no longer able to act on its own, needing international support in the Gulf War. It has lost its monopoly over the atomic bomb and the financial system (its currency is no longer the standard of reserve and exchange); it is confronting competition in the production sector from the Japanese and Europeans; and it depends on external creditors. The entrenchment of the political and strategic union of the EC creates an autonomous and powerful supranational state. A plural and multipolar geopolitics fragments the present American system from a virtually worldwide ambit into various, more rigid sub-regional systems, reliving the pan-regions of Hauschofer. In this context, the manoeuvring space of the semi-periphery is drastically reduced – a visible tendency within the proposal for the creation of a Latin American free market under the regency of the US (the Bush Plan), a new version of the Monroe Doctrine. This alternative, however, does not seem viable for the twenty-first century since the world-economy cannot operate without a semi-peripheral sector.

Another and opposite alternative would be the strengthening of US hegemony which, after the victories of the Cold War and the Gulf War, would extend itself to include Europe, a merely economic confederation though vast, including the Eastern European countries. The economic and technological power of the EC and Japan are not sufficiently strong to surpass that of the USA, which retains the greatest military power, the greatest dominance over transnational networks and information, and whose continental extension and internal unity are crucial advantages in the face of a Balkanization threat in the other groupings. The velocity of economic and political integration will be less in the EC, where there will be difficulty in defining a common geopolitical strategy due to fragmented interests and to tensions as to the reunification of Germany, which revives the English nightmare of Mackinder's heartland. Japan will have political difficulties with its neighbours due to residual resentments from the Second World War and to its commercial policy. In this context, the semi-peripheries remain but with limited autonomy, being reduced to regional powers with a small radius of action. But would not this alternative present the risk of the formation of a world empire?

A third, more viable alternative corresponds to a defence reaction of the world-economy preventing its conversion into a world-empire. The USA loses elements of its power, but continues to be the most powerful state of the system: its economic hegemony declines, but it strengthens itself as a strategic and political hegemonic power. The division of economic power follows the division of political power, signifying not a break from the present structure of the American

system, but simply the system's flexibilization. In this case, there is a possibility of the creation of more manoeuvring space for the semi-peripheries, especially those of great 'weight'. China, in particular, will be a great power, but Brazil could also benefit from the dispute between the blocs.

Because of its resources potential and geographic position, the Brazilian Amazon is a probable scenario in the definition of the three alternatives. This immense and polemic region reveals the transitional structure of the Brazilian state. It is the stage for new regional territorialities which directly articulate themselves with the transnational space, and also the instrument of external pressures for the adhesion of Brazil to the 'North'. These pressures, in turn, reflect the contradictions amongst the dominant powers regarding the redefinition of their zones of influence.

Globalization put forth the ecological challenge as a question of survival for humanity, and the Amazon became the symbol of this challenge. But globalization also attributed biodiversity with a crucial value for science and technology, transforming the ecological question into a techno(eco)logical one, manifest on two contradictory fronts of expansion. On one side, one has the mineral/energy front which gives continuity to the exploitation of resources, particularly through state enterprises. On the other side, there is the biotechnological front linked to the new form of techno-scientific production which values nature as capital and advocates its preservation in order to preserve the largest genetic reserve on the planet (Becker, 1991).

Within the proposal for a debt-for-nature swap, the most diverse interests, and the strangest coalitions converge – Indians, rubber-tappers, environmentalists, Transnational Corporations (TNCs), large banks, and the governments of hegemonic powers. The proposal corresponds to the creation of ecological reserves which could also be 'experimental paradises' or 'free territories' for rapid and direct articulation with the transnational space. The demarcation by the Brazilian government (15 November 1991) of a Yanomami reserve on the border with Venezuela illustrates the conflict: apparently, the

Indians, environmentalists, NGOs and the Catholic Church won the war against the military, regional elites and miners. The conflict resulted in the creation of a 9.4 million-hectare territory; three times the size of Belgium, for the 12,000 survivors of this nation. Add this to the 8.3 million hectares already recognized by the Venezuelan government as 'patrimony of humanity' and one has the Yanomami Indigenous Land of 17.7 million hectares – a contiguous 'free territory' right on the border between two countries.

The dispute for hegemony between powers is revealed in the issue over the paving of the BR-364 highway which, linking the state of Acre and Peru, would accelerate the connection with the Pacific, particularly with Japan – today the third largest investor in Brazil. The USA is pressuring Japan for it not to liberate resources for such a project, seeking to maintain the traditional Amazonian door open to the Atlantic while the UK, Germany and France dispute hegemony with the USA over the region.

In conclusion, the Amazon is the contradictory synthesis of national/transnational articulation, and of the industrialist/ecodevelopmental models dominant at the end of this century, and also of the new levels and forms of spatial globalization/fragmentation that will dominate in the next century.

The capitalist world-economy and the interstate system tend to adjust themselves in order to survive, but the new 'order' is not predetermined; it is a political process whose trajectory will depend on social practices. With the end of the internationalizing pretension of authoritarian socialism, the only universality of the left will be in the concretization of its acts in a precise political and economic environment, and with the recognition of otherness. Ethics, culture, religion – the existential – acting within the states, will be the decisive external forces capable of promoting the demise of the capitalist world-economy.

References

Becker, B. K. 1982. The political use of territory: a Third World perspective. In 'Simposia and Round Tables', *IGU Latin American Regional Conference*, IBGE: Rio de Janeiro.

—— 1988. A Geografia e o resgate da Geopolítica. *Rev. Bras. Geografia*, Ano 50, vol. 2, IBGE.

—— 1991. A Amazônia Brasileira. Uma Área Crítica no Contexto Geopolítico Mundial. *II Simpósio Internacional sobre a América Latina*, Univ. de Varsóvia, Mimeo.

Becker, B. K. and Egler, C. A., 1992, *Brazil: A New Regional Power in the World-Economy*, Cambridge University Press.

Castells, M. 1985. High technology, economic restructuring, and the urban-regional process in the US. In 'High Technology, Space. and Society'. Edited by M. Castells, *Urban Affairs Annual Review*, vol. 28, Sage: Beverly Hills, Ca.

Harvey, D. 1989. *The Condition of Post-Modernity*, Basil Blackwell: Oxford.

Lefebvre, H. 1978. *De l'Etat*, vol. 4. Union Générale: Paris.

Taylor, P. 1993. Contra Political Geography. *Tijdschrift Voor Economische en Sociale Geografie*.

Virilio, P. 1976. *Vitesse et Politique*. Galilée: Paris.

Wallerstein, I. 1983. La Crisis como Transitión. In: *Dinámica de la Crisis Global*. Ed. Amin, S., Arrighi, G., Frank A. G. e Wallerstein, Siglo Veintiuno: Madrid.

Democracy and privatization: a view from India

Chandra Pal Singh

The last decade of the twentieth century has begun with revolutionary changes in the total international environment which will have far-reaching consequences. The biggest event has been the collapse of communist ideology in a number of East European countries and the USSR. *Glasnost* and *perestroika* had initiated the process of democratization in these countries which culminated in the failed coup in the USSR in August 1991. In other parts of the world there were also significant changes. The Iran–Iraq war was over; the sovereignty of Kuwait was restored after the humiliating defeat of Iraq; Russian troops had been withdrawn from Afghanistan, though arms supplies to both the warring groups had continued; democratically elected governments came to power in all the South Asian countries for the first time; relative peace descended in South East Asia with the solution of the Kampuchean problem being in sight; and a shaken China was trying to maintain a calm countenance after the pro-democracy demonstrations and their violent repression. There was also a sea-change in Ethiopia, Angola, the whole of Southern Africa, and more recently the turmoil in Yugoslavia.

The world also became different in terms of increased rapid means of communications, which had their impact on international politics. Supercomputers, satellites, instruments of better transmission and reception enabled signals and images to be received instantly in all parts of the world. Today a TV generation is growing up with a different perspective which is the result of their being able to see events taking place live before their own eyes in the comforts of their drawing room. It was modern technology and fast communications which gave a decisive edge to the United States and its allies during the Gulf War, and the TV media played a crucial role in the collapse of the coup in the USSR. North America, Western Europe and Japan could become part of a single interdependent region, despite their physical separation by the two largest oceans, because of rapid communications and means of transportation. This closeness gave them a decisive say, or power, which will have a decisive affect in reshaping the world in their image in future.

In such an environment, three major processes can be identified which are changing the world today: a universal shift towards capitalism and the globalization of the economy; democratization and, sadly the rise of fundamentalism. The failure of the communist ideology was perceived by Mahatma Gandhi about sixty years ago because it deprived the individual of his liberties, leading to the tyranny of a few elites, and discouraged individual initiative. That was the reason for the Mahatma to advocate a minimal or no role for the state in the economic affairs of the people – advice that was not heeded (Sethi, 1978). It was under Nehru, who was deeply influenced by the egalitarian aspects of communism in Soviet Russia, that a commanding role was assigned to the public (government) sector in the economy of India, with the state exercising complete control over private enterprise through a complex system of permits and quotas. This lofty idealism of a mixed economy is the bane of the Indian economy today. From India one can eas-

ily understand and sympathize with the plight of people in the communist countries. The difference, however, is that in India the state controls were introduced by liberal democrats with the 'sanction' of the people through a democratic system. Like the communist countries, the government controls resulted in a drag on economic growth, and India could achieve what is termed the 'Hindu rate of growth', and it encouraged inefficiency, wastage and corruption at a large scale (Charan, 1991). The admission by Gorbachev of the failure of communist ideology was an honest and brave act, which Indian politicians could not match to this day with regard to the failure of socialism in India. This does not mean that capitalism is the panacea for all ills. This is an ideology based on human greed for higher and higher production by more and more exploitation of human and natural resources. But it is the better of two evils, as it provides a better standard of living with safeguards for the liberties of the individual, since man does not live by bread alone.

Associated with a shift towards capitalism is the effort of all countries to become integrated with the global economy. All the developing countries are making efforts to raise their productivity to levels where it is possible for them to exchange their products with the wealthier countries of North America, Western Europe and Japan. In the process they are opening up their economies and resources to global industries, and inviting foreign capital and knowhow by making their political systems suitable and attractive for such investment. Many times it means the undermining of sovereignty and an invitation to economic colonialism. But wealth cannot be generated in a closed system, and a price has to be paid for raising the standard of living of the people.

Wealth is the basis of knowledge and power. Wealth is also the product of knowledge, and the two together create power, which is the capacity to influence and control others. The terms 'wealth' and 'knowledge' include resources, technical and scientific attainments (including the acquisition of instruments of violence and coercion), human organization, and the cohesiveness of society. To understand the relative importance of a country and its capability of influencing

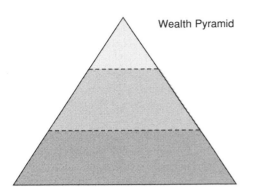
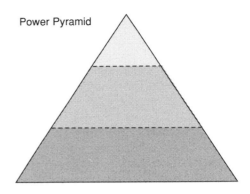

Figure Epil.2 *Relative location of importance of countries in world affairs.*

others at a particular point in time, the countries of the world can be arranged into two pyramids of wealth and power (Figure Epil. 2). In such an arrangement the poor and relatively weak countries will be located towards the base, and the wealthy and powerful towards the apex. A country like Japan can be rich, but may not be powerful without the acquisition of the instruments of violence. A country that is both powerful and wealthy will get located towards the apex on both the pyramids and would dominate the world, like the United States today. But should it slide down on any of the pyramids the situation would change for the worse for that country. The USSR will be located high on the power pyramid but low in the wealth pyramid; this will eventually pull it down on the power pyramid. India will get located slightly higher in the power pyramid but will be low in the wealth pyramid; consequently, it can be marginalized in international affairs except when it concerns the region in which it is located. Germany, on the other hand, will be located in the middle of both the pyramids, which would show its importance in world affairs. Since the position of a country depends on its level of wealth and power, which are variable according to the changing times, the relative position of countries keeps changing within the total galaxy of countries. For this reason all countries strive to improve their wealth, or their economy, by integrating with the global economy, and they scramble for arms and weapons to attain power, even at the cost of essential items for the people.

The third process that is noticed today is the rise of religious fundamentalism in many countries. The activities of fanatics committed to the theocratic control of the lives of people in some countries has assumed serious proportions. It is a paradox that with better means of transportation and communication which enable greater interaction, the differences of race and religion have accentuated. The categorization of the world into religious regions carried out by Mackinder at the beginning of the present century have not become irrelevant, and those regions may continue for a long time to come. It is therefore necessary to divide the world into broad religious/racial regions. They are: the circular Christian belt extending from the Americas through Europe, including most of the Russian Federation, into Australia and New Zealand, and parts of Southern and Central Africa; the Islamic region extending from Northern and Central Africa, through the Middle East, reaching right up to Central Asia comprising the southern parts of the former USSR and western parts of China, and another fork going into Iran, Pakistan, India, Bangladesh, Malaysia and Indonesia; Buddhist Asia in the East and South East; Hindu Asia in the Indian sub-continent which overlaps Buddhism in South–East Asia; and black sub-Saharan Africa with a mix of Christianity, Islam and tribal religions. Thus the Islamic region is surrounded by Christianity in the north and, with some exceptions, the non-Christian and non-Muslim religions in the south extending from East Asia to West Africa. In India, the superimposition of Islam and later Christianity over Hinduism forms the basis of secularism enshrined in the constitution, which has come into conflict with fundamentalism both of Hinduism and Islam.

World prospects till 2000 AD

There are apprehensions that after the decline of Soviet Union as an effective countervailing force at global level, the United States will bring about a new world order in its image. But those fears may be unfounded, as it is impractical and expensive for any single nation to maintain a network of forces required to maintain control throughout the world. The cost–benefit analysis militates against such a role by any nation. In fact, it is doubtful if the United States would have waged the war in the Gulf without the financial and military support of its allies. It is worth noting that the United States is the largest debtor country in the world today and owes more than US $400 billion to the rest of the world. The strategy of the United States could be to develop a network of alliances with regional powers covering the entire globe. It already has reliable partners in Western Europe where NATO has not been disbanded despite the Warsaw Pact countries having wound up their organization. It is making efforts to bring peace in the Middle East and would ensure its presence there through some kind of loose alliance with some of the Arabian countries and possibly Iran.

South Asia has been considered as an independent geopolitical region where India has followed the policy of non-alignment (Cohen, 1982). There are strong possibilities of India and the United States coming closer before the end of the present century (Padgaonkar, 1991). There are several reasons for such optimism. First, India is viewed by the US as becoming one of the six power centres of the world by the end of the present decade. The other five are: the two military superpowers – the former Soviet Union or perhaps Russia, and the US itself; Japan for its economic dominance, and China in East Asia; and Germany for its economic and military power. Second, with the decline of the Soviet Union as an effective counterpole to the US, the *raison d'être* of non-alignment has disappeared

and the Non-Aligned Movement (NAM) may be heading towards its natural death. Third, India is the largest liberal democracy in the world, and with the process of democratization in the newly-emerging countries, Indian democracy may be seen to be more experienced than others. It would be natural for the two countries to come closer. Finally, the internal economic compulsions of India will force it to abandon the earlier policies, some of which were not to the liking of the US. India would require the help of the US in securing loans from international lending institutions and industrialized nations. The process of coming closer has already begun quietly, with the international financial institutions and Germany and Japan pledging substantial monetary help to India. The recent visit of the Indian Chief of Army Staff to the US is a pointer to the possible military co-operation between the two countries.

The former USSR and India had very good economic relations with each other; the USSR was a source of crucial support to India in various international forums. The two traded with each other in Indian rupees, which saved valuable foreign exchange for both of them. The USSR was also a source of more than 70 per cent of military imports and spares. The situation in the Soviet Union is not yet clear. But it can be speculated that India would continue to have good relations with most of the Soviet Republics, as it had had traditional contacts with many of the Muslim Republics even during the pre-Islamic days. The good-will in other republics will also remain, and it would be easy for India to establish good economic and political relations with all the Republics, whether or not they become part of the proposed Economic Union.

It seems that the Russian successor state to the Soviet Union will not regain its self-confidence before the beginning of the twenty-first century. The problem of the former Soviet Union is that communism is dead there, but capitalism is unborn. The coup threw the nation into confusion, and with several Republics becoming independent, no-one knows what the final outcome will be. As compared with the former Soviet Union, India always had some basic structures of capitalism, such as petty traders who were the product of the caste system which had existed in the country for three millennia; a pool of innovative mechanics ingenious in repairing all kinds of machinery; tough and rough truck operators; and experts in finance, insurance, banking and other market institutions. India only needs the removal of controls and restrictions by the state. But the Republics of the former Soviet Union would take time in experimenting with both democracy and capitalism. Instability – economic, political and social – in the vicinity of India to the north would not be in its interest, and for that reason also there will be close cooperation of India with the Soviet Republics. Such relations, however, would not be viewed by the United States with suspicion any more.

The other major contiguous region of geopolitical importance to India is China. Though China has similar diversities as in the USSR and has adopted the communist ideology, there are several differences between the two. China has already allowed selective private ownership of agricultural land and trade. There, production and distribution systems are changing universally, from 'commune command' and 'state control' to largely 'household control'. Foreign investment is expanding in China (a large part, hitherto, by the overseas Chinese) because of the favourable policies of the government in providing infrastructure, services and other facilities. Production and trade linkages are developing with the outside world directly and through Hong Kong, which has very well-developed linkages with large overseas markets. China has no balance-of-payments problem; in 1990 its trade surplus account was nearly US $9 billion. But these factors may prove to be the undoing of the communist system in the country in the not-too-distant future. Selective privatization will produce disparities among the people. With some people becoming more affluent than others, demand for more privatization and the removal of state controls may become strong. The affluent Chinese from Hong Kong and elsewhere could also be a source of dissatisfaction. Even if the Chinese remain insulated from developments in other communist countries (which actually is doubtful), demands for more privatization and democracy are likely to grow further. The events of Tienanmen Square in 1989, and the overwhelming victory of pro-democracy candidates in the first-ever elections in Hong Kong in nearly 150 years, provide evidence of the subterranean processes of shift towards capitalism and democracy which are under way in China. The passing away of the old leadership in China, which does not seem to be very far away, could bring about the demise of communism in the country before the end of the present decade. The only question is whether it would be as peaceful as the transition in the USSR.

Demands for democracy could also bring demands of greater regional autonomy. Areas like Sinkiang, which has more in common in culture and history with the Central Asian Republics of the Soviet Union than with China, and Tibet, where there is deep dissatisfaction over what is perceived by Tibetans as an occupation of their country by China, may even try to become independent. China may also find it expensive to maintain control over them for a long time.

India is also a multi-ethnic, multi-religious, multi-lingual and multi-national state, and it is faced with demands for greater autonomy in the states and for secession in Kashmir, Punjab and Assam. But it is different from the USSR and China with respect to democratization. If the parting of the ways of any part of India became inevitable, then it would be done within an existing democratic set-up. The break-up would not be because of change from one system to another.

World perspective until 2025 AD

Notwithstanding the imponderables, the process of democratization should be complete by the beginning of the twenty-first century. The last bastions of communism should have disappeared by then. The world bodies should have made suitable changes to accommodate new countries emerging out of the demise of communism. Privatization and the globalization of the economy should be the major processes taking over the whole world within the first quarter of the twenty-first century. This is likely to make heavy demands on the environment, and this will further accelerate green movements, and green parties based on green politics may emerge throughout the world – or the existing parties might accommodate these demands. The conflict between the human greed to produce more for higher productivity, and the human need to preserve and protect the environment, should produce cleavages in society that would be different from the cleavages of the twentieth century. Scientific and technological developments that are in the offing could reach their application stage, and this could change production, consumption, communications and transportation, bringing the world closer and making it even more interdependent.

The United States is likely to continue to dominate the world, though Japan and Germany with better economic and military muscle may be in a position to challenge. With the movement of Japan and Germany upwards in the wealth and power pyramids, their relative position with all countries may change. And other countries in the EEC could also move upwards, and the European Community as a whole could be able to influence events in other parts of the world. It should be remembered that although the countries are interdependent and co-operate with each other, they are also competitors for wealth and power. Others like Russia, the East European countries, China and India could be joining this mad race by the beginning of the next century.

The dependence of the nations emerging out of the former USSR (and possibly China) on the more advanced and developed countries is likely to continue during this period, with the exception of the Russian Federation, which has abundant resources, a vast expanse of territory, and is strongest in terms of military and nuclear power, and the Ukraine, which has been the granary of the USSR. A China with more privatization could become even stronger and would have a higher position in the wealth and power pyramids. India is likely to move from one crisis to another till the beginning of the next century, because no individual or political party even today has sufficient political will or courage to drop the ideological baggage of socialism. The capitulation of the present government before the threat of a strike by employees of nationalized banks against alleged privatization as recently as September 1991 is an instance of political expediency adopted by it. But at the

same time some steps have been taken by the government to move away from socialism of Nehru brand. Once socialism goes, the pace of economic growth will increase and India, like China, then could be moving higher in the wealth and power pyramids, and would be in a stronger position to play an important role in world politics. India would still continue to be a leader of the Third World countries, with which it would have better economic and political relations.

The United Nations cannot remain unaffected by the changes in the global environment. There will be many more member countries, whose complex interrelationships and demands on the world body may bring about structural changes within the UN. There are already demands for the restructuring of the Security Council, which is playing a very important role in world affairs, and these may increase further by the beginning of the twenty-first century.

With the rise of a number of countries in wealth and power, the new world order will be multi-polar and even more interdependent in the next century. Standards of living will rise universally, though inequalities will remain. Armed with new powers, man should be in a better position to explore the universe, on the one hand, and destroy his environment, and himself, on the other.

References

Charan, Ram (1991), 'India should become globally competitive', *Times of India*, 19 August.
Cohen, Saul B. (1932), 'A new map of global geopolitical equilibrium: a developmental approach', *Political Geography Quarterly*, 1(3), 223–42.
Padgaonkar, Dileep (1991), 'Foreign policy options: coming to terms with new realities', *Times of India*, 8 July.
Sethi, J. D. (1978), *Gandhi Today*, New Delhi: Vikas.

Coercion and instability: a view from Nigeria/Africa

C.O. Ikporukpo

The past and the present as guide

'All the world's a stage. And all the men and women only players. They have their exits and their entrances.' Thus declared William Shakespeare (1564–1616) in his play, *As You Like It*, published in 1598. This dictum, when applied to nations, aptly describes the changing circumstances which have shaped and will continue to shape the political geography of the world. The 'exits' and the 'entrances' of nations – in war (including cold ones), in peace and in the economic power arena – to a large extent capture the essence of changes in the world's political geography.

The 'players' in the world's political scene have never been equals. Whereas some nations have been actively involved in shaping the political geography of the world, others have been passive actors. The spatial distribution of the active and the passive nations changes over time, determined by the relative economic prosperity of nations.

Whereas such changes in themselves may determine the potential patterns of war and peace, the actual patterns of war and peace are often set by an earlier configuration of victors and of the vanquished. Germany's role in the outbreak of World War II after her defeat in the First World War, and Egypt's initiation of the Yom Kippur War of 1973 after her defeat in the Six Day War of 1967, are good examples.

The changing spatial patterns of power and of powerlessness, and the patterns of war and peace, are to an extent both cause and effect of the tendency towards regional groupings. The character of such groupings has, however, changed over time. Whereas most of the earlier groupings were dictated by circumstances of war or its threat, later ones have been based mainly on the need for peace and economic power.

Given these generalized past and contemporary trends, what is to be expected on the world's political geographic scene? This issue is analysed on the basis of two time-frames: a short term (up to the year 2000) and a medium term (up to 2025).

A short-term view

The most significant likely development in the political geography of the world up to the year 2000 is a change in the spatial pattern of zones of political tension. One factor in this change is the former USSR's economic problems, which had forced her to adopt the policies of *glasnost* and *perestroika*. One consequence of this is a change in the number of nations in the first tier of political actors and a possible increase in the number of tiers. Unlike in the past, when the USA and the USSR made up the first tier, the USA will be the only nation in this tier, with Russia constituting a weak second, other major developed countries such as France and Britain the third, and so on.

This uni-polar pattern, with the USA as the dominant actor and a consequent drastic reduction in the arms race, will result in the end of the threat of a nuclear war in Europe. Thus, the possibility of major inter-state wars such as the two World Wars may be considerably reduced. However, the fact that there will now be no moderating influence on the USA may result in her increasing action as the international policeman of the world, with a consequent coercing of powerless countries perceived as not conforming to her ideals of political and economic behaviour. Within the short term, the collapse of the USSR will also result in more internal strife and tension in the communist/socialist world, such as in China, because of the possibility of pressure from the western capitalist world for social and economic reforms. It is likely that internal dissidents will be the tools for such pressure.

Another development which is likely to have far-reaching effects on the political geography of the world will be the continued economic problems and indebtedness of the less developed countries. The situation approximates an 'economic-slave–economic-master' relationship between the less developed countries and the developed capitalist world. Debt-servicing, which results in a flow of capital from the less developed to the developed countries, is a re-enactment of the colonial situation marked by the flow of all raw materials produced in a given colony to the metropolis. Many of the countries use more than half of their export earnings for debt servicing. These economic problems will continue to breed instability in these countries. With a wide spatial spread of such countries in much of Africa, South America and Asia, this state of affairs will be a threat to world peace.

In sum, the scenario in the short term suggests a change in the centres of political tension and the potential likelihood of war. Whereas the developed western world will be freer from political tension and war, this may not be true of the less developed world and the socialist countries.

A medium-term view

Here we speculate on possible scenarios up to the year 2025. As in our short-term projections, changes in the former USSR and the powerful western countries are of fundamental importance. The scenarios of interest are: (1) the former USSR Republics continue to be under economic stress and operate with a pro-democracy government; (2) the former USSR reverts back to the status quo ante; and (3) other arrangements come into play on the basis of some axioms put forward in the introduction.

In the first scenario, the USA will continue to act as the policeman of the world, ensuring through a 'carrot-and-stick' approach that the politically weak countries conform to some principles of behaviour defined by the western world. In doing this, she may have ready allies in some European countries such as Britain. China may challenge this dominance of the West, since she may attempt to emerge in place of the USSR as the defender and policeman of the socialist world. The USA's reaction to this – an attempt to dislodge the last bastion of socialist ideals – may be a threat to world peace. In general, the major centres of tension and wars will continue to be the weak countries and those of socialist orientation.

This rather simplistic scenario may, however, be complicated by competition amongst continent-based political units, which could be characteristic of the twenty-first century. One example, which will be very important in the political equation, is that of a United Europe. There is the possibility that a United Europe may become an alternative to the United States. The implication is that a United Europe, led by such countries as Germany and France, could assert herself as another international policeman as before the Second World War. This would have two general implications. In the first place,

a world may emerge controlled almost entirely by the West, for even if the might of China and other smaller socialist countries is further developed, this may not be comparable to the combined forces of Europe and the USA. The imbalance of power would mean more and more coercion of weak nations. This implies more tension and perhaps wars in the powerless nations – a setting similar to that of the short term. The second implication is that, after 'conquering' the whole world, the USA and Europe could compete for dominance. This may not necessarily result in a war, although a subtle 'cold war' is a possibility.

The implications of a reversion to socialism by the USSR (scenario 2) may not be so far-reaching that there would be a reactivation of the Cold War. This is because it may not be feasible to enforce many of the repressive principles which underlay the earlier regimes. Furthermore, it could be difficult to operate a closed system, for once the democratization process has been introduced, individuals and some Republics might go to great lengths to resist any reversal. However, a new USSR may attempt to ensure that she re-establishes a sphere of influence made up of weaker countries sympathetic to the socialist cause, but she may never have a strong hold on such countries, as was the case in the past, for she could only be tolerated as a first among equals. The resurgence of the USSR may not necessarily result in a threat to world peace, for the ideological differences between the capitalist and the socialist world may by then be much more subtle.

The re-emergence of the USSR could result in three regions of power – the USSR, the USA and a United Europe. Since none of these is likely to be overwhelmingly dominant, the result could be a better balance of power and hence a much less war-prone world. None of the present developing countries would be economically or politically powerful enough to join this league of dominant nations. Furthermore, whereas a country such as Japan would continue to be economically strong, it may not have the political will to join the competition for strategic dominance. The same may be true of most of the newly industrializing countries.

Nevertheless, some newly-emerging, strategically important countries, particularly those that have had a recent experience of defeat in war, could disturb this balance of peace. It is this possibility that scenario 3 reflects. One example is Iraq, whose experience in defeat is likely to be revenged. This is in spite of attempts to disarm her or the possibility of changes in the leadership. This reasoning is based on the postulate, noted in the introduction, that a nation which suffers a humiliating defeat is usually tempted to revenge. Argentina, with her disagreement with Britain over the Falkland Islands, is another candidate in this regard. Moreover, in spite of current peace efforts the Middle East, with its various histories of war, will continue to maintain its position as the world's leading region of localized political tension and war.

Conclusion

The world is a stage for international rivalry, for the major patterns of political geography have been determined by competition amongst nations. It is to be expected that the future political geography of the world will be consequent on the patterns of future competition.

Index

2560